Invasive Alien Plants

Impacts on Development and Options for Management

CABI INVASIVES SERIES

Invasive species are plants, animals or microorganisms not native to an ecosystem, whose introduction has threatened biodiversity, food security, health or economic development. Many ecosystems are affected by invasive species and they pose one of the biggest threats to biodiversity worldwide. Globalization through increased trade, transport, travel and tourism will inevitably increase the intentional or accidental introduction of organisms to new environments, and it is widely predicted that climate change will further increase the threat posed by invasive species. To help control and mitigate the effects of invasive species, scientists need access to information that not only provides an overview of and background to the field, but also keeps them up to date with the latest research findings.

This series addresses all topics relating to invasive species, including biosecurity surveillance, mapping and modelling, economics of invasive species and species interactions in plant invasions. Aimed at researchers, upper-level students and policy makers, titles in the series provide international coverage of topics related to invasive species, including both a synthesis of facts and discussions of future research perspectives and possible solutions.

Titles Available

1. *Invasive Alien Plants: An Ecological Appraisal for the Indian Subcontinent*
Edited by J.R. Bhatt, J.S. Singh, S.P. Singh, R.S. Tripathi and R.K. Kohli

2. *Invasive Plant Ecology and Management: Linking Processes to Practice*
Edited by T.A. Monaco and R.L. Sheley

3. *Potential Invasive Pests of Agricultural Crops*
Edited by J.E. Peña

4. *Invasive Species and Global Climate Change*
Edited by L.H. Ziska and J.S. Dukes

5. *Bioenergy and Biological Invasions: Ecological, Agronomic and Policy Perspectives on Minimizing Risk*
Edited by L.D. Quinn, D.P. Matlaga and J.N. Barney

6. *Biosecurity Surveillance: Quantitative Approaches*
Edited by F. Jarrad, S. Low-Choy and K. Mengersen

7. *Pest Risk Modelling and Mapping for Invasive Alien Species*
Edited by Robert C. Venette

8. *Invasive Alien Plants: Impacts on Development and Options for Management*
Edited by Carol A. Ellison, K.V. Sankaran and Sean T. Murphy

Invasive Alien Plants

Impacts on Development and Options for Management

Edited by

Carol A. Ellison

CABI, UK

K.V. Sankaran

Kerala Forest Research Institute, India

and

Sean T. Murphy

CABI, UK

CABI

CABI is a trading name of CAB International

CABI
Nosworthy Way
Wallingford
Oxfordshire OX10 8DE
UK

Tel: +44 (0)1491 832111
Fax: +44 (0)1491 833508
E-mail: info@cabi.org
Website: www.cabi.org

CABI
745 Atlantic Avenue
8th Floor
Boston, MA 02111
USA

T: +1 (617)682-9015
E-mail: cabi-nao@cabi.org

The designations employed and the presentation of the material in this publication do not imply the expression of any opinion whatsoever on the part of CABI concerning the legal status of any country, territory, city of area or of its authorities, or concerning the delimitation of its frontiers or boundaries. Lines on maps represent approximate border lines for which there may not yet be full agreement.

A catalogue record for this book is available from the British Library, London, UK.

Library of Congress Cataloging-in-Publication Data

Names: Ellison, Carol A., editor. | Sankaran, K. V., editor. | Murphy, Sean T., editor.
Title: Invasive alien plants : impacts on development and options for management / edited by Carol A. Ellison, K.V. Sankaran, and Sean T. Murphy.
Description: Boston, MA : CABI, [2017] | Series: CABI invasives series ; 8 | Includes bibliographical references and index.
Identifiers: LCCN 2017023563 (print) | LCCN 2017024619 (ebook) | ISBN 9781786391353 (ePDF) | ISBN 9781786391346 (ePub) | ISBN 9781780646275 (hbk : alk. paper)
Subjects: LCSH: Invasive plants--Biological control--Case studies. | Introduced organisms--Case studies.
Classification: LCC SB613.5 (ebook) | LCC SB613.5 .I5513 2017 (print) | DDC 581.6/2--dc23
LC record available at https://lccn.loc.gov/2017023563

ISBN-13: 978 1 78064 627 5

Commissioning editor: David Hemming
Editorial assistant: Emma McCann
Production editor: James Bishop

Typeset by AMA DataSet Ltd, Preston, UK.
Printed and bound in the UK by Antony Rowe, CPI Group (UK) Ltd.

Contents

Contributors

Bhaskar Adhikari, Royal Botanic Garden Edinburgh, 20a Inverleith Row, Edinburgh, EH3 5LR, UK. E-mail: B.Adhikari@rbge.ac.uk

V. Anitha, Kerala Forest Research Institute, Peechi – 680653, India. E-mail: anitha@kfri.org

Hem Sagar Baral, Zoological Society of London – Nepal Office, PO Box 5867, Kathmandu, Nepal and School of Environmental Sciences, Albury-Wodonga Campus, Charles Sturt University, Albury, New South Wales 2640, Australia. E-mail: hem.baral@zsl.org or hem.baral@gmail.com

Matthew J.W. Cock, CABI, Bakeham Lane, Egham, Surrey, TW20 9TY, UK. E-mail: m.cock@cabi.org

Kavya Dashora, Indian Institute of Technology, Hauz Khas, New Delhi-110016, India. E-mail: dashorakavya1@gmail.com

Usha Dev, Division of Plant Quarantine, ICAR (Indian Council of Agricultural Research)-National Bureau of Plant Genetic Resources, Pusa Campus, New Delhi – 110012, India. E-mail: devusha26@rediffmail.com

Jianqing Ding, School of Life Sciences, Henan University, Kaifeng, Henan 475004, China. E-mail: dingjianqing@yahoo.com

Carol A. Ellison, CABI, Bakeham Lane, Egham, Surrey, TW20 9TY, UK. E-mail: c.ellison@cabi.org

Kavita Gupta, Division of Plant Quarantine, ICAR – National Bureau of Plant Genetic Resources, Pusa Campus, New Delhi – 110012, India. E-mail: kavita.gupta@icar.gov.in

Ravi Khetarpal, CABI South Asia, 2nd Floor, CG Block, NASC Complex, DP Shastri Marg, Opp. Todapur Village, Pusa, New Delhi – 110012, India. E-mail: r.khetarpal@cabi.org

Jyotsna Krishnakumar, School of Environmental and Forest Sciences, University of Washington, 4000 15th Avenue NE, Seattle, WA 98195-2100, USA. E-mail: jyotsnak@uw.edu

Sean T. Murphy, CABI, Bakeham Lane, Egham, Surrey TW20 9TY, UK. E-mail: s.murphy@cabi.org

Warea Orapa, (Formerly, Plant Protection Advisor, Secretariat of the Pacific Community), Assistant General Manager, Operations and Inspections Division, National Agriculture Quarantine and Inspection Authority (NAQIA), PO Box 741, Port Moresby, Papua New Guinea. E-mail: warea.orapa@gmail.com or worapa@naqia.gov.pg

R.J. Rabindra, Former Director, ICAR – National Bureau of Agricultural Insect Resources, PO Box 2491, H.A. Farm Post, Hebbal, Bengaluru – 560024, Karnataka, India.

P.S. Ramakrishnan, INSA (Indian National Science Academy) Honorary Senior Scientist, School of Environmental Sciences, Jawaharlal Nehru University, New Delhi – 110067, India. E-mail: psr@mail.jnu.ac.in

K.V. Sankaran, Former Director, Kerala Forest Research Institute, Peechi – 680653, India. E-mail: sankarankv@gmail.com

K.V. Santheep, Kerala Forest Research Institute, Peechi – 680653, India. E-mail: santheep_kv@rediffmail.com

Soetikno S. Sastroutomo, CABI, PO Box 210, 43400 UPM Serdang, Selangor, Malaysia E-mail: s.soetikno@cabi.org

P. Sreerama Kumar, ICAR–National Bureau of Agricultural Insect Resources, PO Box 2491, H.A. Farm Post, Hebbal, Bengaluru – 560024, Karnataka, India. E-mail: psreeramakumar@yahoo.co.in

Soekisman Tjitrosemito, SEAMEO-BIOTROP, Jalan Raya Tajur Km 6, Bogor, West Java, Indonesia. E-mail: s.tjitrosemito@biotrop.org

Abraham Verghese, Former Director, ICAR – National Bureau of Agricultural Insect Resources, Hebbal, Bengaluru, Karnataka, India. Current address: GPS Institute of Agricultural Management, #1, Techno Industrial Complex, Peenya Industrial Estate, Peenya, Bengaluru – 560058, Karnataka, India. E-mail: abraham.avergis@gmail.com

Foreword

We live in a time of global environmental change. One or more of the major and interrelated issues of climate change, desertification and habitat and biodiversity loss, and the human factors driving these, are covered by national and/or world media almost daily. However, one major agent of mostly dramatic negative change – invasive alien species – remains invisible to the majority. These species have been creeping around the world at an ever increasing pace, largely unabated and with their movement facilitated (usually unintentionally) by international human activity. The destruction that some of these species can cause can be life changing; for example, when an invasive alien species almost wipes out a crop or livestock species on a national scale. These cases do reach headlines in national newspapers but they are then frequently forgotten. In other cases, invasive alien species, notably plants, have, over time, come to dominate some landscapes, so much so that the local communities 'assume' that their biodiversity inheritance consists of a plant for which they have little use. Further, and unbeknown to those communities, these aggressive invaders are one of the prime factors driving the local extinction of native biodiversity.

Invasive alien species, like other environmental issues are, however, captured in international frameworks for action to mitigate impacts, and these calls for action have been reiterated by international conventions. Nonetheless, actions against invasive alien species in many countries have to date been low key or non-existent. In the developing world, this places the livelihoods of some of the most vulnerable, rural communities (which form a significant proportion of the world's poor) at high risk from biological invasions. The reasons underlying the general paucity of action at a national level are complex, but significant overriding factors are the lack of awareness of invasive alien species and lack of knowledge or appreciation of the sheer size and impact of (what is now) a rapidly growing and equal problem for agricultural, trade, environmental and other sectors.

The Asia–Pacific region has a fair share of the world's invasive alien species, and some would argue more than a fair share given the number of island states and the particular susceptibility of islands to biological invasions. Some of the more dramatic and widespread biological invaders of these regions are plants. Although these may form only a small percentage of a total national flora, this statistic pales into insignificance given the vast scale of invasion of some of the more aggressive species.

In this book, a number of experienced scientists who have been involved with supporting actions to address invasive alien plant species issues in countries in the region have come together to share knowledge and their experiences of catalysing actions to halt or mitigate impacts. Their aim is to illustrate to us all, by using invasive alien plants as an example, the sheer dimensions of the problems caused and thus the critical need to act now. They also show, on the one hand, the complexities of developing solutions and appropriate policy frameworks but, on the other hand, how barriers to actions can be overcome and how some solutions have worked, or could be made to work; in this context, the particular value of classical biological control as a tool suitable for managing widespread invasive plants is discussed. Many of the authors focus on the specific case of the highly aggressive invasive plant *Mikania micrantha*, which originates from Neotropical regions, as the species demonstrates so many of the issues involved. This plant invader is now widespread in Asia and the Pacific and negatively affects agriculture and natural ecosystems alike.

This is not a book on the biology of invasive alien plants, nor a book about straight technical solutions; it is a book that puts the reality of the negative impacts of invasive alien species into all our laps and clearly illustrates the urgent need for us all – governments, policy makers, agriculturalists, natural resource advisors, researchers, extension workers, industry, funding agencies, farmers and the public – to work together to deliver sustainable solutions.

Fang-Hao Wan
Head of Biological Invasion Research Innovation Group
Institute of Plant Protection
Chinese Academy of Agricultural Sciences
#2, West Road of Yuan-Ming-Yuan, Beijing, 100193
People's Republic of China

Preface

Those people involved wholly or in part in the task of managing biological invasions will know that it is a complex subject – scientifically, technically, socially and politically. The consequences of most biological invasions can be dire, and have far-reaching impacts, yet few beyond those involved in the management of invasive alien species understand the problems and/or realize that they may well be an affected stakeholder. Thus, practitioners of management usually have a maze of issues to address to reach sustainable solutions. Yet there is a wealth of knowledge and experience held by many of those practitioners who have been working on solutions – data for awareness raising, the engagement of communities, practical management tools, policy frameworks, etc. – that needs to be captured and shared with the wider community of stakeholders so that they can reach a broader understanding and be shown that solutions are achievable. One such case is the scientists who have been working on the escalating problems from invasive alien plants in the Asia–Pacific region. This was the inspiration behind the book. Work on the book has been in gestation for a long time but has benefited from the circumstance that problems resulting from invasive alien plants have become more prominent in recent years and hence the need for more urgent action.

The chapters cover a broad range of topics: economic, social and biodiversity impacts, rural community perceptions, policy framework responses, prevention and control experiences, and the net result of all of this on livelihoods and development as a whole. One or more of these topics are covered in each chapter, which are written from a country and/or sector perspective and where, and based on the authors' experiences, provide recommendations for moving forward to reach sustainable solutions.

Carol A. Ellison, K.V. Sankaran and Sean T. Murphy

Acknowledgements

This book is an output from a research project funded by the UK Department for International Development (DFID) for the benefit of developing countries and administered through the Natural Resources International Limited (NRIL) Crop Protection Research Programme (Project No. R8502). The views expressed are not necessarily those of DFID. Support for the completion of this book was kindly provided by the CABI Development Fund (CDF), this fund is supported by contributions from the Australian Centre for International Agricultural Research (ACIAR), DFID and others.

The three editors of this book are indebted to tireless inputs from the substantive editor, Rebecca Murphy, without whose professionalism, knowledge and unwavering determination, this book would not have reached fruition.

1 Invasive Alien Plants as a Constraint to Development in Tropical Asia: Is There a Crisis in the Making?

Sean T. Murphy*
CABI, Egham, UK

Introduction

Poverty, the welfare of the poor and related issues such as natural resource management, particularly in the poorest countries of the world, are now centre-stage concerns of the global community of nations. Even though poverty rates have been falling over recent decades, approximately 896 million people still lived on less than US$1.90 a day in 2012, with concomitant severe problems of undernourishment, high child mortality and lack of opportunities for education (World Bank, 2016). A series of 'Earth Summits' and assessments of progress, which started in the early 1990s, have enabled nations to gradually redefine and clarify the key issues driving poverty and environmental degradation, and the links between the two. Furthermore, the global community has drawn up a set of 17 'Sustainable Development Goals' (https://sustainabledevelopment.un.org), which have provided some focus for the discussions on actions and the orientation of aid budgets for development for the foreseeable future. These points are particularly relevant for tropical Asia, where many of the countries are still developing.

There is of course a multitude of issues that the world's poorest nations are trying to grapple with. In attempts to improve the livelihoods of the poor, most governments have identified, and are trying to address, key problems related to trade, access to land and credit, education, health and gender issues. However, the highest proportion of poor people in developing countries still lives in rural areas. Issues driving poverty in what are essentially agricultural communities are therefore of major concern. Consequently, there is an additional struggle for governments to bed solutions within the development of infrastructure in rural areas that are compatible with sustainable natural resource management and the conservation of biodiversity.

Thus, rural livelihoods and their sustainability require a multidimensional analysis. In order to understand more clearly ways in which interventions can be made to improve livelihoods, previous analyses (e.g. see Carney, 1998) have placed emphasis on the concept of 'capital assets' (natural, social, human, etc.) upon which individuals draw to build their livelihoods and what combinations of these lead (depending on the social group) to sustainability. Two main elements have been identified as facilitating change (both positive and negative): one is the structures and processes that define people's livelihood options (e.g. government policies); the other is referred to as the 'vulnerability context' in which the assets exist

* E-mail: s.murphy@cabi.org

– the trends, shocks and local cultural practices that affect livelihoods.

Such approaches to the livelihoods issue have allowed a clearer understanding of the parameters that affect the spectrum of assets on which particular social groups depend. There has been much discussion and debate on the 'macro-factors' effecting change (e.g. trade policies), but the topic of this chapter is the controversial issue of invasive alien species as a factor that is reported to undermine development and thereby make rural livelihoods more vulnerable to the poverty trap (e.g. see McWilliam, 2000; Rai and Scarborough, 2015).

Invasive alien pathogens, plants and arthropods have been a threat to agriculture since early colonial times; these species were introduced, either deliberately or accidentally, when the ranges of economically important crops and livestock were expanded into new lands to support the migration of human populations, and this tide of invasive alien species that affect agriculture continues (Bebber *et al.*, 2014). Moreover, on a global basis, invasive alien species have come increasingly into the public domain over the last 40 years: (i) because of the increase in both the diversity of taxa that include invasive species and the number of species within these taxa reported as having become invasive; and (ii) as a consequence of the recognition and ensuing concern that economic sectors other than agriculture are now affected (Mooney *et al.*, 2005).

There are published studies – albeit broad scale – that attempt to illustrate the enormousness of the quantitative negative impact of invasive alien species in a few selected countries (e.g. Pimentel *et al.*, 2002) but these studies are focused mostly on the developed world. There have also been a number of international responses to the issue of invasive species and requests for action from countries; an example is the Strategic Plan for Biodiversity, 2011–2020 under the Convention on Biological Diversity (CBD) (www.cbd.int/sp). None the less, there remains a wide spectrum of views on how much of a threat, if any, invasive alien species are to national economies, livelihoods and the environment; there is also a lack of clarity on what national and regional actions are needed. In addition, few countries have invasive alien species as a priority policy issue. Sustainable and cost-effective methods of managing many invasive alien species both exist in principle – for example, biological control through the introduction of host-specific natural enemies – and are available in practice, so technology per se is not a major limiting factor at this point in time.

As invasive alien species have now been reported to affect most developing countries, it is important to take stock of the current situation so as to be able to understand both the likely scale of the threats that they may pose to poor people who rely on agriculture and the environment, and the actions needed to address the threats should they prove real. This chapter examines the reality of the threat of invasive species, with a focus on invasive alien plants in the developing countries of tropical Asia to form a 'case study' of what type of information is available on invasive alien plants, the magnitude of the threats involved and whether or not these plants have any benefits. The extent to which country and other stakeholder responses and actions to the reported threats match this reality is then reviewed. This is followed by a discussion of what factors have led to the paucity of action that has been found and whether this is leading to a crisis of the impacts of invasive plants and other invasive alien species.

Tropical Asia is roughly defined for purposes of this chapter as the region stretching from Afghanistan eastward to Vietnam, and from Nepal southward to Indonesia. The focus is on invasive alien plants as these form a disproportionately large number of the world's worst invasive alien species in terms of numbers and reported impact: Of the 100 of the world's 'most invasive alien species' listed by the Invasive Species Specialist Group (ISSG) under the Species Survival Commission (SSC) of the World Conservation Union (IUCN), 36 species are plants (ISSG, 2000) and many of these are reported as a common problem in tropical Asia (e.g. Saxena, 1991; Waterhouse, 1994).

The Reality of Invasive Alien Plants: What Do We Know?

This section provides a review of the published literature on invasive alien plants in tropical Asia to help understand the real scale of the problem and also identify any situations where invasive plants provide benefits. However, let us first look at the history of the issue as this helps to set the context of the problem.

Where does the issue of invasive alien plants stem from and what size is the problem?

As in other parts of the world, alien floras in tropical Asia have developed as a result of trade and colonialism. Although herbaceous and woody plants have been introduced for centuries into new areas around the globe for economic, aesthetic and other purposes, the major movements of species began in the European mid-colonial times from the 1700s onward. It was then that plants began to be translocated in vast numbers to colonial regions, largely by European concerns such as the East India Companies, for agricultural and ornamental purposes (Heywood, 1989). Botanic gardens established in these regions played a central part in this by acting as 'ports' where new species could be propagated for wider distribution (Heywood, 1989; Hulme, 2011). The number of plants moved during this period was astronomical and resulted in many species becoming first naturalized and then notorious invaders (Cronk and Fuller, 1995).

Trade (e.g. by private seed companies) and other factors such as tourism and transport have developed since those times, particularly from the last part of the 20th century, and many authors identify this rise in human activities as a main driver of new species invasions (Burgiel *et al.*, 2006). Thus, intentional introductions of plant species continue, resulting in the wider dispersal of existing invasive species as well as introductions of other species that are now becoming invasive. It has also long been recognized that accidental plant introductions have been taking place as well (Myers and Bazely, 2003). Once again, the evidence in the published literature as a whole suggests that these are largely because of the current growing volume of trade (Jenkins, 1999; Levine and D'Antonio, 2003); evidence on other invasive species taxa suggests that accidental introductions of those are likewise on the increase (Elmer, 2001). Common pathways of accidental introduction include imports of agricultural products such as crop grains. Nevertheless, up to the early 2000s, intentional introductions were still the most common set of pathways for the spread of invasive alien plant species (Kowarik, 2003; Mack, 2003).

Information on the proportion of *alien* plant species in national floras is very patchy as national audits are, unfortunately, lacking in many developing countries. Further, where data are available, figures are very variable, but sometimes high (Myers and Bazely, 2003). Knowledge about the proportion of *invasive* plant species is equally sketchy, although Waterhouse (1993) estimated that 44% of significant weeds in South-east Asia are non-native. Where national data have been reported, they often lack the definition needed to characterize alien plant species at different stages of naturalization and invasion as suggested by Pyšek *et al.* (2004). However, Khuroo *et al.* (2012) did adopt a systematic approach in assessing the alien and invasive flora for the whole of India for the period 1890–2010. They found that 1599 species are alien and that this constitutes 8.5% of the Indian vascular flora (lower figures than previously reported). Only a small proportion of this alien flora is classed as invasive but, of these species, many are now widespread in the country.

Much research has been aimed at elucidating the species characteristics that lead to invasiveness. Evidence suggests that biological factors are partially responsible, including propagule pressure, both natural and human-mediated (Dehnen-Schmutz *et al.*, 2007), and polyploidy (e.g. through hybridization processes), as plants with this genetic make-up appear to be able to survive in a wider range of ecological conditions than the original diploid plants. In Singapore, for example, where many invasive plant species now occur over wide areas, polyploidy has been recorded in all of the major alien plant taxa studied (Pandit *et al.*, 2006). None the less, extraneous factors, largely anthropomorphically driven, are now exacerbating plant invasions. Dominant among these are land-use change, changes in land-management practices (e.g. increased use of fire in grassland ecosystems) and climate change (Chapin *et al.*, 2000; Masters and Norgrove, 2010).

Overview of the impacts of invasive alien plants

On a global basis, the reported impacts of invasive alien plants are diverse (Weber, 2003; Mooney, 2005) and many agricultural, semi-natural and natural terrestrial and aquatic ecosystems are now affected in some way by them. There are many invasive alien plant species reported from tropical Asia and a large proportion of these are present in more than one country (Waterhouse, 1994; Weber, 2003); the list is also growing steadily. The main problem species are all dicotyledons; in contrast, alien grasses do not seem to be as problematic in Asia as they are in the Americas. It has been hypothesized that this is because the long history of human presence in Asia and Africa may have affected the evolution of the native grasses (D'Antonio and Vitousek, 1992): Old World grasses have evolved under intense grazing pressure from large ungulates and thus have developed adaptions that include features such as perennating organs at ground level, rapid growth in response to defoliation and adaptation to fire, while grasses from the Neotropics have not evolved these features and so are inferior competitors in Asia and Africa.

This review of information about the impacts of alien plants in tropical Asia is not exhaustive, but the aim has been to include a representative set of papers with quantitative information from the available literature. There are many published papers that describe the presence, spread and status (on the last, these are mostly negative, though some are positive) of invasive plants in agricultural and natural ecosystems. There are also studies, albeit not many, on the extent of awareness of local communities about invasive plants (e.g. Rai *et al.*, 2012). Furthermore, experimental studies on impacts, such as actual crop losses in specific farming systems, impacts on rural livelihoods, costs of control and losses to native biota or ecosystem services are quite scarce and much information that is given tends to be anecdotal (Peh, 2010). None the less, some data are available, and an up-to-date review of the literature on all types of impact of invasive plants in the Indian subcontinent is given in Bhatt *et al.* (2012); the situation that these authors depict reflects that in other parts of the developing world.

Some species of invasive alien plants have been reported as particularly widespread and also problematic in tropical Asia through their various impacts on agriculture, natural ecosystems and human and animal health (e.g. Kohli *et al.*, 2006; Peh, 2010), and thus are commonly cited in this review. Examples include *Ageratum conyzoides*, *Chromolaena odorata*, *Mikania micrantha*, *Mimosa pigra* and *Parthenium hysterophorus* in the Asteraceae, and *Lantana camara* in the Verbenaceae. These particular species all originate from the Neotropical region and were either introduced as ornamentals (e.g. *L. camara*) or for agricultural purposes (e.g. *M. micrantha*) or introduced accidentally (e.g. *P. hysterophorus*). The predominance of South American species in an alien Asian flora has been confirmed by Khuroo *et al.* (2012) for India.

The negative impacts of invasive alien plant species have been reported in the literature in the following systems: agriculture and plantation forestry; native biota (e.g. resulting from competition); ecosystems (with impacts on nutrient cycles, water tables, etc.); rural communities and human and livestock health; and industrial processes. Recent work (see Blackburn *et al.*, 2014) has developed a system to categorize impacts on the basis of type; in this review, however, we adhere to the traditional sector approach, as this is also how national governments in the developing world commonly make assessments of invasive species. Apart from a few brief comments, positive impacts are not covered here as studies tend to be anecdotal.

Human and livestock health

In many countries, beside invasions in crops and natural habitats, invasive alien plants are commonly reported occurring in fallows and wastelands, along roadsides and in other disturbed habitats. Globally, the highest proportion of invasive tree and woody shrub species, for example, is found in wastelands and disturbed habitats (Haysom and Murphy, 2003). In such situations, both positive and negative impacts have been reported. In the former case, some species have been reported as useful for soil stabilization and/or local fuelwood supply in degraded habitats (Parthasarathy *et al.*, 2012), largely because the species concerned have a high rate of biomass production.

Nevertheless, rural communities and their livestock experience high exposure to many of the invasive alien plant species through daily livelihood activities. Given the abundance of these species and the lack of advice on how to control them, rural communities have little choice but to incorporate invasive alien plants into their livelihood schedules (Rai *et al.*, 2012). A number of these species are toxic to humans and animals. *L. camara* has been reported as being toxic to livestock and wild ungulates (Ambika *et al.*, 2003). *P. hysterophorus* is a major problem in the farmlands and wastelands of several countries in tropical Asia, and is still spreading. In India, where this plant has invaded over 14 million hectares of farmland alone, it is a major health concern to humans and livestock. In humans, symptoms include allergic eczematous contact dermatitis from prolonged contact, and allergic rhinitis (hay fever) and allergic bronchitis as a reaction to the pollen (Towers and Subba Rao, 1992; Kaur *et al.*, 2014). Livestock tend to avoid the plant, but will feed on it in pure stands, and the death of animals has been reported (Kohli *et al.*, 2006). Likewise, *M. micrantha* has a wide distribution in agricultural areas and wastelands in wetter regions (e.g. north-east India). In these regions, a significant proportion of rural communities allow goats to graze on the plant and they also use it as fodder for goats in winter months, although it is reported to affect animal health (Siwakoti, 2007).

Agricultural, forestry and fisheries systems and rural livelihoods

As part of a global assessment of the impacts of invasive species, the costs of these impacts by invasive alien plants in agroecosystems were estimated at a national level in India by Pimentel *et al.* (2002). Annual losses to crop production were estimated to be US$37.8 billion and losses to pastures at US$0.92 billion; this was 42.5% of the total loss (US$91.02 billion) to crops, pastures and forests from all invasive species, i.e. not only plants.

In annual field crops in the Indian subcontinent, *L. camara* is commonly reported invading wasteland and around agricultural fields (Kohli *et al.*, 2006, 2009; Peh, 2010); it also invades pastures and reduces their productivity (Sharma *et al.*, 2005; Love *et al.*, 2009). *A. conyzoides* is found in arable land, competing with maize, wheat and rice (Batish *et al.*, 2009; Kohli *et al.*, 2009), and it is also found in pastures. This species is particularly widespread in the Himalayan hill ranges and the Western Ghats of India (Kaur *et al.*, 2012a). In Nepal, *A. conyzoides* causes reductions in rice production, with grain yield reduced by 25–47% and straw yield 13–18% (Manandhar *et al.*, 2007). *P. hysterophorus* is reported as a common aggressive invader of wasteland, overgrazed pastures and the borders of agricultural land (Kohli *et al.*, 2009), and in some older studies on crops it was shown to cause yield losses of up to 40% (Khosla and Sobti, 1981) and to reduce forage production by 90% (Nath, 1981). Likewise, in experimental fields of sorghum, the species has been estimated to cause losses in grain weight of up to 30% (Channappagouder *et al.*, 1990).

L. camara and *C. odorata* have frequently been reported from across tropical Asia as affecting tree crops and forest plantation trees; *L. camara* affects coffee, coconut, oil palm, rubber, banana and sugarcane (Sharma and Raghubanshi, 2012), while *C. odorata* seems to be especially troublesome in plantations of coconut, rubber, coffee, cashew, bamboo, teak, *Dalbergia* and *Eucalyptus* (Kohli *et al.*, 2009). In West Timor (Indonesia), *C. odorata* severely affects pastures and thus feed for the cattle that are important to the semi-subsistence farmers (McWilliam, 2000).

M. micrantha is an invasive plant of high rainfall regions in tropical Asia, where it is reported as a major problem in agricultural systems. In India, Abraham *et al.* (2002) showed through experimental studies that *M. micrantha* significantly decreases the growth and dry-matter production of pineapple, banana, cocoa, coconut and rubber. In addition, Kaur *et al.* (2012b) studied the effects of *M. micrantha* on plant richness and rice seedling growth in the Western Ghats. Average plant species richness in sites with *M. micrantha* was significantly lower (by 30%) than in sites without *M. micrantha*. In non-sterile soils treated with *M. micrantha* leaf leachate, these authors also demonstrated that rice seeding germination and growth was retarded by increasing dose levels of the leachate. The plant has been also reported as a problem in home gardens in Kerala, India (Murphy, 2001) and orchards in China (Zhang *et al.*, 2004) where it smothers crops and retards growth. Shen *et al.* (2013) report that in south-western China, dense *M. micrantha* infestations cause 60% reductions in crop yields and make cultivation and harvesting difficult. In a socio-economic study of agroforestry farming systems in Kerala, Muraleedharan and Anitha (2000) found that farmers ranging from marginal (cultivating less than 1.5 ha) to large (with more than 5 ha) landholders all ranked weeding as the major constraint to cultivation; in this region, *M. micrantha* is a predominant weed. The cost of weeding varied between 24 and 35% of the total cost of cultivation, with costs higher in marginal and smallholder farms. On average, farms of all sizes infested with *M. micrantha* were earning Rs. 4000 (= approx. US$75)/ha p.a. less than equivalent farms where the species was absent. Clearly, this is a serious impact on marginal and smallholder farmers who are already poor. In earlier studies in Malaysia, *M. micrantha* was described as a problem in tree crop plantations such as rubber, oil palm and cocoa (Teoh *et al.*, 1985), where yield reductions of 20–27.5% were reported. Traditional manual control of weeds in these crops proved to be unsustainable and so herbicides were used to reduce the *M. micrantha* infestations (Anwar and Sivapragasam, 2001).

Invasive alien plants also cause problems in tribal communities. In the Western Ghats, harvesting of non-wood forest products has become difficult for tribal communities because of the invasion by *M. micrantha* into the forests. The collection of reed bundles has been reduced by almost half in areas with thick *M. micrantha* infestations (Kerala Forest Research Institute – KFRI, unpublished data). Similarly, in the buffer zone of Chitwan National Park in Nepal, *M. micrantha* and other invasive plant species have invaded the community forests and grasslands and are having a major negative effect on community livelihood activities in buffer zones so that, for example, communities are driven to use more resources from the national park (Rai and Scarborough, 2015). Farming households have tried to adapt to the invaded landscape and accommodate the plants into their livelihoods, but they provide few uses (Rai *et al.*, 2012).

Some classic earlier studies showed how invasive plants contribute significantly to the declining productivity of shifting agricultural (slash-and-burn or 'jhum') regimes in the high-rainfall hilly ranges of north-eastern India (Saxena and Ramakrishnan, 1984; Swamy and Ramakrishnan, 1987) (see Chapter 7, this volume). In this region, traditional shifting agricultural cycles of greater than 25 years have become shortened because of human population pressure. This, in turn, has allowed invasive alien plant species – predominantly *M. micrantha* and *C. odorata* – to dominate the flora in shortened regimes because of their superior ability to survive fire (which is used to clear the land for cultivation) and to rapidly utilize soil nutrients, which are abundant after fire (Ramakrishnan, 2001). Even though the invasive plants do play a role in reducing losses of nutrients from runoff, under these conditions, farmers of the region were no longer able to operate sustainably. However, these studies also showed that natural forest succession occurred when farmed plots were left uncultivated as in the traditional longer term cycles of >25 years; in this scenario, the invasive plants are outcompeted by native species because light and nutrients become limiting under the developing forest canopy.

The tropical American shrub *Mimosa pigra* has invaded wetter areas in South-east Asia. The species affects both rice and oil palm cultivation in Thailand (Peh, 2010), and impacts have also been reported in oil palm and fruit orchards in Malaysia (Anwar and Sivapragasam, 2001). In aquatic systems, water hyacinth (*Eichhornia crassipes*) is one of the most widespread invasive plant species in tropical Asia and has been reported choking lakes and rivers and thereby disrupting boat transport and local fishing (MacKinnon, 2002).

Conversely, in agroforestry systems in drier areas of the region, there have been a number of reports of positive impacts by trees and woody shrubs that have become invasive. Most species introduced for agroforestry have been selected for a number of traits, including their suitability for fodder and fuelwood. In many areas of the developing world, the very high demand for these and other forest products has meant that local communities have exhausted natural supplies. In some cases, the aggressive, spreading nature of introduced trees has resulted in reforestation, albeit with an introduced species. This, in turn, has provided rural communities with a 'free' and continuous supply of forest products. For example, in India, the South American tree *Prosopis juliflora* provides an important fuelwood supply for some local communities in arid and semi-arid areas (Sharma, 1981; Pandey *et al.*, 2012), but it is also highly invasive in many areas, where it grows on waste ground, along roadsides and in pastures, and also threatens the environment (Pandey *et al.*, 2012).

In forest plantations, *M. micrantha* was reported as a problem in the 1980s although this only became apparent when regular weeding was stopped because of labour shortages (Palit, 1981). In studies on the impact of the plant on teak plantation production, Muraleedharan and Anitha (2000) showed that the net profit per hectare over an 8-year production period in an infested plantation was reduced by Rs. 6274 (approx. US$118) compared with an *M. micrantha*-free plantation.

One significant feature common to several of the major invasive plant species in tropical Asia is their allelopathic properties, whereby combinations of root and shoot leachates and/or root exudates can reduce the growth of crop or other wild species. *P. hysterophorus*, for example, is a well-known allelopathic plant (Kohli *et al.*, 2009). It releases inhibitors into the soil through leaching from the leaves, shoots and roots, and also through the further decomposition of the leachates in the soil. Likewise, the allelopathic effects of *L. camara* on a wide range of plant species have been well documented (Ambika *et al.*, 2003).

Natural ecosystems and biota

There are several studies in the literature that report the negative impacts of invasive alien plants on natural ecosystems and their constituent biota. The most common impacts include ecosystem-level processes; for example, changes in fire regimes, water tables and nutrient cycles, and the displacement of native species. Very little information exists for tropical Asia on the impacts of hybridization on the genetic diversity of native plant species.

In northern India, *L. camara* has replaced or invaded large areas of natural stands of oak (*Quercus leucotrichophora*) and pine (*Pinus roxburghii*) (Bhatt *et al.*, 1994). In a comparison of ecosystem functioning between *L. camara*-dominated shrubland and oak and pine woodlands, these authors showed that the total net primary production (measured in tonnes/hectare p.a) of *L. camara* ecosystems was similar to that of both the oak and pine woodlands, but that the soil nitrogen and phosphorus levels in the *L. camara* ecosystem were significantly lower than in either type of woodland. Rawat and Singh (1988) showed that *L. camara* thickets have a high litter turnover rate and nutrient cycling, and they suggest that *L. camara*, which is fast growing, can utilize nutrients more efficiently than slow-growing native species; this, in turn, contributes to the further spread of *L. camara* (Sharma and Raghubanshi, 2012).

The effects of invasive plant can be directly or indirectly exacerbated by fire, which is common in grass or grass–forest ecosystems. For example, Murphy *et al.* (2013) suggest that a major factor driving the expansion of *M. micrantha* in the Chitwan area of southern Nepal is the increase in frequency and area of burning of the grasslands every year, because *M. micrantha* is a fire-adapted species. Likewise, Witkowski and Wilson (2001) point to the increased risk of fire in *C. odorata*-invaded areas with a pronounced dry season; this is because the plants become highly flammable owing to the high oil content that develops during dieback after flowering.

Invasive plants can even disrupt tree pollination. *C. odorata* is a major invasive species in the tropical dry forests of Thailand in areas where *Shorea siamensis* is logged on a regular basis. In these areas, the butterfly pollinators of the tree *Dipterocarpus obtusifolius* change their foraging location to *C. odorata* (Ghazoul, 2004).

Studies on the displacement of native species tend to be mostly descriptive, although a few quantitative assessments have been made. Evidence to date suggests that disturbed forests with an open canopy are more prone to invasion than forest fragments with a closed canopy (Teo *et al.*, 2003). Gaps in the forests of the Western Ghats created by selective logging operations are dominated by *C. odorata*, which is absent in undisturbed forests (Chandrashekara and Ramakrishnan, 1994). In the tropical forests of Peninsular Malaysia, the invasive shrub *Clidemia hirta* has been inferred to be a major agent driving changes in the regeneration of forest tree species in tree fall gaps by supressing the normal growth of new native canopy trees in these gaps (Peters, 2001). Dogra *et al.* (2009) have reported that at lower elevations of the Shivalik Hills of the Himalayas (in Himachal Pradesh, India), *A. conyzoides* and other major invasive plant species occupy approximately 20% of the natural ecosystem areas, but that *A. conyzoides* accounts for approximately 30% of the alien plant species.

Other invasive plant species have invaded riparian systems. *M. pigra* has invaded the Mekong River delta in Southeast Asia, where it has been reported to replace native vegetation (Peh, 2010) and form large monocultural stands. The shrub has a prolific seed capacity and this has been suggested as one of the reasons for its rapid spread in the region. It has been shown to cause local reductions in the abundance of native plants and tree seedlings and, thus, bird species diversity in Australia (Braithwaite *et al.*, 1989); these effects are also likely to be happening in the Mekong Delta.

More specific studies include work in the Chitwan National Park, situated in the grass–forest region of southern Nepal, where it has been shown that *M. micrantha* is outcompeting native grasses and other plant species (Sapkota, 2007). Here, the mortality of some dominant grasses (*Saccharum spontaneum* and *Imperata cylindrica*) was high in highly invaded grass–riverine habitat, with little or no regeneration of new plants; likewise, in subtropical hardwood forests, saplings of *Bombax ceiba*, *Dalbergia sissoo* and *Acacia catechu* had died and no regeneration was observed. *S. spontaneum*, *I. cylindrica* and other grasses and shrubs are important fodder and browse species for herbivorous mammals of conservation importance – in particular the one-horned rhinoceros (*Rhinoceros unicornis*) (Laurie, 1982; Hazarika and Saikia, 2012) and deer species. Indeed, recent in-depth studies have confirmed that in *M. micrantha*-infested grass–riverine forest habitats, a number of demographic parameters (e.g. home-range size) of the one-horned rhinoceros are negatively affected and this has been shown to be a direct consequence of the presence of *M. micrantha* (Naresh Subedi, National Trust for Nature Conservation, Nepal, unpublished data). By implication, it is likely that deer populations will also be negatively affected, and this will have serious effects on Bengal tiger (*Panthera tigris tigris*) numbers, as it has been established that these are directly proportional to the abundance of the tiger prey (Karanth and Nichols, 2002).

I. cylindrica is important for thatching in some communities in South-east Asia (FAO, 2016). However, the grass is also reported as being seriously invasive in this and other regions, where it affects crops and the environment (CABI, 2015).

Studies in the tropical dry deciduous forests of the Vindhyan hill ranges of west-central India that are invaded by *L. camara* suggest that the presence of the shrub reduces the recruitment of new native tree seedlings, with up to 60% reduction in species number recorded in these forests (Sharma and Raghubanshi, 2007). In the Western Ghats region of southern India, studies by Ramaswami and Sukumar (2011) on *L. camara* invasion in the seasonally dry and dry–moist forests of Mudumalai have shown that the frequency of occurrence and abundance of several key mammal-dispersed and mechanically dispersed native tree seedlings were significantly less under *L. camara* thickets compared with forest free of the shrub, whereas bird-dispersed species were not affected. The authors infer that lantana thickets create barriers to tree species dispersed by mammals or mechanical means and that, in time, the community composition of these forests will change. Work by Prasad (2010) in the dry deciduous forests of Bandipur Tiger Reserve in Karnataka in India has indicated that *L. camara* contributes significantly to tree species death on the borders of wide clearings in the forests. Like the previous authors, Prasad concludes that *L. camara* invasion is slowly changing the composition of plant communities and that this may be negatively affecting the wild ungulate populations.

In South-east Asia, Osunkoya *et al.* (2005) report that the Australian agroforestry trees *Acacia auriculiformis* and *A. cincinnata* have extensively replaced native shrubs (e.g. *Alphitonia* spp., *Commersonia* spp.) in heath forests in Brunei because the introduced trees are nitrogen-fixing species and the soils of the heaths are nutrient poor.

Inferences from the existing data

Taken as a whole, the inescapable conclusion from the studies reviewed in this section is that invasive alien plants have a dramatic negative impact on crops, pastures, human and livestock health, and ecosystem function and biodiversity in tropical Asia. The positive impacts are few; the main one reported is that some woody invasive plants now form an important supply of fuelwood for local communities, but several of those communities are now deeply concerned about the invasive nature of these species because they have invaded croplands. Although there is a paucity of quantitative information on the total losses to crop yields in specific farming systems, it seems likely that this will be a significant proportion of the total production in any one year. The findings of studies on the impacts on ecosystem function and biodiversity are likewise alarming. However, even though there is clear evidence that invasive plants can cause the local extinction of native plant species, there is no evidence from the literature that any specific native species in tropical Asia is under threat of extinction purely through the action of invasive plants. Peh (2010) reaches the same conclusion in a review of the impacts of invasive species in South-east Asia.

Organizations and Frameworks that Address Invasive Plants at the Regional and National Levels in Tropical Asia

Most of the world's most serious invasive plant species are still spreading unchecked and this is also true in tropical Asia. This region faces new threats as well, with trade and transport opening up new areas to the import and export of agricultural products. Regional and national responses and actions to biological invasions have been varied, but no country in the region has a comprehensive response and management framework in place. Historically, the main frameworks and organizations were developed to protect agricultural interests, and this is still the main concern of these bodies; moreover, they cover invasive species of agricultural importance – with the emphasis tending to be on invasive alien pathogens, insects and mites, rather than plants, and the threats to other sectors still largely neglected.

There are several regional organizations in tropical Asia that partly deal with threats from invasive alien species, but not all cover invasive plants. The two principal organizations cover plant and animal health respectively: the Asia and Pacific Plant Protection Commission (APPPC) and the World Organisation for Animal Health (OIE).

The APPPC is the regional plant protection organization (RPPO), and it takes guidance from the International Plant Protection Convention (IPPC). The IPPC is the international standard-setting body for International Standards for Phytosanitary Measures (ISPMs), and the APPPC, like other RPPOs, uses these ISPMs to set phytosanitary standards for the Asia–Pacific region. The APPPC utilizes the IPPC ISPMs and has developed a number of standards for specific invasive species threats in the region; for example *Guidelines for Protection Against South American Leaf Blight of Rubber* (APPPC, 2009). Currently, there are no specific regional standards that cover procedures for addressing the risks from invasive plants, but the IPPC ISPMs do include relevant guidance, e.g. *Pest Risk Analysis for Quarantine Pests* (IPPC, 2013). In the early 2000s, the IPPC and the CBD recognized the relevance of the IPPC to the CBD's Article 8(h), which calls on all parties to the CDB to 'prevent the introduction of, control or eradicate those alien species which threaten ecosystems, habitats or species', and since 2005 both conventions have been collaborating within the framework of a Memorandum of Cooperation. In 2014, the IPPC joined the Liaison Group of Biodiversity-related Conventions to enhance regional and national capacities to protect agricultural and wild plants, including threats from invasive alien species. It remains to be seen how far these international initiatives will be taken advantage of by RPPOs.

Livestock are central to the livelihoods of most rural communities in tropical Asia. In a similar vein to agricultural plant protection, the OIE sets international standards for animal welfare and has representation in the Asia–Pacific region. The OIE forms collaborations with appropriate bodies in the region. For example, in 2008, the OIE and the Association of Southeast Asian Nations (ASEAN) entered into a Memorandum of Understanding on technical cooperation for the protection of livestock against major animal diseases and zoonoses. The OIE is concerned with alien animal diseases and threats from invasive alien animals, although it does not cover threats to livestock or wild animals from invasive alien plants.

On the environmental side, the South Asia Co-operative Environment Programme (SACEP), established in 1982, does cover some aspects of the invasive alien species issue; this is mainly through the inclusion of invasive alien species threats in awareness raising and advocacy for environmental legislation.

The only truly regional framework to be set up in response to the increasing tide and threat of invasive alien species is the Asia-Pacific Forest Invasive Species Network (APFISN), which was established in 2004 by the Asia-Pacific Forestry Commission (APFC). The APFISN is a cooperative alliance of 33 member countries and its main aim is to help member states detect, prevent, monitor and eradicate or control invasive alien species in forest systems. The APFISN stocktakes national activities to identify gaps and needs in capacity building, and also raises awareness; it has developed an information-sharing forum as well.

Current national situations in some countries are reviewed in detail by other authors in this book, so only some general points are made here. At the national level, accountability for plant protection and livestock welfare has historically lain within ministries of agriculture. This responsibility includes prevention, under the remit of quarantine departments, and early detection and control, which are generally within the remits of agricultural research and extension departments. In India, the responsibility for pest management lies jointly with central and state governments, and involves several organizations; this makes actions against invasive species difficult to coordinate (Mandal, 2011). Signatory countries under the CBD have produced National Biodiversity Strategy and Action Plans (NBSAPs) and these have to include plans to address invasive alien species. These plans have, in most cases, been devised by ministries of environment, however, and in most countries in the region they are not well linked into the existing frameworks under the ministries of agriculture.

In some countries, actions have been and continue to be taken against some major invasive alien plants and other pests of agricultural importance, but the efforts remain focused on species already familiar to agricultural sectors. Little effort has been made to address plants and other species that now threaten other sectors, such as forests, and that have become, or are becoming, a threat to agriculture. The exception is India, which does have a strong record in the use of classical biological control against invasive plant species.

Why Have National Responses to Biological Invasions Been Patchy?

It is clear from the existing published data on impacts that, while many gaps remain in our understanding of the nature of their impacts in farming and natural environments, taken as a whole, invasive alien plants do represent a major and growing threat across the agricultural, trade and environmental sectors in tropical Asia. So, why is there a paucity of investment and action in the region? This question has been raised in the past by other authors (e.g. MacKinnon, 2002).

Part of the problem lies in the nature of the invasive species issue. Overall, the issue is complex for various reasons, as the case study presented in this chapter on invasive alien plants in tropical Asia has illustrated:

- Some invasive species have developed from introductions in colonial times and have been spreading slowly in natural areas throughout countries and regions for well over a 100 years in some cases, so they remain 'unnoticed' because they have become part of the background flora and fauna. Hence, their long-term negative impacts are not recognized.
- A rapid expansion in the range of these species is taking place because of dramatic land-use changes and climate change, and many species are now affecting agricultural areas and forests.
- The development of trade has brought an increasing number of new invasive species, and not all trade pathways are monitored for plants or pests of quarantine significance.

Thus, the invasive alien species issue is multidimensional and difficult to address unless a national and regional approach is taken. As we have seen, invasive species have been an issue in agriculture for centuries, even though they were not always termed 'invasive species'. However, concerns about invasive alien species and their impacts did not become manifest in environmental sectors until about the 1980s, first in North America and then, increasingly, in the developing world (e.g. see Mooney *et al.*, 2005). Out of this realization developed much scientific review of the current ecological knowledge of invasive species in particular regions or countries, the process of invasion and the human dimensions of the problem, and the collation of information about experiences with management. Interestingly, some of the knowledge that emerged was new to the agricultural sector, for example the variety of pathways for the movement of invasive species.

None the less, three core and interrelated issues are at the heart of the problem:

- Within and between countries, there is a lack of national and regional awareness of the scale of the problems associated with invasive alien species, especially their economic and ecological impacts. This is largely a result of two factors: historically, developing countries have not had a sufficient skill base to collect these types of data; furthermore, and as already alluded to, the activities of government ministries, research bodies, public extension organization, non-governmental organizations (NGOs), farmer groups and the private sector are disconnected, resulting in weak linkages and poor communication about the problems. For example, several authors (see Griffin, 2003) have noted the weak links that exist between national plant and animal protection agencies, which focus on agricultural issues, and other ministries or agencies that represent other stakeholders also severely affected by invasive species. Overall, this has led to a lack of coordinated responses to implement effective management with appropriate policy development, and insufficient response to international calls and guidelines for action (Butchart *et al.*, 2010).
- Some capacity exists for managing invasive species within countries but, in general, this has lain in and remains within agricultural sectors (and so is not available to other sectors for the reasons discussed above), and focuses narrowly on traditional agricultural pests. As a consequence, it is inadequate for the wide range of invasive species that now exist, especially invasive alien plants.
- Although there is much knowledge about invasive species globally, some in web-based databases and tools, this knowledge is not available in forms usable by all the stakeholders that most need it. Also, most web-based tools lack detail on up-to-date best management practices for prevention, early detection and sustainable control.

In addition, and as already discussed, many of the reported impacts of invasive alien species in environmental sectors are qualitative, and this is particularly true of older reports. This has resulted in mixed perceptions about the true impact of invasive plants. For instance, there are situations where the impacts of invasions will 'fade way' over time, but these are few.

As a result of the above points, many involved in natural resource strategies have concluded that invasive species are a secondary problem, and the discussion about this issue still lies mainly within scientific circles.

Conclusions: Does a Lack of Response Matter, Is There a Crisis in the Making and Can Issues Be Resolved?

Invasive alien plants pose a major challenge for most of the countries in tropical Asia and across the rest of the developing world. They are a cross-cutting issue and thereby affect all major economic sectors of countries – agriculture, environment and trade. Most significantly, humans are both 'drivers' and victims of invasive plants.

A lack of comprehensive management responses to the issue has created and is perpetuating the unprecedented introduction of new invasive plant species into countries and the rapid expansion of existing invasive species. What is more, the problems continue to escalate in rural areas, with serious negative impacts on the livelihoods of the people. So it is clear that there is a crisis in the making. In some countries, communities try to control or utilize invasive species, including some invasive plants, but this rarely works and has yet to be demonstrated to be sustainable (Rai *et al.*, 2012). In some instances, invasive species drive communities to abandon land and even villages. To take a case from another region, in East Africa, invasive plants have caused conflict between communities (Mwangi and

Swallow, 2008). Moreover, the worst affected are always the poorest people who have little in the way of resources.

What can we learn from experiences to date? Crop losses and threats to livestock and natural resources will continue and are likely to escalate in the absence of national and regional actions. This will threaten food security directly. Invasive plant species are also threatening local biodiversity, and through this, rural livelihoods too. In badly invaded areas though, invasive plants contribute a large proportion of the total biomass and so they play a significant role in reducing losses of nutrient from runoff; they can also provide supplies of local firewood. Notwithstanding, these positive impacts are at the expense of the almost complete monopolization of the environment by the invasive alien plant and, consequently, a severe reduction in local biodiversity, which may not be able to recover. None the less, sustainable solutions do exist. Globally, much is known about many of the serious invasive species and their impacts on crops, livestock and natural resources. For example, a key 'action' for consideration at national and regional levels is to regulate *all* pathways of trade plant movement and, although this may be unwelcome to some trade sectors, it must be remembered that the outcome of unregulated trade in plant material can cause loss of livelihood and severe impacts on other sectors. There are existing control technologies – biological control and integrated pest management (IPM) – that are suitable for invasive plant species, are cost effective and environmentally compatible, and can achieve large-scale geographical impact (Murphy and Evans, 2009). When successful, such interventions against biological invasions are almost always overwhelmingly beneficial, with biological control, for instance, frequently giving benefit–cost figures of well into the hundreds, e.g. the control of water hyacinth in southern Benin resulted in a benefit–cost ratio of 124:1 (De Groote *et al.*, 2003).

Thus, for tropical Asia, as for other parts of the globe, the opportunities are there to address the ongoing crisis of invasions of alien plant species. However, a core urgent need is for coordinated regional and national actions that involve all principal stakeholders in new partnerships to catalyse actions.

Acknowledgements

I wish to thank many colleagues in CABI and in partner institutions in China, India, Indonesia, Nepal, the UK and the USA for helpful discussions over the last few years.

References

Abraham, M., Abraham, C.T. and George, M. (2002) Competition of *Mikania* with common crops of Kerala. *Indian Journal of Weed Science* 34, 96–99.

Ambika, S.R., Poornima, S., Palaniraj, R., Sati, S.C. and Narwal, S.S. (2003) Allelopathic plants 10. *Lantana camara* L. *Allelopathy Journal* 1, 147–161.

Anwar, A.I. and Sivapragasam, A. (2001) Impact and management of selected alien and invasive weeds and insect pests: a Malaysian perspective. In: Balakrishna, P. (ed.) *Report of Workshop on Alien Invasive Species, Global Biodiversity Forum – South and Southeast Asia Session, Colombo, Sri Lanka, 25–26 October 1999.* IUCN Regional Biodiversity Programme, Colombo, pp. 47–60.

APPPC (2009) *Guidelines for Protection Against South American Leaf Blight of Rubber.* Regional Standards for Phytosanitary Measures No. 7, APPC (Asia-Pacific Plant Protection Commission), FAO Regional Office for Asia and the Pacific, Bangkok, Thailand.

Batish, D.R., Kaur, S., Singh, H.P. and Kohli, R.K. (2009) Role of root mediated interactions in phytotoxic interference of *Ageratum conyzoides* with rice (*Oryza sativa*). *Flora* 204, 388–395.

Bebber, D.P., Holmes, T. and Gurr, S.J. (2014) The global spread of crop pests and pathogens. *Global Ecology and Biogeography* 23, 1398–1407.

Bhatt, J.R., Singh, J.S., Singh, S.P., Tripathi, R.S. and Kohli, R.K. (eds) (2012) *Invasive Alien Plants: An Ecological Appraisal for the Indian Subcontinent.* CAB International, Wallingford, UK.

Bhatt, Y.D., Rawat, Y.S. and Singh, S.P. (1994) Changes in ecosystem functioning after replacement of forest by lantana shrubland in

Kumaun Himalaya. *Journal of Vegetation Science* 5, 67–70.
Blackburn, T.M., Essl, F., Evans, T., Hulme, P.E., Jeschke, J.M., Kühn, I., Kumscschick, S., Markova, Z., Mrugala, A., Nentwig, W. *et al.* (2014) A unified classification of alien species based on the magnitude of their environmental impacts. *PLoS Biology* 12(5): e1001850. Available at: http://dx.doi.org/10.1371/journal.pbio.1001850.
Braithwaite, R.W., Lonsdale, W.M. and Estberg, J.A. (1989) Alien vegetation and native biota in tropical Australia: the impact of *Mimosa pigra*. *Biological Conservation* 48, 189–210.
Burgiel, S., Foote, G., Orellana, M. and Perrault, A. (2006) *Invasive Alien Species and Trade: Integrating Prevention Measures and International Trade Rules.* Centre for International Environmental Law and Defenders of Wildlife, Washington, DC. Available at: http://cleantrade.typepad.com/clean_trade/files/invasives_trade_paper_0106.pdf (accessed 18 May 2015).
Butchart, S.H.M., Walpole, M., Collen, B., van Strien, A., Scharlemann, J.P.W., Almond, R.E.A., Baillie, J.E.M., Bomhard, B., Brown, C., Bruno, J. *et al.* (2010) Global biodiversity: indicators of recent declines. *Science* 328, 1164–1168.
CABI (2015) *Imperata cylindrica* (cogon grass). Invasive Species Compendium Datasheet, CAB International, Wallingford, UK. Available at: www.cabi.org/isc/datasheet/28580 (accessed 10 June 2016).
Carney, D. (ed.) (1998) *Sustainable Rural Livelihoods: What Contribution Can We Make*? Department for International Development, London.
Chandrashekara, U.M. and Ramakrishnan, P.S. (1994) Successional patterns and gap phase dynamics of a humid tropical forest of the Western Ghats of Kerala, India: ground vegetation, biomass, productivity and nutrient cycling. *Forest Ecology and Management* 70, 23–40.
Channappagouder, B.B., Panchal, Y.C., Manjunath, S. and Koti, R.V. (1990) Studies on the influence of parthenium on sorghum growth under irrigated conditions. *Farming Systems* 6, 102–104.
Chapin, F.S. III, Zavaleta, E.S., Eviner, V.T., Naylor, R.L., Vitousek, P.M., Reynolds, H.L., Hooper, D.U., Lavorel, S., Sala, O.E., Hobbie, S.E. *et al.* (2000) Consequences of changing biodiversity. *Nature* 405, 234–242.
Cronk, Q. and Fuller, J. (1995) *Plant Invaders.* Chapman & Hall, London.
D'Antonio, C.M. and Vitousek, P.M. (1992) Biological invasions by exotic grasses, the grass/fire cycle, and global change. *Annual Review of Ecology and Systematics* 3, 63–87.
De Groote, H., Ajuonu, O., Attignon, S., Djessou, R. and Neuenschwander, P. (2003) Economic impact of biological control of water hyacinth in southern Benin. *Ecological Economics* 45, 105–117.
Dehnen-Schmutz, K., Touza, J., Perrings, C. and Williamson, M. (2007) A century of the ornamental plant trade and its impact on invasion success. *Diversity and Distributions* 13, 527–534.
Dogra, K.S., Kohli, R.K. and Sood, S.K. (2009) Impact of *Ageratum conyzoides* L. on the diversity and composition of vegetation in the Shivalik hills of Himachal Pradesh (northwestern Himalaya), India. *International Journal of Biodiversity and Conservation* 1, 135–145.
Elmer, W.H. (2001) Seeds as vehicles for pathogen importation. *Biological Invasions* 3, 263–271.
FAO (2016) *Imperata cylindrica* (L.) Beauv. Grassland Species Profiles, Food and Agriculture Organization of the United Nations, Rome. Available at: www.fao.org/ag/agp/AGPC/doc/gbase/data/pf000261.htm (accessed 10 June 2016).
Ghazoul, J. (2004) Alien abduction: disruption of native plant–pollinator interactions by invasive species. *Biotropica* 36, 156–164.
Griffin, B. (2003) The role of the International Plant Protection Convention in the prevention and management of invasive alien species. In: Pallewatta, N., Reaser, J.K. and Gutierrez, A.T. (eds) *Prevention and Management of Invasive Alien Species: Proceedings of a Workshop on Forging Cooperation throughout South and Southeast Asia, Bangkok, Thailand, 14–16 August 2002.* Global Invasive Species Programme, Cape Town, pp. 65–67.
Haysom, K. and Murphy, S.T. (2003) *The Status of Invasiveness of Forest Tree Species outside their Natural Habitat: A Global Review and Discussion Paper.* FAO Forest Health and Biosecurity Working Paper FBS/3E. Forestry Department, Food and Agriculture Organization of the United Nations, Rome.
Hazarika, B.C. and Saikia, P.K. (2012) Food habit and feeding patterns of great Indian one-horned rhinoceros (*Rhinoceros unicornis*) in Rajiv Gandhi Orang National Park, Assam, India. *ISRN (International Scholarly Research Notices) Zoology* 2012, Article ID 259695. Available at: http://dx.doi.org/10.5402/2012/259695.
Heywood, V.H. (1989) Patterns, extents and modes of invasions by terrestrial plants. In: Drake, J.A., Mooney, H.A., di Castri, F., Groves, R.H., Kruger, F.J., Rejmánek, M. and Williamson, M. (eds) *Biological Invasions: A Global Perspective. SCOPE 37.* Wiley, New York, pp. 31–60.

Hulme, P.E. (2011) Addressing the threat to biodiversity from botanic gardens. *Trends in Ecology and Evolution* 26, 168–174.

IPPC (2013) *ISPM 11. Pest Risk Analysis for Quarantine Pests.* International Standards for Phytosanitary Measures (ISPM), Secretariat of the International Plant Protection Convention, Food and Agriculture Organization of the United Nations (FAO), Rome. Available at: https://www.ippc.int/en/core-activities/standards-setting/ispms/ (accessed 2 March 2017).

ISSG (2000) *100 of the World's Worst Invasive Species: A Selection from the Global Invasive Species Database.* Invasive Species Specialist Group, University of Auckland, Auckland, New Zealand.

Jenkins, P.T. (1999) Trade and exotic species introductions. In: Sandland, O.T., Schei, P.J. and Viken, Å. (1999) *Invasive Species and Biodiversity Management.* Kluwer (of Springer), Dordrecht, Netherlands, pp. 229–235.

Karanth, K.U. and Nichols, J.D. (eds) (2002) *Monitoring Tigers and Their Prey: A Manual for Researchers, Managers and Conservationists in Tropical Asia.* Centre for Wildlife Studies, Bangaluru, India.

Kaur, M., Aggarwal, N.J., Kumar, V. and Dhiman, R. (2014) Effects and management of *Parthenium hysterophorus*: a weed of global significance. *International Scholarly Research Notices* 2014, Article ID 368647. Available at: http://dx.doi.org/10.1155/2014/368647.

Kaur, R., Malhotra, S. and Inderjit (2012b) Effects of invasion of *Mikania micrantha* on germination of rice seedlings, plant richness, chemical properties and respiration of soil. *Biology and Fertility of Soils* 48, 481–488.

Kaur, S., Batish, D.R., Kohli, R.K. and Singh, H.P. (2012a) *Ageratum conyzoides*: an alien invasive weed in India. In: Bhatt, J.R., Singh, J.S., Singh, S.P., Tripathi, R.S. and Kohli, R.K. (eds) *Invasive Alien Plants: An Ecological Appraisal for the Indian Subcontinent.* CAB International, Wallingford, UK, pp. 57–76.

Khosla, S.N. and Sobti, S.N. (1981) Effective control of *Parthenium hysterophorus* L. *Pesticides* 15, 18–19.

Khuroo, A.A., Reshi, Z.A., Malik, A.H., Weber, E., Rashid, I. and Dar, G.H. (2012) Alien flora of India: taxonomic composition, invasion status and biogeographic affiliations. *Biological Invasions* 14, 99–113.

Kohli, R.K., Batish, D.R., Singh, H.P. and Dogra, K.S. (2006) Status, invasiveness and environmental threats of three tropical American invasive weeds (*Parthenium hysterophorus* L., *Ageratum conyzoides* L., *Lantana camara* L.) in India. *Biological Invasions* 8, 1501–1510.

Kohli, R.K., Singh, H.P., Batish, D.R. and Dogra, K.S. (2009) Ecological status of some invasive plants of Shiwalik Himalayas in northwestern India. In: Kohli, R.K., Jose, S., Singh, H.P. and Batish, D.R. (eds) *Invasive Plants and Forest Ecosystems.* CRC Press, Boca Raton, Florida, pp. 143–155.

Kowarik, I. (2003) Human agency in biological invasions: secondary releases foster naturalisation and population expansion of alien plant species. *Biological Invasions* 5, 293–312.

Laurie, A. (1982) Behavioural ecology of the greater one-horned rhinoceros (*Rhinoceros unicornis*). *Journal of Zoology* 196, 307–341.

Levine, J.M. and D'Antonio, C.A. (2003) Forecasting biological invasions with increasing international trade. *Conservation Biology* 17, 322–326.

Love, A., Babu, S. and Babu, C. (2009) Management of lantana, an invasive alien weed, in forest ecosystems of India. *Current Science* 97, 1421–1429.

Mack, R.N. (2003) Global plant dispersal, naturalization, and invasion: pathways, modes, and circumstances. In: Ruiz, G.M. and Carlton, J.T. (eds) *Invasive Species: Vectors and Management Strategies.* Island Press, Washington, DC, pp. 3–30.

MacKinnon, J.R. (2002) Invasive alien species in Southeast Asia. *ASEAN Review of Biodiversity and Environmental Conservation* 2, 9–11.

Manandhar, S., Shrestha, B.B. and Lekhak, H.D. (2007) Weeds of a paddy field at Kirkipur, Kathmandu. *Scientific World* 5, 100–106.

Mandal, F.B. (2011) The management of alien species in India. *International Journal of Biodiversity and Conservation* 3, 467–473.

Masters, G. and Norgrove, L. (2010) *Climate Change and Invasive Alien Species.* CABI Working Paper 1, CAB International, Wallingford, UK. Available at: www.cabi.org/Uploads/CABI/expertise/invasive-alien-species-working-paper.pdf (accessed 18 May 2015).

McWilliam, A. (2000) A plague on your house? Some impacts of *Chromolaena odorata* on Timorese livelihoods. *Human Ecology* 28, 431–469.

Mooney, H.A. (2005) Invasive alien species: the nature of the problem. In: Mooney, H.A., Mack, R.N., McNeely, J.A., Neville, L.E., Schei, P.J. and Waage, J.K. (eds) *Invasive Alien Species: A New Synthesis. SCOPE 63.* Island Press, Washington, DC, pp. 1–15.

Mooney, H.A., Mack, R.N., McNeely, J.A., Neville, L.E., Schei, P. J. and Waage, J.K. (eds) (2005) *Invasive Alien Species: A New Synthesis. SCOPE 63.* Island Press, Washington, DC.

Muraleedharan, P.K. and Anitha, V. (2000) The economic impact of *Mikania micrantha* on teak plantations in Kerala. *Indian Journal of Forestry* 23, 248–251.

Murphy, S.T. (2001) Alien weeds in moist forest zones of India: population characteristics, ecology and implications for impact and management. In: Sankaran, K.V., Murphy, S.T. and Evans, H.C. (eds) *Alien Weeds in Moist Tropical Zones: Banes and Benefits. Proceedings of a Workshop, Kerala Forest Research Institute, Peechi, India, 2–4 November 1999.* Kerala Forest Research Institute, Peechi, India and CABI Bioscience, UK Centre (Ascot), Ascot, UK, pp. 20–27.

Murphy, S.T. and Evans, H. (2009) Biological control. In: Clout, M. and Williams, P.A (eds) *Invasive Species Management: A Handbook of Principles and Techniques.* Oxford University Press, Oxford, UK, pp. 77–92.

Murphy, S.T., Subedi, N., Jnawali, S.R., Lamichhane, B.R., Upadhyay, G.P., Kock, R. and Amin, R. (2013) Invasive mikania in Chitwan National Park, Nepal: the threat to the greater one-horned rhinoceros *Rhinoceros unicornis* and factors driving the invasion. *Oryx* 47, 361–368.

Mwangi, E. and Swallow, B. (2008) *Prosopis juliflora* invasion and rural livelihoods in the Lake Baringo area of Kenya. *Conservation and Society* 6, 130–140.

Myers, J.H. and Bazely, D.R. (2003) *Ecology and Control of Introduced Plants.* Cambridge University Press, Cambridge, UK.

Nath, R. (1981) Note on the effect of parthenium extract on seed germination and seedling growth in crops. *Indian Journal of Agricultural Sciences* 51, 601–603.

Osunkoya, O.O., Othman, F.E. and Kahar, R.S. (2005) Growth and competition between seedlings of an invasive plantation tree, *Acacia beccarianum. Ecological Research* 20, 205–214.

Palit, S. (1981) *Mikania* – a growing menace in plantation forestry in West Bengal. *Indian Forester* 107, 96–101.

Pandey, C.N., Pandey, R. and Bhatt, J.R. (2012) Woody, alien and invasive *Prosopis juliflora* (Swartz) DC: management dilemmas and regulatory issues in Gujarat. In: Bhatt, J.R., Singh, J.S., Singh, S.P., Tripathi, R.S. and Kohli, R.K. (eds) *Invasive Alien Plants: An Ecological Appraisal for the Indian Subcontinent.* CAB International, Wallingford, UK, pp. 299–303.

Pandit, M.K., Tan, H.T.W. and Bisht, M.S. (2006) Polyploidy in invasive plant species of Singapore. *Botanical Journal of the Linnean Society* 15, 395–403.

Parthasarathy, N., Pragasan, L.A. and Muthumperumel, C. (2012) Invasive alien plants in tropical forests of the south-eastern Ghats, India: ecology and management. In: Bhatt, J.R., Singh, J.S., Singh, S.P., Tripathi, R.S. and Kohli, R.K. (eds) *Invasive Alien Plants: An Ecological Appraisal for the Indian Subcontinent.* CAB International, Wallingford, UK, pp. 162–173.

Peh, K.S.H. (2010) Invasive species in Southeast Asia. *Biodiversity Conservation* 19, 1083–1099.

Peters, H.A. (2001) *Clidemia hirta* invasion at the Pasoh Forest Reserve: an unexpected plant invasion in an undisturbed tropical forest. *Biotropica* 33, 60–68.

Pimentel, D., McNair, S., Janecka, J., Wightman, C., Simmonds, C., O'Connell, E., Wong, L., Russel, J., Zern, T., Aquino, T. and Tsomondo, T. (2002) Economic and environmental threats of alien plant, animal, and microbe invasions. In: Pimentel, D. (ed.) *Biological Invasions: Economic and Environmental Costs of Alien Plant, Animal and Microbe Species.* CRC Press, Boca Raton, Florida, pp. 307–329.

Prasad, A.E. (2010) Impact of *Lantana camara,* a major invasive plant, on wildlife habitat in Bandipur Tiger Reserve, southern India. Available at: www.rufford.org/rsg/projects/ayesha_prasad (accessed 18 May 2015).

Pyšek, P., Richardson, D.M., Rejmanek, M., Webster, G.L., Williamson, M. and Kirschner, M. (2004) Alien plants in checklists and flora: towards better communication between taxonomists and ecologists. *Taxon* 53, 131–143.

Rai, R.K. and Scarborough, H. (2015) Understanding the effects of the invasive plants on rural forest-dependent communities. *Small-scale Forestry* 14, 59–72.

Rai, R.K., Scarborough, H., Subedi, N. and Lamichhane, B. (2012) Invasive plants – do they devastate or diversify rural livelihoods? Rural farmers' perception of three invasive plants in Nepal. *Journal for Nature Conservation* 20, 170–176.

Ramakrishnan, P.S. (2001) Biological invasion as a component of global change: the Indian context. In: Sankaran, K.V., Murphy, S.T. and Evans, H.C. (eds) *Alien Weeds in Moist Tropical Zones: Banes and Benefits. Proceedings of a Workshop, Kerala Forest Research Institute, Peechi, India, 2–4 November 1999.* Kerala Forest Research Institute, Peechi, India, and CABI Bioscience, UK Centre (Ascot), Ascot, UK, pp. 28–34.

Ramaswami, G. and Sukumar, R. (2011) Woody plant seedling distribution under invasive *Lantana camara* thickets in a dry-forest plot in Mudumalai, southern India. *Journal of Tropical Ecology* 27, 365–373.

Rawat, Y.S. and Singh, J.S. (1988) Structure and function of oak forests in Central Himalaya. 1. Dry matter dynamics. *Annals of Botany* 6, 397–411.

Sapkota, L. (2007) Ecology and management issues of *Mikania micrantha* in Chitwan National Park, Nepal. *Banko Janakara* 17, 27–39.

Saxena, K.G. (1991) Biological invasions in the Indian subcontinent: review of invasion by plants. In: Ramakrishnan, P.S. (ed.) *Ecology of Biological Invasion in the Tropics*. International Scientific Publications, New Delhi, pp. 53–73.

Saxena, K.G. and Ramakrishnan, P.S. (1984) Herbaceous vegetation development and weed potential in slash and burn agriculture (jhum) in northeastern India. *Weed Research* 24, 135–142.

Sharma, G.P. and Raghubanshi, A.S. (2007) Effect of *Lantana camara* L. cover on plant species depletion in the Vindhyan tropical dry deciduous forest of India. *Applied Ecology and Environmental Research* 5, 109–121.

Sharma, G.P. and Raghubanshi, A.S. (2012) Invasive species: ecology and impact of *Lantana camara* invasions. In: Bhatt, J.R., Singh, J.S., Singh, S.P., Tripathi, R.S. and Kohli, R.K. (eds) *Invasive Alien Plants: An Ecological Appraisal for the Indian Subcontinent*. CAB International, Wallingford, UK, pp. 19–42.

Sharma, G.P., Raghubanshi, A.S. and Singh, J.S. (2005) Lantana invasion: an overview. *Weed Biology and Management* 5, 157–165.

Sharma, I.K. (1981) Ecological and economic importance of *Prosopis juliflora* in the Indian Thar Desert. *Journal of Taxonomy and Botany* 2, 245–248.

Shen, S-C., Xu, G.-F., Zhang, F.-D., Jin, G.-M., Liu, S.-F., Liu, M.-Y., Chen, A.-D. and Zhang, Y.-H. (2013) Harmful effects and chemical control study of *Mikania micrantha* H.B.K. in Yunnan, southwest China. *African Journal of Agricultural Research* 8, 5554–5561.

Siwakoti, M. (2007) Mikania weed: a challenge for conservationists. *Our Nature* 5, 70–74.

Swamy, P.S. and Ramakrishnan, P.S. (1987) Weed potential of *Mikania micrantha* H.B.K. and its control in fallows after shifting agriculture (jhum) in north-east India. *Agriculture, Ecosystems and Environment* 18, 195–204.

Teo, D.H.L., Tan, H.T.W., Corlett, R.T., Wong, C.M. and Lum, S.K.Y. (2003) Continental rain forest fragments in Singapore resist invasion by exotic plants. *Journal of Biogeography* 30, 305–310.

Teoh, C.H., Chung, G.F., Liau, S.S., Ghani Ibrahim, A., Tan, A.M., Lee, S.A. and Mariati, M. (1985) Prospects for biological control of *Mikania micrantha* H.B.K. in Malaysia. *Planter* 61, 515–530.

Towers, G.H.N. and Subba Rao, P.V. (1992) Impact of the pan-tropical weed, *Parthenium hysterophorus* L., on human affairs. In: Combellack, J.H., Levick, K.J., Parsons, J. and Richardson R.G. (eds) *Proceedings of the First International Weed Control Congress, Vol. 1, Melbourne, Australia, 17–21 February 1992.* Weed Science Society of Victoria, Frankston, Australia, pp. 134–138.

Waterhouse, D.F. (1993) *The Major Arthropod and Weeds of Agriculture in Southeast Asia: Distribution, Importance and Origin.* Australian Centre for International Agricultural Research (ACIAR), Canberra.

Waterhouse, D.F. (1994) *Biological Control of Weeds: Southeast Asian Prospects.* Australian Centre for International Agricultural Research (ACIAR), Canberra.

Weber, E. (2003) *Invasive Plant Species of the World: A Reference Guide to Environmental Weeds.* CAB International, Wallingford, UK.

Witkowski, E.T.F. and Wilson, M. (2001) Changes in density, biomass, seed production and soil seed banks of the non-native invasive *Chromolaena odorata*, along a 15 year chronosequence. *Plant Ecology* 15, 13–27.

World Bank (2016) Poverty overview. Available at: www.worldbank.org/en/topic/poverty/overview (accessed 10 June 2016).

Zhang, L.Y., Ye, W.H., Cao, H.L. and Feng, H.L. (2004) *Mikania micrantha* H.B.K. in China – an overview. *Weed Research* 44, 42–49.

2 Profile of an Invasive Plant: *Mikania micrantha*

Carol A. Ellison[1]* and K.V. Sankaran[2]

[1]CABI, Egham, UK; [2]Kerala Forest Research Institute (KFRI), Peechi, India

Introduction

Despite the long history of biological invasions in the Asia–Pacific region, little has been achieved in terms of effective management actions for such species. This situation needs be remedied given that: (i) many crop species are affected, thereby threatening the food security of all countries in the region; (ii) the region has a great diversity of habitats, rich biodiversity and a high level of species endemism; (iii) many endemic species and their habitats in the region are threatened; and (iv) the degradation and fragmentation of natural habitats are common. Thus, damage due to invasive alien species is arguably higher in this region than in other parts of the world.

Although the threat posed by invasive species to livelihoods and biodiversity is now high profile, the mechanisms by which an invasive species succeeds are largely ill-defined. A fundamental understanding of these mechanisms might eventually allow the prediction of which species will become invasive and, therefore, the development of more effective management strategies (Deng *et al.*, 2004). Nevertheless, at present it remains difficult to predict whether a species that establishes in a new environment will become invasive or not. The best predictors of invasiveness for plants are whether the species has such a history elsewhere and whether it reproduces vegetatively (Kolar and Lodge, 2001), but neither of these predictors is appropriate with newly invasive species. However, there are some general factors that are either likely to operate or known to be important in facilitating an exotic species to become invasive: for example, similarities in climate and soil between the original habitat of the exotic species and its new habitat (Nye and Greenland, 1960; Holgate, 1986); and the absence of all or most of its co-evolved, host-specific natural enemies (which were present in the original habitat) from the adventive range (as mentioned in the next paragraph).

Some of the most damaging invasive alien species are plants. Plant species have been moved around the world by humans for centuries – either deliberately as crops and ornamental species, or accidently as casual passengers. Most of these species do not become invasive in their introduced range. Moreover, plants are usually moved without most or all of the natural enemies referred to above, and, often aided by human disruption of natural habitats, population explosions of some species occur, with the subsequent development of plant invasions (Mack *et al.*, 2000). The Neotropical plant *Mikania micrantha* Kunth illustrates the latter situation very well, and it is used as a core example by most of the authors in this volume. In its native range, the species generally has a cryptic riparian non-weedy habit, and grows along riverbanks and among reed-like vegetation around standing water (Cock, 1982; Barreto and Evans,

* Corresponding author. E-mail: c.ellison@cabi.org

 Invasive Alien Plants (eds C.A. Ellison, K.V. Sankaran and S.T. Murphy)

1995). None the less, since the plant was introduced into Asia at various times in the 20th century as a cover crop, and for various other purposes (see next section: Taxonomy, description and distribution), the environments in Asia and the Pacific have allowed it to express a full, or near-full, capacity for reproduction, growth and spread across many types of ecosystems. As a result, *M. micrantha* has become a serious threat to a range of crop plants and to native plants, particularly in India in the high-biodiversity areas of the Western Ghats and in the state of Meghalaya (Bhatt *et al.*, 2012; and Chapter 1, this volume).

Here, current information on the taxonomy and distribution of *M. micrantha* are summarized, followed by a review of relevant details on its life cycle and physiological ecology. The last of these topics has received much attention over recent years and advances have been made in particular on the allelopathic properties and photosynthetic strategy of the plant. Additional information about its physiology has been elucidated from a comparison with *M. cordata*, which is native to Asia (Deng *et al.*, 2004). These topics form an important backdrop for the book, because the theme of how some of the plant's biological traits have facilitated population outbreaks in agro-ecosystems and disturbed native ecosystems owing to the nature of these habitat types is pursued by authors in a number of chapters in this volume. The topic of the role of the natural enemies in suppressing plant invasions is discussed separately in Chapter 10, this volume.

Lastly, another consequence of the domination of *M. micrantha* in many habitats in the Asia–Pacific region is that several rural communities have been driven to seek ways of exploiting the plant, and these have been largely based on its perceived traits and properties. The reported benefits are few, but they are reviewed at the end of this chapter in order to complete the species profile.

A general data sheet on *M. micrantha* can be found in the open-access CABI Invasive Species Compendium (available at: www.cabi.org/isc/datasheet/34095).

Taxonomy, Description and Distribution

M. micrantha was described in 1818 by Kunth (Humboldt *et al.*, 1818). Common English-language names for the species include mikania weed, American rope, bittervine, climbing hemp vine and Chinese creeper. Although it is also commonly called mile-a-minute weed, this name is best avoided as it is also used for the for the unrelated weed *Persicaria perfoliata* in the USA. The classification of the species is as follows:

Class:	Dicotyledonae
Subclass:	Asteridae
Order:	Asterales
Family:	Asteraceae (Compositae)
Subfamily:	Asteroideae
Tribe:	Eupatorieae
Species:	*Mikania micrantha* Kunth

Mikania is a species-rich genus with over 425 described species (King and Robinson, 1987). Almost all are New World native species, with only nine being considered native to the Old World, and only one of these, *M. cordata* (Burm. f.) Robinson, being native to Asia (Holmes, 1982). There has been considerable confusion over the correct identification of *M. micrantha* in Asia, with early literature mistakenly referring to it as *M. scandens* (L.) Willd. (a North American species) or *M. cordata*. However, Parker (1972) concluded that *M. micrantha* is the only species that causes a significant weed problem, and this was confirmed by the work of Holmes (1982).

M. micrantha is a fast-growing, perennial, creeping or twining vine. A full taxonomic description from Holm *et al.* (1991) is as follows:

> *stems* branched, pubescent to glabrous, ribbed; *leaves* opposite, thin, cordate, triangular, or ovate, blade 4 to 13 cm long, 2 to 9 cm wide, on a petiole 2 to 8 cm long, base cordate or somewhat hastate, tip acuminate, margins coarsely dentate, crenate, or subentire, both surfaces glabrous, three- to seven-nerved from base; *flowers* in heads 4.5 to 6 mm long, in terminal and lateral openly rounded, corymbous panicles; *involucral bracts* four,

oblong to obovate, 2 to 4 mm long, acute, green, and with one additional smaller bract 1 to 2 mm long; four flowers per head; *corollas* white, 3 to 4 mm long; *fruit* in an achene, linear-oblong, 1.5 to 2 mm long, black, five-angled, glabrous; *pappus* of 32 to 38 mm long, soft white bristles 2 to 3 mm long.

The species possesses semi-translucent enations (small scaly leaf-like structures without vascular tissue) between the petioles at the nodes of young vegetative shoots, which are a rare feature in the Asteraceae. These are not seen on flowering branches and they wither on older shoots. Differences in the form of the enations can help to distinguish *M. micrantha* from *M. cordata*. The enations of *M. micrantha* are membranous flaps with incised lobes, whereas *M. cordata* produces ear-like enations with furry ridges, which are not membranous (see Fig. 2.1a). The length of the pappus (the fine feathery hairs that surround the fruit in the Asteraceae) bristles on the seed is also a distinguishing feature between the two species. Fig. 2.1b shows the structure of *M. micrantha* seed, which has pappus bristles that are 2–3 mm in length, while those of *M. cordata* are significantly longer (4–5 mm) and the seed is significantly larger.

M. micrantha is a plant of Neotropical origin that has become invasive in most countries within the humid tropical zones of Asia and the Pacific (Waterhouse, 1994). It occurs from sea level up to 1500 m. The first record of *M. micrantha* in Asia dates back to 1884, and is from the Hong Kong Zoological and Botanical Gardens (Li *et al.*, 2003). However, in Malaysia and Indonesia, evidence suggests that the plant was introduced on a number of occasions, both to botanical gardens and as a plantation cover crop (Wirjahardja, 1976; Holmes, 1982; Cock *et al.*, 2000). It was recorded to have been brought from Paraguay to Bogor Botanical Garden in West Java, where it was planted for medicinal purposes in 1949 (Cock *et al.*, 2000). Some scientists support the view that *M. micrantha* was introduced as a non-leguminous ground cover for crops in India (Borthakur, 1977; Palit, 1981), and Parker (1972) reported that the plant was intentionally introduced into north-eastern India during the Second World War as ground

Fig. 2.1. (a) *Mikania micrantha* nodes showing the structure of the enations, found at the nodes of young shoots (arrows). Inset: line drawing of enations of *M. cordata*. Photo courtesy C.A. Ellison. (b) *Mikania micrantha* seed showing pappus bristles (vertical line = 3 mm). Photo courtesy B. Adhikari.

cover in tea plantations. In addition, anecdotal evidence suggests that it was used for airfield camouflage in Assam during World War II (A.C. Barbora, Assam Agricultural University, 1999, personal communication).

The species is native to the tropical and subtropical zones of Central and South America, from Mexico to Argentina. It is widespread, but is not recorded as a significant weed in its native range. Its status in the Asia–Pacific region is shown in Fig. 2.2; further more detailed country-specific distribution is covered in some of the chapters in this volume. *M. micrantha* is currently under an 'eradication programme' in Queensland, Australia (Brooks *et al.*, 2008), and has been reported as a noxious weed in Florida, USA (Derksen and Dixon, 2009).

Life Cycle

M. micrantha has a vigorous capacity for both vegetative and sexual reproduction (Swamy and Ramakrishnan, 1987a), but cannot tolerate dense shade (Holm *et al.*, 1991). It occurs largely in humid environments and at temperatures between 13 and 35°C. Seed production by a single plant can reach 20,000–40,000 achenes (one-seeded fruits) in one season (Dutta, 1977) and these are dispersed long distances by wind (Holm *et al.*, 1991). The species is also able to produce perennating organs (rosette roots) for survival during less favourable conditions and ramets form from these rosettes when conditions allow active growth. In addition, the plant can grow vegetatively from very small sections of stem (Holm *et al.*, 1991). The growth of young plants is extremely fast – an increase of 8–9 cm has been recorded in 24 h (Choudhury, 1972) – and they will use trees and crops to support their growth. In the Asia–Pacific region, these features result in the plant rapidly forming a dense cover of entangled stems bearing many leaves (Holm *et al.*, 1991). Areas freed from the plant by slashing can be recolonized within a fortnight (Choudhury, 1972). Damage is caused to crops because the weed can smother, penetrate crowns, choke and even pull over other plants.

Ramakrishnan and Vitousek (1989) reported that within its native and adventive ranges, *M. micrantha* occurs in humid tropical areas with highly leached soils. More recent work in China, however, shows that

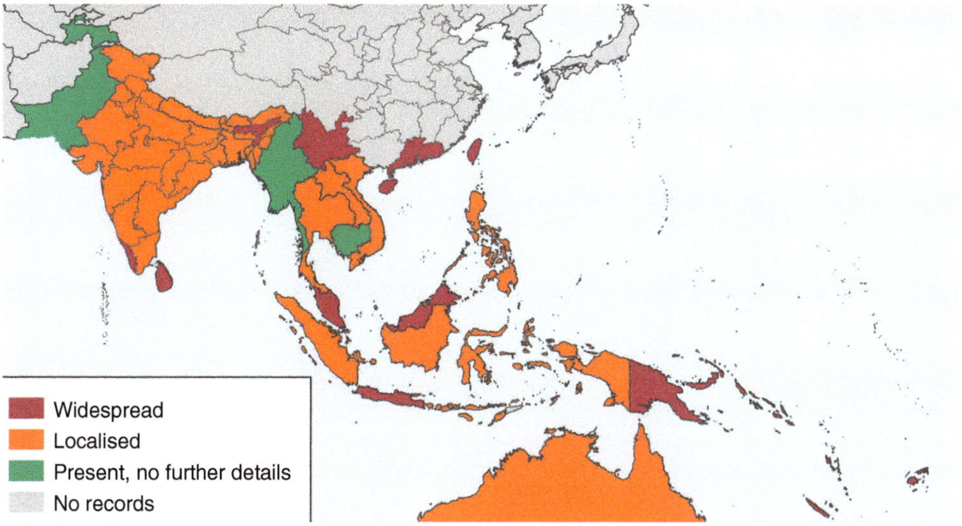

Fig. 2.2. Distribution of *Mikania micrantha* in South-east Asia. Adapted from the CABI Invasive Species Compendium distribution map in South East Asia and the Pacific. Available at: http://www.cabi.org/isc/datasheet/34095.

the plant commonly grows on a wide range of soils – from acidic to alkaline (pH 4.15–8.35) and with a broad range of relative proportions of organic matter (Zhang *et al.*, 2003). Under some circumstances, the plant also successfully competes for soil nutrients, while there is evidence from several studies that the weed can retard plant growth through the production of allelopathic substances (see next section).

Physiological Ecology

Allelopathic properties

Allelopathy is considered to be an important mechanism in successful exotic plant invasion. There is evidence from several studies that *M. micrantha* can retard plant growth through the production of allelopathic substances (Holm *et al.*, 1991; Cronk and Fuller, 1995) and this may be an important factor driving its invasion success in the Asia–Pacific region. The evidence falls into two categories: the properties of the *M. micrantha* rhizosphere soil; and the effects of the application of extracts of *M. micrantha* tissue on other plants.

Chen *et al.* (2009) found that soil in which *M. micrantha* had been growing inhibited the seed germination and growth of test plants and, in addition, had higher nutrient levels. *M. micrantha* has also been shown to have a disproportionately larger impact on the growth of young cocoa than weedy *Cyperus* spp. or *Imperata cylindrica* (Zaenuddin *et al.*, 1986). Gray and Hew (1968) reported a similar effect in oil palm; they found a 20% reduction in yield due to *M. micrantha* when it was used as a cover crop, and even when fertilizer was applied there was still an 11% yield reduction. This suggests that *M. micrantha* has an impact on growth through both competition for nutrients and direct allelopathic effects. Kaur *et al.* (2012) found that rice seedling growth is suppressed in *M. micrantha* rhizosphere soil; they found higher levels of organic matter, organic nitrogen and water-soluble phenolics in soil in which the plant had been growing compared with soil from which the plant was absent. Further, these authors found lower soil respiration in the *M. micrantha* rhizosphere and suggested that this may be due to a higher ratio of fungal/bacterial biomass in the soil.

Even as mulch, leachates of *M. micrantha* can adversely affect the growth of other plants. Ismail and Mah (1993) found that the leachates caused a significant reduction in the growth of three weed species, and Abraham and Abraham (2006) attributed the inhibition of rubber seedling growth by *M. micrantha* mulch and extracts to allelopathy. In a pot experiment with young rubber seedlings, Shao *et al.* (2005) isolated and identified three sesquiterpenoids from the aerial parts of *M. micrantha* which inhibited germination and growth in *Acacia mangium*, *Eucalyptus robusta* and *Pinus massoniana* in natural habitats in southern China.

Clearly, more work is needed to better understand the mechanisms involved, and the scale of impact and range of plants affected, as this would facilitate the improvement of management options for *M. micrantha*.

Photosynthetic strategy

The mechanisms of photosynthesis differ between plants. The two main types of photosynthesis are usually referred to as C_3 (*c.*90% of plant species) and C_4 (*c.*10%) (see Box 2.1); this is because the CO_2 is fixed by initially incorporating it into either a 3- or 4-carbon compound, respectively. In tropical weed communities in a variety of systems – such as smallholdings, plantations and shifting agriculture – both C_3 and C_4 plants are found. For example, of the predominant weeds in the early fallow stage in the shifting agriculture of north-eastern India, the most important exotic species, including *M. micrantha*, are C_3 plants, while all of the important native plants are C_4 species (Saxena and Ramakrishnan, 1984a).

Although C_4 plants are generally regarded as better adapted to hot environments with high light intensity,

Box 2.1. Photosynthesis and C_3 and C_4 plants.

Photosynthesis, the ability to transform light energy, carbon dioxide and water into sugar (releasing oxygen in the process) characterizes green plants. Greater efficiency at doing this may give a plant a competitive advantage over other plants growing alongside it.

The principal function of a leaf is photosynthesis, and its structure and function reflect this. CO_2 is taken up from the atmosphere through pores (stomata) on the leaf surface. Water is lost through these same stomata. Water has specific functions in plants beside photosynthesis. It allows plants to maintain turgor (structural rigidity created by water pressure in cells) and to keep cool via transpiration (evaporation through the stomata). Stomata can be closed partly or fully to conserve water – typically during the hotter daylight hours – but need to be open to allow CO_2 to enter for photosynthesis, which also takes place only during the daylight. Plants must balance these opposing needs. The main (and contrasting) features of C_3 and C_4 photosynthesis are as follows:

- C_3 photosynthesis is more efficient under cool and moist conditions with normal light levels. C_4 photosynthesis is more efficient under hot conditions with high light levels.
- C_3 photosynthesis is less water efficient than C_4 photosynthesis.
- C_3 plants exhibit higher nutrient uptake than C_4 plants and store more nutrients in their tissues; conversely, C_4 plants make more efficient use of nitrogen and other nutrients, and can thrive where soil nutrients are in short supply.
- C_3 plants tend to rely on heavy seed production as a primary reproductive strategy, while C_4 plants rely more on vegetative propagation.

M. micrantha and many other exotic invasive tropical C_3 plants are also able to predominate. At least part of the explanation for this apparent paradox is that C_4 photosynthesis comes at a cost in terms of the investment of resources in the cell structures and biochemical pathways involved. In some circumstances, this can give C_3 photosynthesis a competitive edge. Furthermore, unlike other exotic C_3 weeds such as *Chromolaena odorata*, which rely largely on heavy seed production (Kushwaha *et. al.*, 1981), *M. micrantha* also reproduces easily by vegetative propagation (Swamy and Ramakrishnan, 1987a).

Research illustrating how the properties of *M. micrantha* and other C_3 plants can confer competitive advantages over native plants under certain circumstances was carried out in the shifting agriculture systems in the hill regions of north-eastern India by Ramakrishnan and several co-workers. In summary, three main important situations were identified:

- *Variation in light intensity over the year.* C_4 photosynthesis is at its peak in the higher light intensity and thus temperatures in the middle of the year in the region, coinciding with the start of the growing season (April–June), and growth by C_4 species is strongest at this time (Saxena and Ramakrishnan, 1984b). In contrast, with light intensity and temperatures falling during the latter part of the growing season (October–December), C_3 photosynthesis comes into its own and these species grow more strongly (Saxena and Ramakrishnan, 1984a).
- *Variation in nutrient availability.* Leaching and erosion on the steep slopes where shifting agriculture is practised means that soil nutrients are not evenly distributed (Toky and Ramakrishnan, 1981). The nutrient-poor higher slopes are occupied by C_4 species, while the lower slopes where nutrients have accumulated are occupied by C_3 species (Saxena and Ramakrishnan, 1983, 1984a).
- *Fire.* Soil fertility status may increase after burning (which is done to prepare land for planting) and the increase would be to a greater extent in older fallows because of their increased fuel loads. This would tend to favour the growth of the C_3 species, which are more efficient at nutrient uptake, when the light intensity and temperature also favour the growth of these species (Swamy and Ramakrishnan, 1987a,b, 1988a,b).

These and other dimensions of C_3 plant invasions in shifting agriculture are reviewed in detail in Chapter 7, this volume.

The Photosynthetic Strategy of *Mikania micrantha* Compared With *M. cordata*

In southern China, the non-invasive *M. cordata*, which is native to the Asia–Pacific region, has been displaced by *M. micrantha* (Deng *et al*., 2004). Superficially, *M. micrantha* and *M. cordata* appear to be very similar in their morphologies and life histories (which has led to them being confused). However, *M. cordata* grows more slowly and only in shaded habitats, and does not achieve weed status; *M. micrantha*, in contrast, is highly invasive in open, disturbed habitats, including home gardens, plantations, abandoned cultivated land and forest gaps.

In seeking the mechanisms underlying these differences between the two species, Deng *at al.* (2004) looked at whether *M. micrantha* had a greater capacity for photosynthesis or used resources more efficiently than its non-invasive relative. These authors demonstrated a number of key differences between the two plants:

- At high light levels, *M. micrantha* photosynthesizes at a higher rate than *M. cordata*. Conversely, the two species photosynthesize at similar and lower rates at lower light levels. Thus, at high light levels, *M. micrantha* assimilates CO_2 at a higher rate, with associated higher photosynthetic enzyme activity. *M. micrantha* also fixes CO_2 at higher concentrations inside the leaf than does *M. cordata*, so it creates a larger pool of organic carbon for use in growth and reproduction.
- *M. micrantha* uses fewer leaf resources for photosynthesis than *M. cordata*. 'Leaf nitrogen' is a common proxy for investment in photosynthesis (as enzymes, cell structures), and *M. micrantha* photosynthesized more rapidly than *M. cordata* across a range of leaf nitrogen levels, indicating that it made more efficient use of this investment (and so was left with more for growth and reproduction).
- *M. micrantha* has greater CO_2 uptake for less water loss through the stomata than *M. cordata*. This greater 'water-use efficiency' indicates that faster photosynthesis by *M. micrantha* is not achieved by keeping its stomata open to take in more CO_2. The authors highlighted that *M. micrantha* combines this characteristic with high resource (photosynthetic nitrogen) use efficiency, rather than needing a trade-off, most likely because of its highly efficient photosynthesis.
- *M. micrantha* has a larger area of leaf per unit weight (specific leaf area) than *M. cordata* – in other words, it has relatively large thin leaves – a trait that has been linked by other authors to greater light capture and CO_2 uptake.
- The 'construction costs' are lower for *M. micrantha* leaves. So even though it creates more photosynthetic products, it uses fewer of them in photosynthesis itself.

In summary, Deng *et al.* (2004) showed how *M. micrantha*'s leaf physiology and morphology allow it to exploit more open, drier habitats than *M. cordata*; the structure and function of its leaves facilitate faster and more efficient photosynthesis at higher light levels, allowing it to assimilate more CO_2 without compromising its water balance.

Beneficial Effects of *Mikania micrantha*

Although widely introduced into Southeast Asia and the Pacific as a cover crop, *M. micrantha* is considered to be of restricted value, particularly when compared with nitrogen-fixing legumes such as *Mucuna bracteata* (Teoh *et al*., 1985). Cattle eat it, but its nutritional value is considered to be inferior to that of the pasture plants it is able to smother; and it is known to cause hepatotoxicity and liver damage in dairy cattle. Owing to its sprawling nature, the species has been used to prevent soil erosion and to serve as mulch. It is considered by some to be less noxious than those weeds that might occupy a particular niche if it were not present (Waterhouse, 1994).

In villages in Indonesia, besides being slashed as herbage, *M. micrantha* is used as a poultice for swellings, itching and wounds. A similar practice is reported from some Pacific island countries (see Chapter 6, this volume); reports from Fiji indicated that the leaves were applied as a poultice for ant bites and hornet and bee stings as early as the 1950s (Parham, 1958). Biological control measures in Western Samoa are being delayed while a perceived conflict of interest is addressed: *M. micrantha* leaves are used medicinally, and there is a perception that the release of a biocontrol agent would have an impact on the amount of plant material available. In addition, there has been a perception among farmers that *M. micrantha* is a good mulch and easier to control than other weeds, as well as being 'good' for the soil and reducing other pests. This view is held despite much time being spent by family members in controlling the weed and preventing it from smothering crops such as cocoa, taro, bananas and papaya, as these types of activities by family members are not considered to be a 'cost' by farmers (Ellison *et al.*, 2014). Day *et al.* (2011), in a socio-economic study in Papua New Guinea, reported that 32% of respondents used *M. micrantha* as a medicinal plant to treat cuts and wounds. Baruah and Dutta Choudhury (2012) included *M. micrantha* among the medicinal plants used in the Barak Valley of Assam in north-eastern India.

Other work has shown the potential of *M. micrantha* extracts as an insect deterrent. Cen *et al.* (2003) found that alcohol extracts of the plant inhibited oviposition by the citrus leaf miner (*Phyllocnistis citrella*), and Zhang *et al.* (2004) showed that essential oils extracted from *M. micrantha* had oviposition deterrent and repellent properties against the diamondback moth (*Plutella xylostella*). Studies have also highlighted insecticidal properties of the plant against several agricultural pest species. Methanol extracts of *M. micrantha* have shown potential as insect growth regulators for the rhinoceros beetle (*Oryctes rhinoceros*) (Zhong *et al.*, 2012); these authors showed that including the extracts in beetle food had a feeding deterrent effect but also delayed development and increased mortality at all developmental stages, as well as leading to deformities. Lu *et al.* (2012) found that alcohol extracts of *M. micrantha* caused developmental delay and increased deformities in the coconut leaf beetle (*Brontispa longissima*) and reported 80% control in semi-field trials.

In its native range, there are unsubstantiated reports of the use of *M. micrantha* in folk medicine as a cure for snakebites (R.W. Barreto, Universidade Federal de Viçosa, Brazil, 1991, personal communication). Evidence has emerged more recently of other significant medicinal potential of the species. Investigations by Bakir *et al.* (2004) led to the isolation of mikanolide compounds from a variety of *M. micrantha* growing in Portland, Jamaica. The mikanolides have been shown by Teng *et al.* (2005) to have potential for treating cancer, and the authors have patented (in the USA) a method of treating proliferative diseases (cancers) in warm-blooded animals by administering an appropriate amount of at least one compound selected from the sesquiterpenoid group consisting of mikanolides and dihydromikanolides.

While these potential benefits should not be discounted and may lead to various uses for *M. micrantha*, the available evidence to date clearly indicates that the damage caused to crop and forest tree production and to native biodiversity far outweighs these benefits.

Conclusions

An increasing amount of research over recent decades has revealed much about the physiology and ecology of *M. micrantha* and this has provided a clearer basis for understanding how and why this plant has become invasive in so many agricultural and native ecosystems. The main biological traits and features of the plant, which are summarized in Box 2.2, may not substantially

Box 2.2. The main biological traits and features of *Mikania micrantha*.

- Native to the New World tropics, *M. micrantha* is a highly invasive weed in the Asia–Pacific region of the Old World, where it smothers other vegetation; it occurs from sea level up to 1500 m, and at temperatures of between 13 and 35°C.
- It is a fast-growing, perennial vine that is able to smother plants in agricultural ecosystems, agroforestry and plantation systems and natural habitats, thereby reducing productivity and biodiversity.
- It grows in a wide range of soil types – from acid to alkaline, but does not tolerate shade.
- Its seed production is prolific, and the seeds are predominantly wind dispersed as they are light in weight. The seeds have no dormancy and are an effective means of propagation in unutilized, cleared or burnt ground. The plant also reproduces vegetatively rather vigorously – new plants can develop from nodes, which produce roots when in contact with damp soil.
- The plant has allelopathic properties, but the mechanism of this action is unclear.
- The rapid photosynthesis exhibited by M. micrantha, courtesy of a large leaf area and efficient biochemical pathways, means that it grows quickly, particularly in unshaded situations and at high light intensities.
- It is efficient at taking up nutrients where plenty are available, and accumulates them in its tissues, but it is not so good at sequestering nutrients when they are in short supply; thus, burning, which releases nutrients from the standing plant biomass and destroys shade, favours growth.

differentiate it from other indigenous sprawling vines. However, though it may have competitive advantages, having evolved in isolation from the indigenous fauna of the Asia–Pacific region, there is also strong evidence that the absence of specialized natural enemies following its introduction from the Neotropical region is also critical (Cock *et al*., 2000). This aspect is further discussed in Chapter 10, this volume.

References

Abraham, M. and Abraham, C.T. (2006) Preliminary investigations on the allelopathic tendency of *Mikania micrantha*, a common weed in young rubber (*Hevea brasiliensis*) plantations. *Natural Rubber Research* 19, 81–83.

Bakir, M., Facey, P.C., Hasan, I, Mulderand, W.H. and Porter, R.B. (2004) Mikanolide from Jamaican *Mikania micrantha*. *Acta Crystallographica Section C: Structural Chemistry* 60, 798–800.

Barreto, R.W. and Evans, H.C. (1995). The mycobiota of the weed *Mikania micrantha* in southern Brazil with particular reference to fungal pathogens for biological control. *Mycological Research* 99, 343–352.

Baruah, M.K. and Dutta Choudhury, M. (2012) Ethno-medicinal uses of Asteraceae in Barak Valley, Assam. *International Journal of Plant Sciences* 7, 220–223.

Bhatt, J.R., Singh, J.S., Singh, S.P., Tripathi, R.S. and Kohli, R.K. (eds) (2012) *Invasive Alien Plants: An Ecological Appraisal for the Indian Subcontinent.* CAB International, Wallingford, UK.

Borthakur, D.N. (1977) *Mikania* and *Eupatorium*: two noxious weeds of N.E. region. *Indian Farming* 26, 25–26.

Brooks, S.J., Panetta, F.D. and Galway, K.E. (2008) Progress towards the eradication of mikania vine (*Mikania micrantha*) and limnocharis (*Limnocharis flava*) in northern Australia. *Invasive Plant Science and Management* 1, 296–303.

Cen, Y.-J., Pang, X.-F., Zhang, M.-X., Deng, Q.-S., Du, Y.-Y. and Wang, B. (2003) Oviposition repellent of alcohol extracts of 26 non-preferable plant species against citrus leafminer. *Journal of South China Agricultural University* 24(3), 27–29. [In Chinese, English abstract.]

Chen, B.-M., Peng, S.-L. and Ni, G.-Y. (2009) Effects of the invasive plant *Mikania micrantha* H.B.K. on soil nitrogen availability through allelopathy in south China. *Biological Invasions* 11, 1291–1299.

Choudhury, A.K. (1972) Controversial *Mikania* (climber) – a threat to the forests and agriculture. *Indian Forester* 98, 178–186.

Cock, M.J.W. (1982) Potential biological control agents for *Mikania micrantha* HBK from the Neotropical region. *Tropical Pest Management* 28, 242–254.

Cock, M.J.W., Ellison, C.A., Evans, H.C. and Ooi, P.A.C. (2000) Can failure be turned into success for biological control of mile-a-minute weed (*Mikania micrantha*)? In: Spencer, N.R. (ed.) *Proceedings of the X International Symposium on Biological Control of Weeds, Bozeman, Montana, 4–14 July 1999.* USDA Forest Service, Forest Health Technology Enterprise Team, Morgantown, West Virginia, pp. 155–167.

Cronk, Q.C.B. and Fuller, J.L. (1995) *Plant Invaders.* Chapman and Hall, London.

Day, M.D., Kawi, A., Tunabuna, A., Fidelis, J., Swamy, B., Ratutuni, J., Saul-Maora, J., Dewhurst, C.F. and Orapa, W. (2011) The distribution and socio-economic impacts of *Mikania micrantha* (Asteraceae) in Papua New Guinea and Fiji and prospects for its biocontrol. In: *23rd Asian-Pacific Weed Science Society Conference, The Sebel Cairns, Australia, 26–29 September 2011: Weed Management in a Changing World. Conference Proceedings, Volume 1.* Asian-Pacific Weed Science Society, pp. 146–153. Available at: http://apwss.org/documents/23rd%20APWSS%20Conference%20Proceedings%20Vol%201.pdf (accessed 14 July 2017).

Deng, X., Ye, W.-H., Peng, H.-L., Yang, Q.-H., Cao, H.-L., Xu, K.-Y. and Zhang, Y. (2004) Gas exchange characteristics of the invasive species *Mikania micrantha* and its indigenous congener *M. cordata* (Asteraceae) in south China. *Botanical Bulletin of Academia Sinica* 45, 213–220.

Derksen, A. and Dixon, W. (2009) Interim report: Florida Cooperative Agricultural Pest Survey for Mile-A-Minute, *Mikania micrantha* Kunth in Miami-Dade County. Florida Department of Agriculture and Consumer Services, Miami, Florida. Available at: https://www.researchgate.net/publication/239531988_Interim_Report_Florida_Cooperative_Agricultural_Pest_Survey_for_Mile-A-Minute_Mikania_micrantha_Kunth_in_Miami-Dade_Co (accessed 14 July 2017).

Dutta, A.C. (1977) Biology and control of *Mikania. Two and a Bud* 24, 17–20.

Ellison, C.A., Day, M.D. and Witt, A. (2014) Overcoming barriers to the successful implementation of a classical biological control strategy for the exotic invasive weed *Mikania micrantha* in the Asia-Pacific region. In: Impson, F.A.C., Kleinjan, C.A. and Hoffmann, J.H. (eds) *Proceedings of the XIV International Symposium on Biological Control of Weeds, Kruger National Park, South Africa, 2–7 March 2014.* University of Cape Town, Rondebosch, South Africa, pp. 135–141.

Gray, B.S. and Hew, C.K. (1968) Cover crops experiments in oil palm on the west coast of Malaysia. In: Turner, P.D. (ed.) *Oil Palm Developments in Malaysia.* Incorporated Society of Planters, Kuala Lumpur, pp. 138–151.

Holgate, M.W. (1986) Summary and conclusions: characteristics and consequences of biological invasions. *Philosophical Transactions of the Royal Society of London. Series B, Biological Sciences* 314, 733–742.

Holm, L.G., Plucknett, D.L., Pancho, J.V. and Herberger, J.P. (1991) *The World's Worst Weeds: Distribution and Biology.* Krieger, Malabar, Florida.

Holmes, W.C. (1982) Revision of the Old World *Mikania* (Compositae). *Botanisches Jahres Beitraegefuer Systematik* 103, 211–246.

Humboldt, A. von, Bonpland, A. and Kunth, C.S. (1818) *Nova Genera et Species Plantarum,* quarto edn. Lutetiae Parisiorum, Paris, France.

Ismail, B.S. and Mah, L.S. (1993) Effects of *Mikania micrantha* H.B.K. on germination and growth of weed species. *Plant and Soil* 157, 107–113.

Kaur, R., Malhotra, S. and Inderjit (2012) Effects of invasion of *Mikania micrantha* on germination of rice seedlings, plant richness, chemical properties and respiration of soil. *Biology and Fertility of Soils* 48, 481-488.

King, R.M. and Robinson, H. (1987) *Genera of Eupatorieae.* Monographs in Systematic Botany No. 22, Missouri Botanical Garden Press, St Louis, Missouri.

Kolar, C.S. and Lodge, D.M. (2001) Progress in invasion biology: predicting invaders. *Trends in Ecology and Evolution* 16, 199–204.

Kushwaha, S.P.S., Ramakrishnan, P.S. and Tripathi, R.S. (1981) Population dynamics of *Eupatorium odoratum* in successional environments following slash and burn agriculture. *Journal of Applied Ecology* 18, 529–535.

Li, M., Wang, B. and Yu, Q. (2003) Controlling a noxious exotic weed *Mikania micrantha* in the field. In: Zhang, R., Zhou, C., Pang, H., Zhang, W., Gu, D. and Pang, Y. (eds) *Proceedings of the International Symposium on Exotic Pests and Their Control, Zhongshan, China, 17–20 November 2002.* Zhongshan (Sun Yat Sen) University Press, Guangzhou, China, pp. 187–192.

Lu, C., Zhong, B., Zhong, G., Weng, Q., Chen, S., Hu, M., Sun, X. and Qin, W. (2012) Four botanical extracts are toxic to the hispine beetle, *Brontispa longissima,* in laboratory and semi-field trials. *Journal of Insect Science* 12(1):58. Available at: http://dx.doi.org/10.1673/031.012.5801.

Mack, R.N., Simberloff, D., Lonsdale, W.M., Evans, H., Clout, M. and Bazzaz, F.A. (2000) Biotic invasions: causes, epidemiology, global consequences and control. *Ecological Applications* 10, 689–710.

Nye, P.H. and Greenland, D.J. (1960) *The Soil under Shifting Cultivation.* Technical Communication No. 51, Commonwealth Bureau of Soils, Harpenden, UK.

Palit, S. (1981) *Mikania* – a growing menace in plantation forestry in West Bengal. *Indian Forester* 107, 96–101.

Parham, J. (1958) *The Weeds of Fiji.* Bulletin 35, Department of Agriculture, Suva, Fiji.

Parker, C. (1972) The *Mikania* problem. *Pest Articles and News Summaries* 18, 312–315.

Ramakrishnan, P.S. and Vitousek, P.M. (1989) Ecosystem-level processes and consequences of biological invasions. In: Drake, J.A., Mooney, H.A., di Castri, F., Groves, R.H., Kruger, F.J., Rejmanek, M. and Williamson, M. (eds) *Biological Invasions: A Global Perspective.* SCOPE 37. Wiley, New York, pp. 281–300.

Saxena, K.G. and Ramakrishnan, P.S. (1983) Growth and allocation strategies of some perennial weeds of slash and burn agriculture (jhum) in north-eastern India. *Canadian Journal of Botany* 61, 1300–1306.

Saxena, K.G. and Ramakrishnan, P.S. (1984a) C_3/C_4 species distribution among successional herbs following slash and burn in north-eastern India. *Acta Oecologia: Oecologia Plantarum* 5, 335–346.

Saxena, K.G. and Ramakrishnan, P.S. (1984b) Herbaceous vegetation development and weed potential in slash and burn agriculture (jhum) in north-eastern India. *Weed Research* 24, 135–142.

Shao, H., Peng, S.-L., Wei, X.-Y, Zhang, D.-Q. and Zhang, C. (2005) Potential allelochemicals from an invasive weed *Mikania micrantha* H.B.K. *Journal of Chemical Ecology* 31, 1657–1668.

Swamy, P.S. and Ramakrishnan, P.S. (1987a) Weed potential of *Mikania micrantha* H.B.K. and its control in fallows after shifting agriculture (jhum) in north-eastern India. *Agriculture, Ecosystems and Environment* 18, 195–204.

Swamy, P.S. and Ramakrishnan, P.S. (1987b) Effect of fire on population dynamics of *Mikania micrantha* H.B.K. during early succession after slash and burn agriculture (jhum) in north-eastern India. *Weed Research* 27, 397–404.

Swamy, P.S. and Ramakrishnan, P.S. (1988a) Growth and allocation patterns of *Mikania micrantha* H.B.K. in successional environments after slash and burn agriculture. *Canadian Journal of Botany* 66, 1465–1469.

Swamy, P.S. and Ramakrishnan, P.S. (1988b) Effect of fire on growth and allocation strategies of *Mikania micrantha* H.B.K. under early successional environments. *Journal of Applied Ecology* 25, 653–658.

Teng, P.B., Pannetier, B., Chistine, M., Moumen, M. and Prevost, G. (2005) USA Patent Application No. 20050208155 Kind Code A1. September 22, 2005. Available at: http://www.google.com/patents/US20050208155 (accessed 21 February 2017).

Teoh, C.H., Chung, G.F., Liau, S.S., Ghani Ibrahim, A., Tan, A.M., Lee, S.A. and Mariati, M. (1985) Prospects for biological control of *Mikania micrantha* H.B.K. in Malaysia. *Planter* 61, 515–530.

Toky, O.P. and Ramakrishnan, P.S. (1981) Run-off and infiltration losses related to shifting agriculture (jhum) in north-eastern India. *Environmental Conservation* 8, 313–321.

Waterhouse, D.F. (1994) *Biological Control of Weeds: Southeast Asian Prospects.* Australian Centre for International Agricultural Research (ACIAR), Canberra.

Wirjahardja, S. (1976) Autecological study of *Mikania* sp. In: *Proceedings of Fifth Asian-Pacific Weed Science Society Conference, Tokyo, Japan, October 5–11, 1975.* Asian-Pacific Weed Science Society, pp. 70–71. Available at: http://apwss.org/documents/5th_APWSSC_P.pdf (accessed 21 February 2017).

Zaenuddin, S., Ronoprawiro and Marzuki, A. (1986) Competition and allelopathy of some weeds towards cacao plant in relation to its control in the field. In: Madkar, O.R., Soedarsan, A. and Sastroutomo, S.S. (eds) *Proceedings of the 8th Indonesian Weed Science Conference, Bandung, 24–26 March 1986.* Indonesian Weed Science Society, pp. 83–89.

Zhang, L.Y., Ye, W.H., Cao, H.L. and Feng, H.L. (2003) *Mikania micrantha* H.B.K. in China – an overview. *Weed Research* 44, 42–49.

Zhang, M.-X., Ling, B., Chen, S.-Y. Liang, G.-W. and Pang, X.-F. (2004) Repellent and oviposition deterrent activities of the essential oil from *Mikania micrantha* and its compounds on *Plutella xylostella. Entomologia Sinica* [now *Insect Science*] 11, 37–45.

Zhong, B.-Z., Lu, C.-J., Wang, D.-M., Li, H. and Qin, W.-Q. (2012) Effects of methanol extracts of *Mikania micrantha* on the growth and development of the rhinoceros beetle, *Oryctes rhinoceros* (Coleoptera: Dynastidae). *Acta Entomologica Sinica* 55, 1062–1068. [In Chinese, English abstract.]

3 Social and Economic Implications of *Mikania micrantha* in the Kerala Western Ghats

V. Anitha,[1]* K.V. Santheep[1] and Jyotsna Krishnakumar[2]

[1]Kerala Forest Research Institute (KFRI), Peechi, India; [2]School of Environmental and Forest Sciences, University of Washington, Seattle, USA

Background

Mikania micrantha, or mikania weed, is posing a serious threat to both agricultural and natural ecosystems in the parts of the Western Ghats that lie in Kerala State. The weed, a perennial climber, affects trees and subsistence crops, either through cost escalation or income reduction, or both, in addition to causing other serious ecological damage (Sankaran *et al*., 2001; Kaur *et al*., 2012; Gopakumar and Motwani, 2013). Economically, the impact of mikania weed infestation leads to an escalation in the costs of agroecosystems that could make them economically unviable (Muraleedharan and Anitha, 2001).

In India, *M. micrantha* was first recorded in 1918 from Bengal, from which a herbarium specimen of the weed was sent to the Royal Botanic Gardens, Kew (UK) for identification (Choudhury, 1972; Dutta, 1977). In Kerala, the mikania weed infestation was first observed in the 1980s (Sankaran *et al*., 2001); in certain parts of the state the weed is used as fodder during the summer season when the availability of grass becomes limited, although its value as fodder has been questioned. The use of mikania as fodder has also been reported for cattle in Indonesia (Soerjani, 1977) and for goats in Nepal (Murphy *et al*., 2013).

Current control measures in Kerala focus on physical and chemical methods, which are expensive, ineffective and unsustainable, and can be environmentally damaging (Palit, 1981; Sen Sarma and Mishra, 1986; Muniappan and Viraktamath, 1993; Ellison and Murphy, 2001; Sankaran *et al*., 2001). Similar drawbacks have been documented recently in Fiji and Papua New Guinea (Day *et al*., 2011; Macanawai *et al*., 2011). Classical biological control involving the introduction of host-specific organisms from the weed's country of origin can sometimes offer a highly effective and environmentally friendly solution to the problem of invading alien weeds (Singh, 2001). However, according to Ramakrishnan (2001), historical analysis of the causative factors of the invasion, a detailed understanding of the original and modified natural ecosystem of the area and the land use dynamics and, arising out of these, a participatory effort in landscape management based on community perceptions and participation, could also contribute to the effective management and, perhaps, the elimination of exotic weeds over a period of time.

Kerala State, situated in south-western India, covers an area of 38,863 km^2; of this, 11,125.59 km^2 are either natural forests or forest plantations. The land of the state is

* Corresponding author. E-mail: anitha@kfri.org

 Invasive Alien Plants (eds C.A. Ellison, K.V. Sankaran and S.T. Murphy) highly diversified in its physical features and agro-ecological conditions. The undulating topography ranges from below mean sea level (msl) to 2694 m above msl. Physiographically, the state is classified into four zones: the lowlands (up to 7.5 m above msl); the midlands (7.5–75 m above msl); the highlands (75–750 m above msl); and high ranges (over 750 m above msl) (KSLUB, 1997). The highlands are mostly forested areas located along the Western Ghats, accounting for 48% of the total area and 15% of the people in the state; the midlands cover 40% of the area and provide shelter to 59% of the population (Yesodharan *et al.*, 2007). The state has a warm, humid monsoon climate.

The agricultural economy of Kerala has been undergoing structural transformation since the mid-1970s, switching over a large proportion of its traditional crop area, which had been devoted to subsistence crops, such as rice and tapioca, to more remunerative crops, such as coconut and rubber. None the less, the contribution of agriculture to the state income has been on the decline as the other sectors have registered higher rates of growth. During the year 2012/13, 8.95% of the gross state domestic product came from the agriculture sector (Kerala State Planning Board, 2013).

A survey conducted in Kerala during 1999–2000 showed that the level of mikania weed infestation was severe and that the percentage of areas infested was greater in the southern and central zones of the state (82.5 and 75.5%, respectively) than in the northern zone (45.3%) (Sankaran *et al.*, 2001). In the agroforestry production system, the severity of infestation was found to be high in seven districts: Thrissur, Ernakulam, Alappuzha, Idukki, Kottayam, Kollam and Pathanamthitta (Fig. 3.1) (Sankaran *et al.*, 2001). Although 92% of the agricultural systems surveyed were infested by mikania, in general, the severity of infestation had been reduced by smallholders in some areas through intensive weed management (Sankaran *et al.*, 2001). *M. micrantha* has since spread from Kerala into neighbouring states in India (Murphy *et al.*, 2000; Ellison and Murphy, 2001; Sankaran *et al.*, 2001; Sreenivasan, 2004; Reddy and Raju, 2009; Jagtap and Bachulkar, 2010).

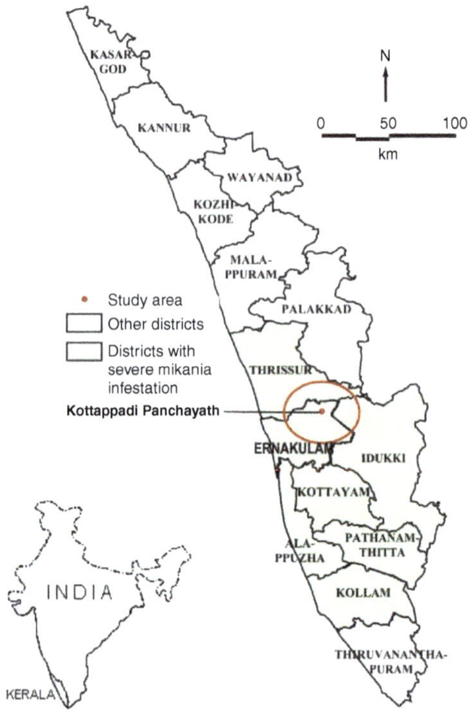

Fig. 3.1. Map of Kerala State, showing study area and other districts with severe mikania weed infestations.

Against this backdrop, the chapter examines and explains the socio-economic and management implications of mikania infestation and suggests a feasible and sustainable solution from the smallholder perspective. It is based on a study that used participatory rural assessments (PRAs) carried out with 50 households in the Kottappadi Grama Panchayath (village local self-governance area) of Ernakulam District in central Kerala (see Fig. 3.1).

Landholders and Lease Cultivation

The study focused on the impact of mikania weed on farmers with a cultivated land area of 0.2–4 ha in the Kerala region of the Western Ghats. The farmers were not a homogenous group. In terms of land ownership, they included traditional smallholders with full ownership, some who were marginal landholders and some who were leaseholders. Though the target group for this study was the traditional smallholder, the existence of leaseholder cultivators in the group made it difficult to distinguish between marginal landholders and smallholders. Hence, for the purposes of the study, the landholder types were broadly classified as smallholders (including lease cultivators), with 0.2–4.0 ha of land, and large landholders, with >4.0 ha of land (who were not included in the study); see Fig.3.2.

It was important to include lease cultivation (see Box 3.1) in the study, as it is an important feature of the agriculture scenario in the study area. Smallholders whose primary income source is agriculture often take land on lease for a period of 3 years and cultivate seasonal crops as a business venture. For example, pineapple cultivators take land planted with young rubber on lease and intercrop it with pineapple. The cost of an annual lease varies from Rs. 10,000 to 20,000 (Rs. 27,400–49,400/ha), depending on the relative bargaining power of the landholder and the lease cultivator. In the case of other crops, lease cultivators take uncultivated land and give one tenth of the produce as a lease payment. These cultivators grow high-yielding varieties, irrigate the land area and use chemical fertilizers to increase their output. They apply various chemicals for controlling mikania weed infestations. Lease cultivation is a short-term activity undertaken mainly for income generation. Thus, it differs in some respects – crop type, inputs, finance – from the long-term cultivation of a smallholder's own land because for lease cultivation the focus is on the financial return. In contrast, in conventional smallholder farming, the primary aim is food security. The different aims of the two types of cultivation lead to different reactions to mikania infestation.

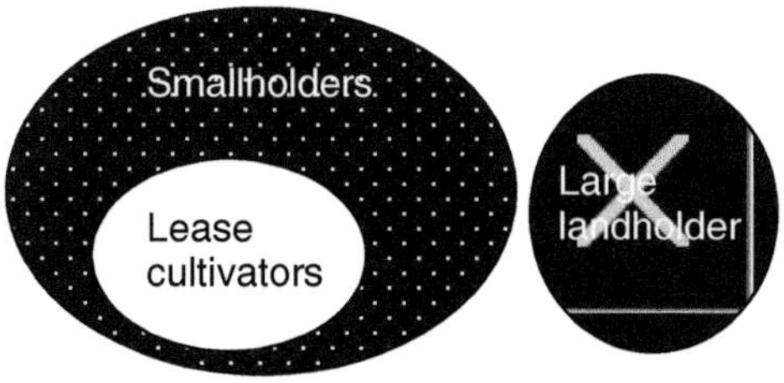

Fig. 3.2. Venn diagram illustrating the sample used in the study.

Cropping Patterns

Smallholders in the study area undertook mixed crop cultivation (Table 3.1). Coconut was the dominant crop, along with rubber

Table 3.1. Cropping patterns of smallholders in the study area.

Least susceptible crops		Highly susceptible crops	
Crop	Area (%)	Crop	Area (%)
Coconut	30.20	Ginger	2.95
Rubber	27.50	Tapioca	2.50
Paddy	24.30	Pineapple	2.40
Mango	2.01	Plantain	2.00
Jack fruit	1.80	Banana	1.50
Cashew	0.67	Pepper	0.22
Cocoa	0.60	Turmeric	0.20
Papaya	0.52		
Tamarind	0.40		
Areca nut	0.23		
Total	88.23	Total	11.77

Box 3.1. Characteristics of lease cultivation.

- Lease cultivators are smallholders who cultivate annual crops.
- Lease cultivation is profit oriented.
- The market mechanism of the system determines the cropping pattern.
- Both scientific and traditional methods of cultivation are used.
- Chemicals are applied for weed control.
- The social and environmental costs of chemical application are not considered.
- Mikania weed is a threat to profitability.

and paddy; together, they occupied 82% of the total cultivated area. During the initial stages of rubber cultivation, the planters intercropped pineapple with rubber. Crops highly susceptible to mikania weed were cultivated on only 11.77% of the total cultivated area in the region. Ginger, tapioca, pineapple, plantain and banana, the most susceptible crops, together constituted 96.4% of the total cultivated area of highly susceptible crops. None the less, farmers continued to invest in these highly susceptible crops for a number of reasons, which are summarized in Box 3.2.

The Infestation

A historical overview of mikania weed infestation in the study area (Table 3.2) was developed, based on the discussions with the sample households. It emerged that the weed was first reported in the area during the 1970s. After that, it underwent a rapid increase, and during the 1990s, the intensity of the infestation peaked, but currently it is on a decline in the area. Interestingly, in the initial stages of infestation, mikania had no positive effects, as it was not put to any use, though over time, with increasing infestation, its perceived positive effects have moved from low to medium.

Box 3.2. Factors motivating the persistent cultivation of crops highly susceptible to mikania weed.

- increasing market price;
- increasing demand;
- for self-consumption; and
- small size of the landholding required.

Table 3.2. History of mikania weed in smallholder crops, 1970–2000.

Year	Infestation	Positive effect	Negative effect
1970	Starting	Nil	Low
1980	Increasing	Low	High
1990	High	Medium	High
2000	Decreasing[a]	Medium	High

[a]Owing to periodic weed management – see text.

Detailed explorations indicated that the perceived severity of mikania infestations varied with the type of agricultural ownership and cultivation. Leasehold cultivators rated infestations as high owing to their adverse impact on production and profitability. Those who were engaged in agriculture merely as an additional source of income perceived the infestation as a small problem. The farmers who were engaged in breeding livestock along with cultivating perennial crops did not perceive the infestation as a problem because the weed was used as feed for their livestock – although there are reports indicating that mikania weed can cause hepatotoxicity and liver damage in dairy cattle (KAU, 1993).

Social and Economic Implications

The study locale covered an area of intensive cultivation. The agricultural output and costs of cultivation were estimated, and the responsiveness of the agricultural system to the employment of capital and labour was calculated. Agricultural output was estimated in monetary terms as the revenue from agriculture. The costs of capital incurred by highly susceptible and least susceptible crops were categorized into irrigation cost and implementation cost. Labour cost was categorized into costs of sowing, weeding, harvesting and 'other purposes'. The expenditure on crops highly susceptible to mikania and that on least susceptible crops was then compared.

It became clear that mikania weed could have variable economic effects, with both positive and negative impacts identified (Table 3.3). In addition, the negative impact of the weed on production and profitability in the agricultural production system was only moderate, as weeding was carried out intensively. On average, a household spent only 10–16% of the total production costs on weeding mikania. However, in forest plantations where weeding was not carried out intensively and regularly, the

Table 3.3. Social and economic impacts of mikania weed infestation.[a]

Social	Economic	
Positive	Positive	Negative
Women's employment (73)	Reduces fodder cost (67)	Reduces profitability (34)
Fodder value (67)	Increases livestock (47)	Absorbs fertilizers (27)
Medicinal value (7)	Increases rural incomes (39)	Increases cultivation costs (20)
		Reduces quality of agricultural produce (41.5)
		Reduces quantity of agricultural produce (14)

[a]Figures in parentheses: % of total sample reporting stated impact.

productivity and profitability were reduced drastically (Muraleedharan and Anitha, 2000). In moist deciduous forests, infestation by mikania weed is on the increase, which makes the harvesting of reeds, bamboos and other non-wood forest products laborious and time-consuming (Sankaran *et al.*, 2001).

Use of mikania weed as fodder

Mikania weed is used as a fodder in many countries. In Kerala, the weed is used in this way in some parts of the state, especially during summer when other fodder is scarce (Muraleedharan and Anitha, 2001). In the initial period of the infestation, mikania had no utility and the livestock rejected it as a fodder, but gradually it became necessary to accept it. Nowadays, it is used as fodder throughout the study area, and some smallholders are even ready to cultivate mikania on their land because of its fodder value. Among the sample households, 67% possessed livestock and all of these used mikania as fodder. Livestock offers an additional source of income, particularly to women, because this is something they take on in their free time, and they are keen to grow mikania on their land. (The ratio of women to men participating in livestock production is 83:17.) Waste from mikania fodder is also used as bio-fertilizer in the study area.

However, 41.5% of the sample households said that even though using mikania weed as fodder increases the quantity of milk production, it reduces its quality, and people only value the weed as fodder because of the scarcity of other fodders in the area. With the shift in the cropping pattern in Kerala, production of food crops has drastically declined in the state from 40% of net sown area in the 1960s to only 9.63% in the year 2012/13. This has simultaneously had an impact on the availability of straw as cattle feed (Kerala State Planning Board, 2013).

Other uses and benefits of mikania weed

Despite its use as fodder, the economic gain attributable to mikania weed is meagre compared with the losses caused by its weediness – though there are reports of various other uses and benefits of the plant (see Table 3.4). In Assam, the Kabi tribes use the juice from the leaf of mikania as an antidote for insect bites and scorpion stings, and the leaves are known to be used for treating stomach ache as well (Bora and Pandey, 1995). In Kerala, the Kadar tribes use juice from mikania leaf as an antidote for skin diseases (peeling skin) locally called *chuduvatham*. The use of

Table 3.4. Uses of and resulting benefits from mikania weed.[a]

Uses of mikania weed	Benefits from use
Fodder (67)	Increases livestock and rural income
Green manure (62)	Reduces cost of fertilizers
Controlling the growth of other weeds (49)	Reduces other weeding costs
Mulch (53)	Maintains humidity of soil
Medicine (7)	Reduces cost of medicine

[a]Figures in parentheses: % of total sample reporting use.

mikania weed juice as a curative agent for itching is reported from Malaysia. The weed has also been used as a cover crop in rubber plantations in Malaysia (Soong, 1976; Soong and Yap, 1976).

Multiplier impact of mikania weed on agricultural profitability

Mikania weed infestation has a multiplier impact on the profitability of agriculture because it affects the system in multidimensional ways. It reduces the agricultural productivity of an area, increases the costs of cultivation, and reduces the quality and quantity of the produce. Furthermore, it has an adverse impact outside the agricultural system by reducing biodiversity in the forest areas. In any production system, mikania infestation results in either cost escalation or income reduction, or both. From data collected during this study, it was calculated that one unit change in expenditure for weeding mikania owing to a high level of infestation leads to a 12.64 unit reduction in the profitability of the susceptible crops (authors' unpublished data).

How the infestation is perceived

In addition to the socio-economic sample of 50 households targeted by the PRAs, a total of 182 landholders with holdings of varying sizes were asked whether or not mikania weed was a problem to them. The size of landholding and the perception of the severity of the weed infestation were positively related. As the size of the holding increases, manual weeding becomes more difficult and more costly. Also, because livestock breeding is not carried out on a large scale and is only a supplementary income source, farmers are not able to effectively use the extra mikania as fodder if the infestation increases at a rapid rate. Among the 50 sample households, 67% rated the weed as good fodder, 47% rated it as an aggressive plant, and 20% rated it as an aggressive plant that increased the cost of agricultural cultivation (see Table 3.3). Moreover, statistical analysis found an exponential relationship between the size of the landholding and the probability of the landholder regarding mikania weed as a problem.

Management Options

Based on the results obtained during the study, we developed a model of the mikania weed infestation and its implications, and from that we proposed a feasible and sustainable solution, which is described below.

Current management systems

The current management system for mikania weed infestations used by smallholders differs from that used by the lease cultivators, though the socio-economic conditions are similar for both groups. There are three reasons for the emergence of this difference between the two groups:

- difference in the purpose of cultivation;
- difference in the method of cultivation; and
- difference in the term of dependence on the cultivable land for income.

Smallholders employ various methods for controlling mikania weed, which are summarized in Table 3.5. However, they tend to focus their efforts on the conventional methods of manual (blanket, circle, strip) weeding, because this ensures that the weed can be used as fodder and/or biofertilizer, and does not create any pollution. Furthermore, family labour is sufficient for manual weeding, and so no additional monetary cost is incurred. None the less, even using this method, the real cost of cultivation is increased. Smallholders state that using chemicals for weed control is costlier than weeding by conventional methods and, indeed, a recent study indicated manual weeding as the most commonly used method, followed by mechanical and chemical methods (Jones, 2012).

In contrast, lease cultivators prefer a combination of conventional manual weeding and chemicals (using herbicide). They carry out weeding by employing labour and thus incur a monetary cost. The use of chemicals for controlling weeds also creates various environmental problems, such as reducing soil fertility, and creating soil, water and air pollution in addition to related social problems. These farmers said that mikania weed poses a great threat to the profitability of their cultivation. Further, the weed is a threat to both large landholders and lease cultivators, as the infestation level is very high before control, which makes control difficult and not always effective; the net result is an adverse effect on the quality, quantity and profitability of agricultural cultivation for them (Fig. 3.3).

Table 3.5. Current mikania weed management methods among farmers.

Method of weeding[a]	% of sample
Blanket	20.0
Blanket and fodder	20.0
Fodder	13.2
Blanket and strip	13.3
Blanket, strip and fodder	6.7
Blanket, circle and fodder	6.7
Circle	6.7
Manual and chemical	6.7
No weeding	6.7

[a]Blanket – complete manual weeding of crop area;
Circle – manual weeding around each crop plant;
Fodder – weeded material fed to livestock;
Manual and chemical – combination of herbicide use and manual weeding;
Strip – manual weeding along crop plant rows.

In the case of smallholders, because of their scarce resources, they make the best use of mikania weed as fodder, green manure and mulch, and this has some positive impact on the rural economy. People's participation is a necessary factor for the effective management and efficient use of

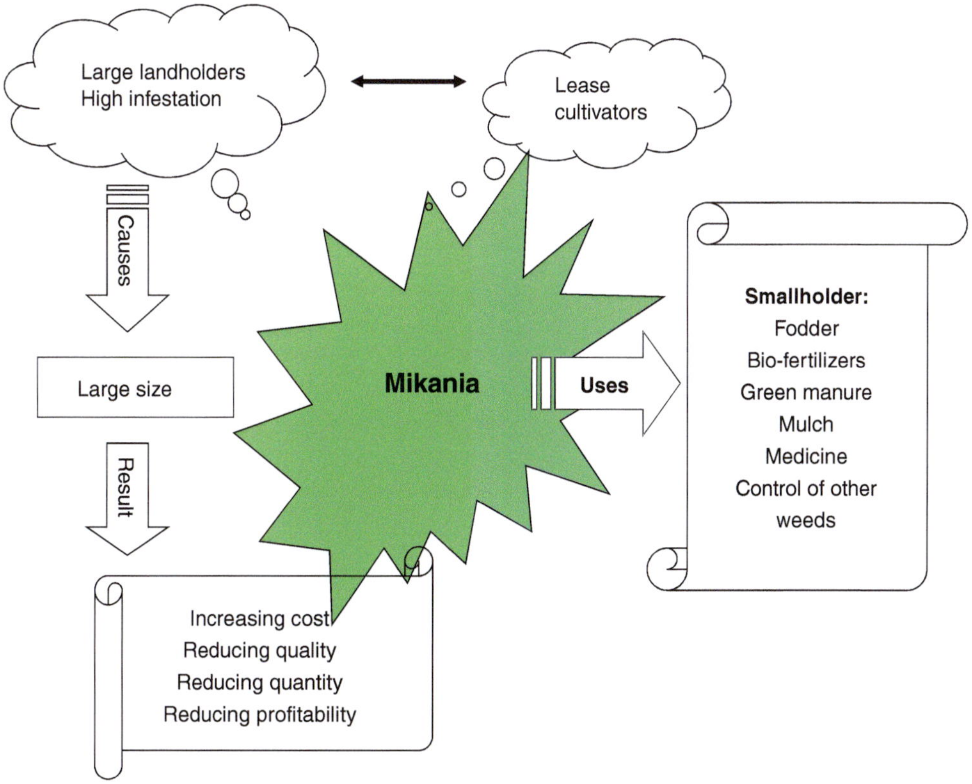

Fig. 3.3. Conceptual model of the mikania weed infestation/implications in the study area.

the weed, and social organizations such as environmental clubs, farmers' organizations and others should be encouraged to participate and contribute to mikania controlling programmes as an alternative integrated management option (Fig. 3.4).

Integrating management with biological control

As mentioned above, setting up mikania weed control groups at the village level is the first step towards effective management through the participation of the concerned stakeholders. These groups collect mikania from the severely affected areas and provide this to smallholders for use as fodder (see Fig. 3.4). The implementation of classical biological control in the Western Ghats using the rust fungus *Puccinia spegazzinii*, which is showing promise in Papua New Guinea (Day *et al.*, 2013), could bring a new dimension to mikania management. In this situation, mikania weed control groups could play a role in encouraging the use of classical biological control in the infested areas. The groups could also encourage people to find substitutes for the weed as fodder by increasing the cultivation of fodder grass in the areas where classical biological control is successfully implemented.

Conclusions

Mikania weed poses a serious threat to agriculturally dependent communities, especially smallholders undertaking lease cultivation in the affected area. The infestation has a multiplier impact on profitability as it reduces the agricultural productivity of the area, increases the cost of cultivation, and reduces the quality and quantity of the product. The size of landholding and the probability of considering the infestation as a problem are positively related. Although mikania reduces agricultural productivity,

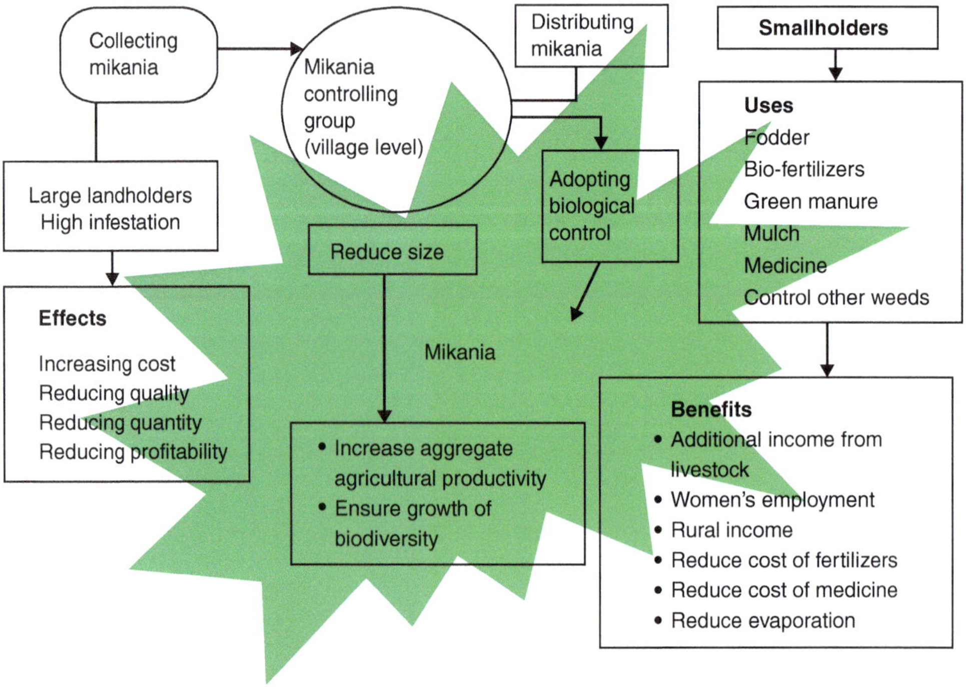

Fig. 3.4. Conceptual model of the mikania weed infestation/implications and a probable feasible and sustainable solution.

smallholders use conventional methods to weed it, and then use it as fodder and/or bio-fertilizer, a process that creates no pollution and incurs no direct costs (though it does increase the real cost of cultivation). While there is no clear knowledge of the total cost incurred from mikania management by the farmers, it is predicted to be high owing to the rapid growth and abundance of the weed. Moreover, decreasing diversity in agriculture as more and more farmers switch to monoculture cash crops is only likely to aggravate the problem. Last but not the least, with the threats from mikania weed predominantly occurring in the Western Ghats region, which is an area recognized as an important biodiversity hotspot, more targeted efforts to check the spread of the invasive plant are urgently required. However, the wider economic picture portrays a severe situation needing governmental intervention. An integrated approach that includes people's active participation, along with the government, as well as the support or intervention of non-governmental organizations (NGOs), is necessary for the effective management and efficient use of mikania weed. The goal should therefore be to consider the societal and sustainable livelihood concerns of the local community in the short run, and the sustainable development of the area in the long run, which will depend on both policy and extension support.

References

Bora, H.R. and Pandey, A.K (1995) Traditional uses of some weeds of Assam. *World Weeds* 2, 223–231.

Choudhury, A.K. (1972) Controversial *Mikania* (climber) – a threat to the forest and agriculture. *Indian Forester* 98, 178–186.

Day, M.D., Kawi, A., Tunabuna, A., Fidelis, J., Swamy, B., Ratutuni, J., Saul-Maora, J., Dewhurst, C.F. and Orapa, W. (2011) The distribution and socio-economic impacts of *Mikania micrantha* (Asteraceae) in Papua New Guinea and Fiji and prospects for its biocontrol. In: *23rd Asian-Pacific Weed Science Society Conference, The Sebel Cairns, Australia, 26–29 September 2011: Weed Management in a Changing World. Conference Proceedings, Volume 1.* Asian-Pacific Weed Science Society, pp. 146–153. Available at: http://apwss.org/apwss-publications.htm (accessed 22 August 2017).

Day, M.D., Kawi, A.P., Fidelis, J., Tunabuna, A., Orapa, W., Swamy, B., Ratutini, J., Saul-Maora, J. and Dewhurst, C.F. (2013) Biology, field release and monitoring of the rust fungus *Puccinia spegazzinii* (Pucciniales: Pucciniaceae), a biological control agent of *Mikania micrantha* (Asteraceae) in Papua New Guinea and Fiji. In: Wu, Y., Johnson, T., Singh, S., Raghu, S., Wheeler, G, Pratt, P., Warner, K., Center, T., Goolsby J. and Reardon, R. (eds) *Proceedings of the XIII International Symposium for Biological Control of Weeds, Waikoloa, Hawaii, 11–16 September 2011.* US Forest Service, Morgantown, West Virginia, pp. 211–217.

Dutta, A.C. (1977) Biology and control of *Mikania. Two and a Bud* 24, 17–20.

Ellison, C.A. and Murphy, S.T. (2001) *Dossier on:* Puccinia spegazzinii *de Toni (Basidiomycetes: Uridinales) a Potential Biological Control Agent for* Mikania micrantha *Kunth. ex H.B.K. (*Asteraceae*) in India.* Report for the Indian plant health authorities. CABI Bioscience UK Centre (Ascot), Ascot, UK.

Gopakumar, B. and Motwani, B. (2013) Factors restraining the natural regeneration of reed bamboo *Ochlandra travancorica* and *O. wightii* in Western Ghats, India. *Journal of Tropical Forest Science* 25, 250–258.

Jagtap, D. and Bachulkar, M. (2010) *Mikania micrantha* Kunth: addition to the *Asteraceae* flora of Maharashtra. *Journal of Economic and Taxonomic Botany* 34, 561–563.

Jones, K. (2012) Home gardens as a form of sustainable agriculture, invasive plants and resilience in social–ecological systems. A survey of home gardens in Kerala, India. MSc thesis, Royal Holloway, University of London, Egham, UK.

KAU (1993) *Annual Report of the AICRP on Weed Control.* Kerala Agriculture University, Thrissur, India, pp. 9–10.

Kaur, R., Malhotra, S. and Inderjit (2012) Effects of invasion of *Mikania micrantha* on germination of rice seedlings, plant richness, chemical properties and respiration of soil. *Biology and Fertility of Soils* 48, 481–488.

Kerala State Planning Board (2013) *Economic Review.* Kerala State Planning Board, Thiruvananthapuram, India, pp. 33–38 and 73–79.

KSLUB (1997) *Kerala State Resource Based Perspective Plan 2020 AD.* Kerala State Land Use Board, Thiruvananthapuram, India.

Macanawai, A.R., Day, M.D., Tumaneng-Diete, T. and Adkins, S.W. (2011) Impact of *Mikania micrantha* on crop production systems in Viti Levu, Fiji. In: *23rd Asian-Pacific Weed Science Society Conference, The Sebel Cairns, Australia, 26–29 September 2011: Weed Management in a Changing World. Conference Proceedings, Volume 1.* Asian-Pacific Weed Science Society, pp. 304–312. Available at: http://apwss.org/apwss-publications.htm (accessed 22 August 2017).

Muniappan, R. and Viraktamath, C.A. (1993) Invasive alien weeds in the Western Ghats. *Current Science* 64, 555–557.

Muraleedharan, P.K. and Anitha, V. (2000) The economic impact of *Mikania micrantha* on teak plantations in Kerala. *Indian Journal of Forestry* 23, 248–251.

Muraleedharan, P.K. and Anitha, V. (2001) The economic impact of *Mikania micrantha* on the agroforestry production system in the Western Ghats of Kerala. In: Sankaran, K.V., Murphy, S.T. and Evans, H.C. (eds) *Alien Weeds in Moist Tropical Zones: Banes and Benefits. Proceedings of a Workshop, Kerala Forest Research Institute, Peechi, India, 2–4 November 1999.* Kerala Forest Research Institute, Peechi, India and CABI Bioscience UK Centre (Ascot), Ascot, UK, pp. 80–85.

Murphy, S.T., Ellison, C.A. and Sankaran, K.V. (2000) *The Development of a Biocontrol Strategy for the Management of the Alien Perennial Weed,* Mikania micrantha *H.B.K. (*Asteraceae*) in the Tree Crop Based Farming Systems in India.* Final technical report for DFID Project No. R 6735. CABI Bioscience UK Centre (Ascot), Ascot, UK.

Murphy, S.T., Subedi, N., Jnawali, S.R., Lamichhane, B.R., Upadhyay, G.P., Kock, R. and Amin, R. (2013) Invasive mikania in Chitwan National Park, Nepal: the threat to the greater one-horned rhinoceros *Rhinoceros unicornis* and factors driving the invasion. *Oryx* 47, 361–368.

Palit, S. (1981) *Mikania* – a growing menace in plantation forestry in West Bengal. *Indian Forester* 107, 96–101.

Ramakrishnan, P.S. (2001) Biological invasion as a component of global change: the Indian context. In: Sankaran, K.V., Murphy, S.T. and Evans, H.C. (eds) *Alien Weeds in Moist Tropical Zones: Banes and Benefits. Proceedings of a Workshop, Kerala Forest Research Institute, Peechi, India, 2–4 November 1999.* Kerala Forest Research Institute, Peechi, India and CABI Bioscience UK Centre (Ascot), Ascot, UK, pp. 28–34.

Reddy, C.S. and Raju, V.S. (2009) *Aeschynomene americana* L. and *Mikania micrantha* Kunth – new invasive weeds in flora of Andhra Pradesh, India. *Journal of Economic and Taxonomic Botany* 33, 540–541.

Sankaran, K.V., Muraleedharan, P.K. and Anitha, V. (2001) *Integrated Management of the Alien Invasive Weed* Mikania micrantha *in the Western Ghats.* KFRI Research Report No. 202, Kerala Forest Research Institute, Peechi, India, pp. 35–45.

Sen Sarma, P.K. and Mishra, S.C. (1986) Biological control of forest weeds in India – retrospect and prospects. *Indian Forester* 112, 1088–1093.

Singh, S.P. (2001) Biological control of invasive weeds in India. In: Sankaran, K.V., Murphy, S.T. and Evans, H.C. (eds) *Alien Weeds in Moist Tropical Zones: Banes and Benefits. Proceedings of a Workshop, Kerala Forest Research Institute, Peechi, India, 2–4 November 1999.* Kerala Forest Research Institute, Peechi, India and CABI Bioscience UK Centre (Ascot), Ascot, UK, pp. 11–19.

Soerjani, M. (1977) Weed management and weed science development in Indonesia. In: Day, M.D. and McFadyen, R.E. (eds) *Chromolaena in the Asia–Pacific Region. Proceedings of the Sixth Asian-Pacific Weed Science Society Conference, Jakarta, Indonesia, 11–17 July 1977, Vol. 1.* Asian-Pacific Weed Science Society, pp. 31–41. Available at: http://apwss.org/apwss-publications.htm (accessed 22 August 2017).

Soong, N.K. (1976) Feeder root development of *Hevea brasiliensis* in relation to clones and environment. *Journal of the Rubber Research Institute of Malaya* 24, 283–298.

Soong, N.K. and Yap, W.C. (1976) Effects of cover management on physical properties of rubber growing soil. *Journal of the Rubber Research Institute of Malaya* 24, 145–159.

Sreenivasan, M.A. (2004) Natural distribution and control of the invasive weed *Mikania micrantha* in the Western Ghats. PhD thesis, Forest Research Institute, Dehra Dun, India.

Yesodharan, E.P., Kokkal, K., Harinarayanan, P. (eds) (2007) *State of Environment Report – Kerala – Land Environment, Wetlands of Kerala and Environmental Health, Vol. 1.* Kerala State Council for Science, Technology and Environment, Government of Kerala, Thiruvananthapuram, India.

4 Impacts and Management Options for *Mikania micrantha* in Plantations

K.V. Sankaran,[1]* Soekisman Tjitrosemito[2] and Soetikno S. Sastroutomo[3]

[1]Kerala Forest Research Institute (KFRI), Peechi, India; [2]SEAMEO-BIOTROP, Bogor, Indonesia; [3]CABI, Serdang, Malaysia

Introduction

Plantations under various commercial crops and forest species across the globe are already extensive in area and are expanding. The benefits from these have been economic, environmental and social, and there are many opportunities and needs for their further development (Nambiar and Brown, 1997). The occasional failure of these ventures can be attributed to a variety of factors, one of the foremost in the tropics and subtropics being invasion by alien plants. Invasion by *Mikania micrantha* (mikania), a perennial climbing vine native to Central and South America, is a good case in point. Mikania is one of the most serious invasive species (one among the world's ten worst weeds) in plantations, agroforestry and agricultural systems, and natural forests in the Asia–Pacific region and in Miami-Dade County in Florida in the USA (Holm *et al.*, 1991; Derksen and Dixon, 2009; Sankaran and Suresh, 2013a). The various impacts of mikania in plantations and commercial crops in the Asia–Pacific region have been well documented, as is summarized in Table 4.1. Plantation crops are particularly vulnerable to mikania in the establishment phase because it can thrive in moist fertile soil with high light intensity. It may cease to be a problem in mature plantations of crops such as oil palm and rubber as the plant cannot grow under a closed canopy. However, in most cases, the long-term productivity of the crops is seriously affected by weed-impaired early growth. This chapter looks at the impact of mikania in plantation crops and reviews options for its management.

The potential of mikania to kill immature plants/young trees by overgrowing and smothering them, by competing for light, soil water and nutrients, and by releasing allelopathic substances that inhibit their growth and regeneration, is well known (Teoh *et al.*, 1985; Sankaran *et al.*, 2001; Matthews, 2004). Mikania also interferes with flowering, natural pollination by insects and the harvesting of crops (Day *et al.*, 2012). The presence of the weed alters soil chemical characteristics and the soil microbial community, possibly creating a favourable condition for its prolific growth and spread (Li *et al.*, 2006). Studies in southern China have shown that there is local adaptation at the genome level in mikania and that this may represent a major evolutionary mechanism for successful invasion wherever the species is introduced (Wang *et al.*, 2008). The weed grows rapidly – up to 9 cm in 24 h – and can produce enormous amounts of seed, which can then be dispersed by wind or by attachment to machinery, animal fur and human clothing or possessions (Waterhouse and Norris 1987; Holm *et al.*, 1991).

Mikania was introduced as a cover crop into rubber and oil palm plantations in

Corresponding author: E-mail: sankarankv@gmail.com

Table 4.1. Plantations and commercial crops affected by *Mikania micrantha*.

Crop	Impact	Location	Reference
Acacia auriculiformis	Smothers young trees	India, Kerala	Sankaran *et al.*, 2001
Aibika (*Abelmoschus manihot*)	Outcompetes	Papua New Guinea	Day *et al.*, 2012
Ailanthus triphysa	Smothers young trees	India, Kerala	Sankaran *et al.*, 2001
Albizia (*Falcataria moluccana*)	Smothers young trees	India, Kerala	Sankaran *et al.*, 2001
Areca nut (*Areca catechu*)	Heavy infestation	India, Kerala	Abraham and Abraham, 2002
Bamboo (*Bambusa arundinacea*)	Smothers and kills entire clumps	India, Kerala	Sankaran *et al.*, 2001
Bamboo	Yield loss (up to 25% when infestations are severe)	China, Yunnan	Shen *et al.*, 2013
Dendrocalamus asper	Smothers and kills young clumps	Indonesia, Sumatra	Widjaja and Tjitrosoedirdjo, 1991
Dendrocalamus strictus	Smothers and kills entire clumps	India, Kerala	Sankaran *et al.*, 2001
Banana/plantain (*Musa* spp.)	Flowering delayed, fruit yield reduced in experimental situation	India, Kerala	Abraham *et al.*, 2002a
	Overgrows and kills young plants; larger plants enshrouded, failed to yield fruit	India, Kerala	Sankaran *et al.*, 2001
		Papua New Guinea	Day *et al.*, 2011, 2012
	Heavy infestation, smothers crop	Fiji, Viti Levu	Macanawai *et al.*, 2011
	Loss in yield (up to 65% when infestations are severe)	China, Yunnan	Shen *et al.*, 2013
Cassava (*Manihot esculenta*)	Overgrows and affects growth	India, Kerala	Sankaran *et al.*, 2001
		Papua New Guinea	Day *et al.*, 2011
		Fiji, Viti Levu	Macanawai *et al.*, 2011
Casuarina (*Casuarina equisetifolia*)	Smothers young trees	India, Kerala	Sankaran *et al.*, 2001
Citrus spp.	Crop infestation frequent	India, north-east	Gogoi, 2001

Citrus (orange, lemon, shaddock/pomelo)	Yield loss (up to 78% when infestation is severe)	China, Yunnan	Shen *et al.*, 2013
Cocoa (*Theobroma cacao*)	Highly competitive in young crops; smothers and retards growth	Malaysia, West	Teoh *et al.*, 1985
		India, Kerala	Sankaran *et al.*, 2001; Abraham *et al.*, 2002a
	Smothers and retards growth; delay in flowering and fruiting; lower yields	Papua New Guinea	Day *et al.*, 2011, 2012
Coconut (*Cocos nucifera*)	Heavy infestation	Indonesia	Wirjahardja, 1975
	Highly competitive in young crops	Malaysia, West	Teoh *et al.*, 1985
	Smothers young palms, retards growth	India, Kerala	Sankaran *et al.*, 2001; Abraham *et al.*, 2002a
	Heavy infestation; plantations abandoned	Samoa	Brooks *et al.*, 2008
	Overgrows and retards growth of young, poorly managed palms; reduced flowering/fruiting; impedes harvesting of fallen nuts	Papua New Guinea	Day *et al.*, 2011, 2012
Coffee (*Coffea arabica*)	Overgrows and reduces crop growth and yield	India, Assam	Gogoi, 2001
	Smothers bushes	India, Kerala	Sankaran *et al.*, 2001
		Papua New Guinea	Jones, 2012
	Outcompetes/smothers plants	Papua New Guinea	Day *et al.*, 2012
Cotton (*Gossypium herbaceum*)	Crop infestation frequent	India, Assam	Gogoi, 2001
Eucalypt (*Eucalyptus tereticornis*)	Serious impact on growth; smothers trees	India, Kerala	Sankaran *et al.*, 2001
Ginger (*Zingiber officinale*)	Smothers crop	India, Assam	Gogoi, 2001
		India, Kerala	Sankaran *et al.*, 2001
Kava (*Piper methysticum*)	Smothers crop	Papua New Guinea	Jones, 2012
Manila copal (*Agathis* spp.)	Competes for nutrients, space, light; mechanical damage	Indonesia	Suharti and Sudjud, 1978
Maize (*Zea mays*)	Infests the crop field	India, Assam	Gogoi, 2001

continued

Table 4.1. *continued*

Crop	Impact	Location	Reference
Oil palm (*Elaeis guineensis*)	Highly competitive in young crops; reduces yield by ≤20% for 5 years post establishment	Malaysia, West	Gray and Hew, 1968; Teoh *et al.*, 1985
	Forms <30% ground cover in young plantations; replaced by other species as shade develops in mature plantations	Malaysia	Teng and Teh, 1990
	Severe infestation in young crops	India, Kerala	Sankaran and Sreenivasan, 2001
	Smothers young trees; escalates weeding costs	Indonesia, Java	Tjitrosemito, 2005
	Depresses growth and yield; hinders agronomic operations	Malaysia	Chung, 2010
	Overgrows young crops and retards growth; reduced flowering/fruiting; interferes with harvesting	Papua New Guinea	Day *et al.*, 2011, 2012
Papaya (*Carica papaya*)	Overgrows and leaves crop in poor condition	Papua New Guinea	Day *et al.*, 2011, 2012
Pine (*Pinus merkusii*)	Competes for nutrients, space and light; mechanical damage	Indonesia	Suharti and Sudjud, 1978
Pineapple (*Ananas comosus*)	Severely affects crop; flowering delayed, fruit yield reduced	India, Kerala	Sankaran *et al.*, 2001; Abraham *et al.*, 2002a
	Reduces crop growth and yield	India, Assam	Gogoi, 2001
Reed (*Ochlandra travancorensis* and *O. wightii*)	Smothers crop; affects regeneration	India, Kerala	Sankaran *et al.*, 2001; Gopakumar and Matwani, 2013
Rubber (*Hevea brasiliensis*)	Retarded rubber growth and delayed tapping by ≥1 year	Indonesia	Mangoensoekarjo and Soewadji, 1973
	Highly competitive in young crops, reduces early growth and yield; invades leguminous cover crops; increases time to tappable age; tolerated in mature crops	Malaysia	Teoh *et al.*, 1985
	Dominant in the pre-tapping phase; replaced by other weeds (including *Chromolaena odorata*) as canopy closes; retards growth in nurseries and young plantations	India, Kerala	Abraham and Abraham, 2000; Sankaran and Sreenivasan, 2001

Sesame (*Sesamum indicum*)	Infests the crop field	India, Assam	Gogoi, 2001
Sugarcane (*Saccharum officinarum*)	Overgrows, causing yield loss	Indonesia, Sumatra	Widyatmoko and Riyanto, 1986
	Reduces crop growth and yield	India, Assam	Baruah and Gogoi, 1995; Gogoi, 2001
	Heavy infestation, smothers crop	Fiji, Viti Levu	Macanawai *et al.*, 2011
	Yield loss (up to70% when infestations are severe)	China, Yunnan	Shen *et al.*, 2013
Sweet potato (*Ipomoea batatas*)	Smothers and kills the crop	Papua New Guinea	Day *et al.*, 2012
Taro (*Colocasia esculenta*)	Smothers and leaves crop in poor condition	Papua New Guinea	Day *et al.*, 2011, 2012
	Heavy infestation	Fiji, Viti Levu	Macanawai *et al.*, 2011
Tea (*Camellia sinensis*)	Smothers the crop	India, Kerala	Sankaran *et al.*, 2001
	Large impact ≤4 years after tea planting; sequesters substantial amounts of nitrogen; long term damage via impacts on branching and frame development; in mature crops, grows into and over surface, hampering harvesting; escalates production costs so threatening viability; host for *Helopeltis* tea pest	India, Assam	Abraham *et al.*, 2002b; Ellison, 2004; Rajkhowa *et al.*, 2005; Puzari *et al.*, 2010
	Yield loss (up to 36% when infestations are severe)	China, Yunnan	Shen *et al.*, 2013
Teak (*Tectona grandis*)	Competes for nutrients, space, light; causes mechanical damage	Indonesia	Suharti and Sudjud, 1978
	75% of plantations infested; seedlings ≤3 years old most damaged and can be smothered; increases planting and maintenance costs and reduces profitability; has impact on state forestry income by impeding harvest	India, Kerala	Muraleedharan and Anitha, 2000; Sankaran and Pandalai, 2004
Turmeric (*Curcuma longa*)	Infests the crop field	India, Assam	Gogoi, 2001
Vanilla (*Vanilla planifolia*)	Smothers and reduces yield	Papua New Guinea	Jones, 2012
Yam (*Dioscorea* sp.)	Outcompetes	Papua New Guinea	Day *et al.*, 2012

Malaysia and Indonesia in the 1950s (Wirjahardja, 1975; Alif, 2001). Evidence from these countries shows that the species may invade and compete with legume cover crops in young rubber and oil palm, thereby adversely affecting the establishment of plantations. Experimental studies in Kerala in south-west India illustrated the potential for damage to plantation species: they showed that the growth of cocoa, rubber, coconut and teak seedlings, and banana and pineapple suckers, was significantly reduced when plants were grown in competition with even a single mikania plant (Abraham *et al.*, 2002a).

Impacts of *Mikania micrantha* in Plantations

Plantation crops in Malaysia and Indonesia

Mikania has been listed as one among the most important weeds that invade plantation crops and forest plantations in Malaysia (Alif, 2001). Teoh *et al.* (1985) reported that it competes with immature rubber, oil palm and cocoa, and that it poses a threat to tea, coffee and coconut plantations and fruit orchards. In spite of this, there have been few studies on the economic impact of mikania in plantations. Teoh *et al.* (1985) summarized the results of a questionnaire-based survey conducted in 1982, by which date mikania had already been classified as Class C vegetation (undesirable or noxious) in rubber and oil palm. The survey covered almost all of the growers of rubber, oil palm and cocoa in both plantation and smallholder sectors, with a few reporting on coconut and 'other' (mostly orchard) crops as well. The growers reported mikania as especially troublesome in young plantations, with over half the respondents considering it highly competitive in young rubber (55%), oil palm (51%), cocoa (53%), coconut (61%) and other crops (50%); at least three-quarters of them considered it to be competitive (or worse) for all crops. However, it was less of a problem in mature plantations, especially those of rubber and oil palm, at which stage many estates ceased to control it.

As mikania control in plantations was undertaken as part of overall weed control, Teoh and co-workers could not be precise about costs but, using two different methods, they estimated that in the early 1980s, 5% of total annual spending on herbicides in the country, and 6% of annual weed control costs in the three plantation crops, were devoted to control mikania. Oil palm and rubber remain Malaysia's most important export crops, with cocoa increasing in significance as world demand expands. It is clear from this study that mikania is therefore not merely a plantation weed, but also a constraint to the exports and economy of Malaysia.

Estimates based on the above survey in rubber and oil palm by Teoh *et al.* (1985) were supported by research data. Large-scale trials in rubber plantations have shown that in plots with mikania, tree girth was reduced by 27% and cumulative yields by 27–29% during the first 32 months of production, compared with trees under leguminous cover (Watson *et al.*, 1964; RRIM, 1965). Moreover, rubber trees grown with mikania took longer to reach tappable age (Watson *et al.*, 1964; Ti *et al.*, 1971).

Chung (2010) noted that mikania significantly depresses growth and yield, and also hinders agronomic operations in oil palm. He reported significant nutrient immobilization by mikania grown as a cover crop under young (1 year old) oil palm, and cites earlier research by Gray and Hew (1968) which showed that mikania ground cover reduced fresh fruit bunch yield in oil palm by almost 20% compared with *Pueraria/Centrosema* leguminous cover; this represented a loss of 23.8 t/ha over the 4.75 years during which data were recorded. Similarly, in West Java, Indonesia, where mikania was one of the four most abundant weeds in a young (≤3 month old) oil palm plantation, controlling it in the first year of planting required six weeding operations immediately under the oil palms and three between the rows. This equated to 36 man-days for weeding rows, or 18 man-days for weeding around each tree on a per hectare basis (assuming 130 plants/ha) (Tjitrosemito, 2005). Indeed, Chung (2010) commented that one of the challenges in weed and cover crop management in oil palm plantations is the shortage of labour, especially of skilled labour.

Commercial/semi-commercial enterprises in Papua New Guinea and Fiji

Mikania is a noxious invader in many cropping systems in the Pacific and field surveys have confirmed that it is widespread in Papua New Guinea (Day *et al.*, 2011). It smothers cash crops such as banana and taro, often killing the plants, and overgrows cocoa, young oil palms and coconut palms, retarding their growth and yield. The weed is also known to reduce flowering and pollination and to interfere with the harvesting of coconut (by smothering fallen nuts), oil palm and cocoa (Day *et al.*, 2011), which happen to be three of the four most important export crops of Papua New Guinea (FAOSTAT, 2011).

A socio-economic survey among farmers in Papua New Guinea revealed that crop losses attributable to mikania were rated as high as 30% for the worst affected land by 40% of the respondents, though less than 30% of these were commercial or semi-commercial producers (Day *et al.*, 2011). Of those respondents, 79% considered mikania to be a serious weed in their cropping systems. The weed has become more invasive in the country over the past decade and is now present in all 15 lowland provinces and all agricultural systems (except for grazing land); it has had severe impacts on production in many crops via a combination of reduced yields and increased weeding costs (Day *et al.*, 2012). Some 75% of commercial/semi-commercial enterprises judged that yield was reduced by mikania. Poorly managed plantations of young cocoa and oil palm were often covered with the weed. A field trial in Papua New Guinea's West New Britain Province looked at the impact of mikania in young banana plantations. Those left unweeded (at 1 m tall) were overgrown by 4 months, and were dead at 6 months, having been smothered by the weed. In the same study, unweeded larger banana plants were shown to be enshrouded by the weed within 6 months and failed to produce fruit (Day *et al.*, 2012).

An implication of this study is that although estates may have the financial resources to cope with mikania, and in many cases are implementing good control strategies, this is at considerable cost because of the scale of the mikania problem. Where weed control in plantations is suboptimal, yield, income and the sustainability of the enterprise are compromised.

Mikania was first reported in Fiji in 1907, and the damage that it causes to root crops (taro and cassava) and in plantations of banana and sugarcane has been noticed for around 50 years (Macanawai *et al.*, 2011). A questionnaire survey conducted among farmers in Fiji showed that the negative impact of mikania has been recognized by a large majority in the root crop and sugarcane growing areas. The majority of these farmers (over 94%) indicated that they control the weed to prevent crop loss, but some do so to prevent the spread of mikania to other areas. This study showed that there is a substantial cost associated with mikania infestations in crop production, with the annual cost to farmers on average US$25/ha in root crops and US$18/ha in sugarcane (Macanawai *et al.*, 2011). In a similar survey conducted by Day *et al.* (2011) among farmers in four islands in Fiji, 60% of respondents considered mikania to be a serious weed, with losses of about 30% of the potential crop yield due to the weed, while 33% had more than 30% of their farms infested. Only 15% of respondents needed to weed fortnightly, with 56% using slashing and/or hand pulling as the main means of control.

Tea in Assam, India

Mikania can be found in tea plantations in Kerala, but the high altitude where tea is grown is not conducive to rapid growth of the weed, and this is also the case in most of the tea-growing regions of Asia. However, the impact of the weed is particularly evident in the tea gardens of Assam in north-eastern India, where it invades from the forests as the rains begin. The visual effect of the fast-growing weed moving relentlessly into the crop and smothering it has been likened to a green tidal wave (Ellison, 2004). This low-lying north-eastern Indian state, criss-crossed by the Brahmaputra River and its tributaries, was originally covered with tropical rainforest, but is now the world's largest contiguous tea-growing region and produces more than half of India's tea crop. Although mikania has been recognized as a weed in Assam since the 1940s, its importance has escalated in recent decades along with the large-scale degradation of the natural forests and their invasion by mikania. From these strongholds, mikania has been able to invade the tea gardens. Mikania smothers tea plants by growing into and over the canopy, and seriously disrupts plucking by growing among new, harvestable shoots. The increasing weed burden since the emergence of mikania means that control costs have escalated, making some practices economically unviable (Puzari *et al.*, 2010).

In Assam, the mikania infestation is at its worst in April–September, which includes the major tea plucking periods (Barbora, 2001). Tea is most susceptible to weed competition in the first 3–4 years after planting (Ellison, 2004; Rajkhowa *et al.*, 2005), and failure to control weeds in tea plants under 4 years old can seriously affect the long-term productivity of the bushes, because the weeds affect frame formation (Barbora, 2001). However, mature tea is also directly affected by the weed interfering with harvesting. Growers may be forced into a trade-off between the cost of control and yield loss. A study assessing the impact of mikania on cultivation and control costs and tea yields (Puzari *et al.*, 2010) found that heavy infestations inflicted losses of 19–42% through a combination of reduced yield and increased weed management (labour) costs. Where weed control is not carried out intensively and regularly, productivity and profitability are drastically reduced.

An indirect threat of mikania weed to tea cultivation was revealed by research conducted in Kerala. As part of initiatives to develop more sustainable control methods, surveys were made for natural insect enemies of mikania, but the pests recorded were polyphagous, and a number of them were pests on some crops (Abraham *et al.* 2002b). Notably, in this context, the mirid bug *Helopeltis theivora*, which was found to be damaging to mikania in Kerala, is a serious pest of tea in Assam and elsewhere (as the tea bug or tea mosquito); it is also a pest on cocoa in tropical Asia (Manabendra and Rudrapal, 2011). This brings new dimensions to the problem, for by acting as an alternate host during the non-cropping season, mikania has the potential to act as a reservoir for the populations of this insect pest. In short, in areas infested by the tea mosquito, both the weed and the insect pests may be poised to move into tea at the start of the growing season. Thus, the dual role of mikania as a reservoir for the tea bug and as a weed could indirectly lead to an increase in the use of insecticide as well as herbicide in tea.

Timber in Kerala, India

A survey to assess the occurrence, spread and severity of mikania infestation in natural forests, forest plantations and agricultural systems of Kerala's Western Ghats showed that the weed has become widespread in the 35 years since it was first reported in the state (Nair, 1968). Among forest plantations, 78% of teak, 38% of eucalypt and 88% of other miscellaneous species were affected, including *Acacia auriculiformis* and *Falcataria molucanna* (Sankaran and Pandalai, 2004).

Teak (*Tectona grandis*) is one of the most valuable timbers in the world and the principal timber tree of peninsular India (CABI, 2005). It comprises over half the plantation area in Kerala (75,258 ha; KFD, 2013). Its seedlings are sensitive to weed competition during the initial 2–3 years of establishment (Tewari, 1992). Young teak is severely affected by mikania more than any other timber species, probably because of the favourable conditions provided by the lighter canopy

(Sankaran *et al.*, 2001). Mikania infestation at this stage affects the growth and productivity of young trees and even smothers them (Muraleedharan and Anitha, 2000).

An assessment of the impact of mikania on teak plantations in Kerala showed that the weed adversely affects the costs of planting and maintenance of plantations, and income from them, with a per hectare overall difference of Rs. 6274 (*c.*US$100 based on current conversion rates) between infested and uninfested plots for the first 8 years after planting (Muraleedharan and Anitha, 2000). Lack of adequate funds and labour shortage frustrate periodic weeding in young plantations, which results in poor growth or death of trees (Muraleedharan and Anitha, 2000, Sankaran *et al.*, 2001). In summary, current cultural practices and thin canopies favour mikania invasion in teak plantations in Kerala, and where the weed is not managed intensively and regularly, productivity and profitability are significantly reduced (Muraleedharan and Anitha, 2000).

Control of *Mikania micrantha*

In most plantation situations, especially young plantations, mikania will form part of a weed flora, and control strategies will vary with weed composition (Barbora, 2001; Tjitrosemito, 2005). Controlling weeds is essential to maintaining crop growth and yield, and represents a substantial proportion of labour and production costs in many plantation systems, especially during the establishment phase (Mangoensoekarjo, 1978; Teoh *et al.*, 1985; Barbora, 2001; Tjitrosemito, 2005). Options for controlling mikania in plantations include manual/mechanical measures and the application of herbicides but, as this section will indicate, there are often significant constraints to implementing these methods.

Physical control

Sickle weeding, uprooting and digging are the main physical control measures for mikania in practice. Hoeing, shovelling, tilling and mowing are also used infrequently. However, all these methods need to be practised before the weed flowers and produces seeds. Sickle weeding/slashing at the base will give temporary control, but quick regrowth may occur from cut stems and the underground stolon. Furthermore, the slashed/uprooted biomass must be collected carefully and dried/burnt to avoid regrowth from leftover stem fragments. Physical removal of mikania after seed set will enhance spread compared with natural means of spreading. Uprooting during the initial stages of growth is the most effective physical control method. None the less, this has its limitations because of the enormous amounts of easily dispersed, viable seeds, which can be carried by wind and can reinvade the weed-cleared areas. Kuo *et al.* (2002) reported that in Taiwan, cutting mikania near the ground once a month for 3 (consecutive) months during the summer and autumn eliminated 90% of the vines, though this method was less effective in the winter and spring. In Fiji, a small group of farmers opted for weeding at fortnightly intervals, whereas others used slashing and/or hand pulling as the main means to control mikania (Day *et al.*, 2011). According to Alif (2001), manual control of mikania in newly infested areas by rolling, drying and burning was unsustainable in Malaysia. In tea in north-eastern India, Puzari *et al.* (2010) described how mikania's aggressive creeping and twining habit through and over the plucking surface of tea bushes makes manual weeding time-consuming and causes damage to the fragile new tea shoots.

Hoeing is used in tea in Assam to remove mikania, especially in nurseries and young plantations (Rajkhowa *et al.*, 2005; Puzari *et al.*, 2010), but this can result in damage to the roots of young tea, which lie close to the surface, as well as causing increased weed growth from the soil seed bank. All of the physical methods of control discussed above are generally labour intensive, costly, time-consuming, unsustainable and inefficient in bringing about effective long-term control. Some of these methods can also disturb the soil, resulting in erosion, especially on steep slopes (Cock *et al.*, 2000). Soil erosion caused by clean weeding in tea estates in the high rainfall areas of north Sumatra, Indonesia, led to herbicide options being explored (Sutedjo and Lubis, 1971). In less fragile environments, and where crop spacing allows (e.g. timber plantations in India), mechanical control can be used, but it is not economically feasible because weeding needs to be done several times a year, and herbicide use may prove cheaper (see Box 4.1 below).

Mikania has been used as herbage for goats and sheep in Java, but it is less valuable as a fodder crop than pasture species (Waterhouse and Norris, 1987). Sheep grazing has been reported to be commonly used in Malaysia to remove weeds from leguminous cover crops in rubber plantations (Chee *et al.*, 1992). The use of mikania as a fodder during the summer season has also been reported from Kerala (Sankaran *et al.*, 2001), although a case of suspected poisoning of cattle after grazing mikania was recorded from northern Sumatra in 1984 (Murdiati and Stoltz, 1987). So the fodder value of the weed still needs to be properly ascertained.

Chemical control

In Indonesia, short-term to seasonal control of mikania in plantations has been achieved with a variety of herbicide combinations and spraying schedules (Mangoensoekarjo, 1978; Hutauruk *et al.*, 1982; Teng and Teh, 1990). The use of herbicide mixtures will allow reduction in costs and improve efficacy. Seeds of mikania do not undergo dormancy so are amenable to pre-emergence herbicides such as diuron. The weed was reported to have brought under control in sugarcane in 4 years using diuron + 2,4-D, followed by paraquat (Widyatmoko and Ryanto, 1986). Foliar herbicides, such as 2,4-D (sodium/dimethylamine salts), glyphosate, paraquat, diuron, oxyflurofen, dalapon and triclopyr, are widely used in plantations. Successful foliar treatment requires precise timing, i.e. the weed needs to be actively growing, but treatment needs to be done before it has an impact on crop growth/yield and in synchrony with fertilizer application; above all, the mikania needs to be controlled before it flowers and sets seed (Barbora, 2001).

Herbicide use also requires well-maintained, appropriate spray equipment (Chung *et al.*, 2000) and appropriate product choice for stage of the crop, weed composition and season (Barbora, 2001; Rajkhowa *et al.*, 2005). Additives may allow doses to be reduced; for instance, in forest plantations in Kerala the efficacy of glyphosate and diuron was increased by adding ammonium sulphate and urea, respectively (Sankaran and Pandalai, 2004). Most importantly, label instructions must be followed to achieve maximum efficacy and cost-effectiveness, and to minimize non-target effects, such as those reported for 2,4-D applied against mikania in rubber and oil palm in Malaysia (Teoh *et al.*, 1985). In northern Sumatra, glyphosate was recommended in place of 2,4-D for the control of mikania in immature oil palm because it was less toxic (Mangoensoekarjo, 1979). Based on intensive herbicidal trials in mikania-infested farmland under various crops in Yunnan, China, Shen *et al.* (2013) recommended that it is much safer to use atrazine in sugarcane and rubber, glyphosate in non-farming land and rubber, and 2,4-D in maize.

Sankaran and Pandalai (2004) developed a herbicide-based strategy for forest plantations in Kerala that was cheaper and more effective than mechanical interventions (see the case study in Box 4.1). This work did highlight an important issue: how the non-availability of herbicide products may hinder control. For example, one of the effective treatments used by these authors for mikania in teak and eucalypt plantations – a mixture of triclopyr and picloram (Grazon DS) – was not available commercially in India during the time of the study.

Following the above recommendations, Brooks and Setter (2014) conducted herbicidal trials on mikania in Queensland, Australia, and suggested that triclopyr-based herbicides, and several rates of fluroxypyr, are cost-effective in controlling the weed. Other studies by Seller *et al.* (2014), in

Box 4.1. Chemical control of *Mikania micrantha* – a case study from Kerala, India.

In 2000–2002, the herbicidal control of mikania was attempted in plantations of teak (5–7 years old) and eucalypt (7 years old), and in natural reed (*Ochlandra* spp.) areas, all of which were heavily infested by mikania. Trials were carried out with six herbicide treatments, viz., glyphosate, triclopyr, triclopyr+picloram, 2,4-D, diuron and paraquat. The results indicated that triclopyr+picloram (1.75 l/ha), triclopyr (0.5 l/ha), glyphosate (2.5–5 l/ha) and diuron (1–1.5 kg/ha) were highly effective in controlling mikania in all three situations, with control apparent after 1 month, and weed biomass reduced by 95% after 3 months (see Fig 4.1 for results in the teak plantation). No significant regrowth was observed in treated plots, even after a period of 8 months.

The total cost of applying one of these herbicides in areas of low to high mikania infestation was in the range Rs. 1150–2000/ha (US$20–33 at the current Rs./US$ conversion rates) This is much less than the annual cost of mechanical weeding in plantations infested by the weed, as this has to be carried out at least three or four times a year to attain satisfactory control, and has an associated cost of Rs. 5200–6000/year (US$85–100), depending on the severity of infestation.

The effects of paraquat and 2,4-D were short-lived and they were therefore unsuitable for mikania control. Moreover, as 2,4-D is toxic to animal life and has a relatively long and persistent residual action, it has been suggested that its use needs to be avoided in all ecosystems (Sankaran and Pandalai, 2004).

Although a single application of one of the effective herbicide treatments (triclopyr+picloram, triclopyr, glyphosate) could keep the weed under control in a given area for more than 8 months (see Fig. 4.1), regrowth was observed from windborne seeds at the beginning of the monsoon each year. Thus, it would be necessary to repeat the applications for several years depending on the severity of reinfestation (Sankaran and Pandalai, 2004).

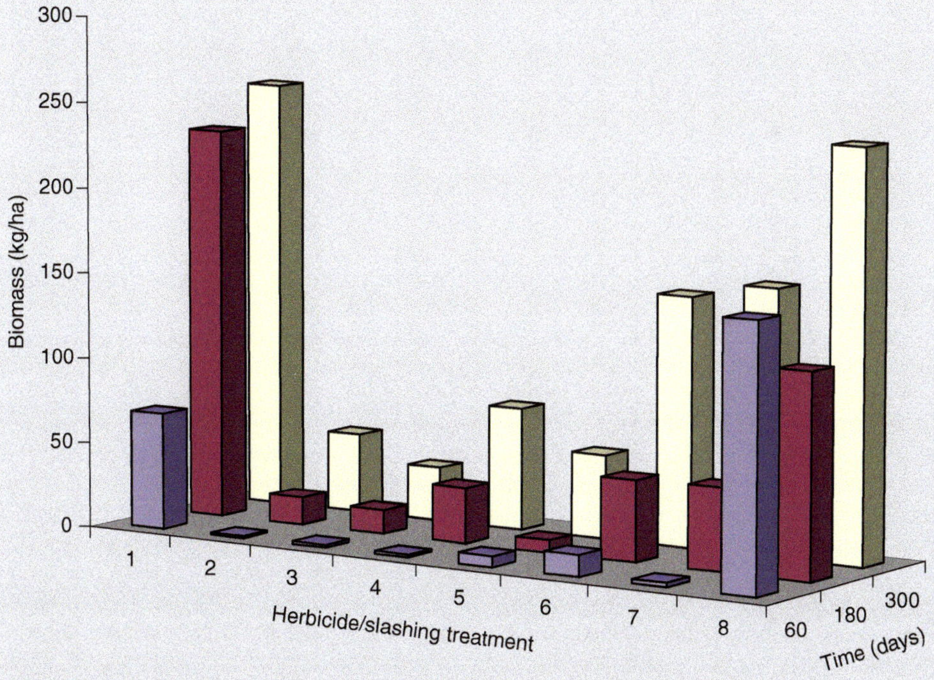

Fig. 4.1. Biomass of mikania (kg/ha dry weight) at various intervals after the application of herbicides or a slashing treatment in a 5-year-old teak plantation in Vazhachal, Kerala, India. Key: 1, slashing; 2, triclopyr+picloram (1.75 l/ha); 3, triclopyr (500 ml/ha); 4, diuron (1 kg/ha); 5, glyphosate (2.5 l/ha); 6, 2,4-D (500 g/ha); 7, paraquat (1 l/ha); 8, control.

The addition of adjuvants – ammonium sulphate (with glyphosate) and urea (with diuron) – improved the efficacy of the herbicides. What is more, in further trials, it was shown that the combination of either glyphosate or diuron with paraquat gave better control than individual applications. It was recommended that herbicide application should be carried out before mikania flowered or set seed (August–October in Kerala). Also, applications should be made on dry days because rainfall within 48 h of spraying will reduce the efficacy of the treatment.

These recommendations were made as a short-term solution for mikania control until alternative cost-effective and eco-friendly methods were developed.

Additionally, Table 4.2 compares the cost and efficacy of the herbicidal control of mikania with that of using either knife weeding or biological control.

Table 4.2. Cost and efficacy of the control of mikania using different methods in 2002. Data from Sankaran and Pandalai, 2004.

Method	No. of applications required each year	Efficacy of control	Cost (US$/ ha p.a.)[a]
Knife weeding	3–4	Short-lived	85–100
Herbicidal	1	Longer-lived than slashing	20–33
Biological	Agents likely to require multiple releases over several years, until established and spreading in the field	Permanent (with effective agents)	No cost to farmer

[a]Based on current conversion rates of the Indian rupee.

Florida, USA, where mikania has been recorded more recently (2009), proved that aminopyralid or aminocyclopyrachlor, as well as fluroxypyr, triclopyr or glyphosate, when they are applied during early vegetative growth, are effective in controlling the weed.

Markets and pesticide regulation

The specific crop and market may limit chemical control options, and tea provides a good example of this. Tea is a beverage crop that is plucked regularly and there are thus additional considerations in selecting herbicides and timing applications (Puzari *et al.*, 2010). A substantial proportion of the herbicides used in India are applied in tea (Rajkhowa *et al.*, 2005), with paraquat, 2,4-D and glyphosate the most commonly used (Barooah, 2011).The high surface-to-mass ratio of tea shoots, coupled with the frequency of both herbicide treatment and tea plucking, make it difficult to keep residues below the accepted levels in harvested tea (Barooah, 2011). Cock *et al.* (2000) argue that herbicide use in tea plantations in Assam is uneconomic and sometimes has adverse impacts in young tea and on new flushes of leaves, aside from problems in terms of residues affecting the marketability of the crop. Also, mikania has been implicated anecdotally as a threat to tea exports in India because of the massive increase in spraying against the weed. According to Gurusubramanian *et al.* (2008), stringent regulations on allowable residue levels in tea for export cannot be met given existing pest management practices. In 2005, the Indian authorities reviewed the use of all pesticides in tea and recommended inter alia fewer than ten herbicides for use in the crop; they issued the *Tea (Distribution and Export) Control Order 2005* to help protect India's place in the world market (Gurusubramanian *et al.*, 2008; Barooah, 2011). More complications came with the introduction of a 'positive list' by Japan in 2006, which effectively meant that 2,4-D could not be used in tea destined for that market (Barooah, 2011). As manual control is particularly problematic in tea, there is a real need for a control option that is both effective and safe.

Chemical versus non-chemical control: efficacy and labour costs

Traditionally, labour costs in plantations are high, with the majority being devoted to weed control as uncontrolled weeds reduce both the quantity and quality of yields, and shorten the production period of the crop (Soedarsan *et al.*, 1977). Mikania is a heavy burden, and where infestations are more established and denser, herbicide use is popular as it is perceived to be more effective and cheaper than manual control. Teoh *et al.* (1985) reported that wage costs and labour shortages led managers of (mostly oil palm rubber and cocoa) plantation estates in Malaysia to prefer chemical control. Sukasman (1979) argued that using glyphosate to control a weed flora including mikania in a 1-year-old tea plantation in Indonesia was economic because it gave longer weed suppression and required less labour and equipment than alternative methods. These views are borne out by recent experiences of the tea industry in Assam and studies in forest plantations in Kerala (see above).

Chemical and non-chemical weed control methods (slashing or livestock grazing) were compared for efficacy in rubber plantations in Malaysia. In a 2-year-old plantation, there was 90% regrowth in plots that had been mechanically slashed or grazed by sheep, but only 10% regrowth in those treated with herbicide (glyphosate + picloram) (Ahmad-Faiz, 1992). However, the use of sheep to graze mikania and other weeds gave an estimated 15–25% saving in overall weed control costs (Arope *et al.*, 1985). Another study, which explored the potential for integrating sheep, poultry rearing and beekeeping into smallholder rubber plantations in Malaysia, described how sheep controlled weeds with cost savings of 21% over customary practices, while the internal rate of return from sheep rearing was as high as 44% (Tajuddin, 1986).

Day *et al.* (2011) commented that chemical control is more effective than manual weeding/slashing against mikania and reduces the frequency of weeding interventions. In Papua New Guinea, 30% of commercial/semi-commercial enterprises used herbicides alone or in combination with other (manual) means, and the commercial (plantation) enterprises were most likely to use herbicides; 32% of these estates needed to weed fortnightly, and 39% hired extra labour for weeding. In contrast, smaller scale operations that did not use herbicides found that weeding could occupy 1–2 days a fortnight or more.

Intensive weeding is needed to reduce the impact of mikania on productivity in tea in north-eastern India, and necessitates employing 55–65 unskilled labourers/ha (Puzari *et al.*, 2010), which may not be economically viable compared with the cost for herbicide applications. Herbicides were introduced by the industry some three decades ago, not only because they were cheaper than manual weeding, but also because labour was scarce during the peak growing/harvesting season (Barbora, 2001). Herbicide applications require only one fifth of the labour needed for manual/mechanical control, so chemical control is seen as more efficient and cost-effective, and is a popular choice (Rajkhowa *et al.*, 2005). Financial and labour constraints mean that manual control remains difficult to implement effectively, especially as the greatest need for labour for weeding coincides both with the peak tea plucking period and the main season for growing paddy; as a result, tea estates tend to have only about half the ideal labour force (Puzari *et al.*, 2010). Thus, it is not only the costs of chemical versus manual control that are at issue, but the availability of labour as well, especially if there is better paid (and perhaps more congenial) work available.

The problem of affording/finding labour is particularly acute in young plantations, where the burden of controlling mikania – and weeds in general – means that a combination of chemical and manual control is often used and weed control costs are high. Teoh *et al.* (1985) reported that even though chemical control was the most common measure for mikania control across oil palm, rubber and cocoa plantations at all stages of development in Malaysia, manual weeding was frequently used during plantation establishment, with or without chemical control. Muraleedharan and Anitha (2000) reported that in forest plantations in Kerala that relied on manual labour for mikania control during establishment, the lack of availability of labour and financial constraints could make weeding operations suboptimal.

A combination of weed control measures is often advised; for example, Rajkhowa *et al.* (2005) outlined the potential for integrated weed management in tea in Assam, especially in young tea. In a review of weed and cover crop management in oil palm, Chung (2010) endorsed an integrated approach to weed management, suggesting that cover crop management is important in the prevention and exclusion of weeds, while weed control aims at a general reduction in the weed flora and, if possible, the eradication of noxious weeds from the cropping areas.

The drive to reduce herbicide use, fuelled by evidence of their detrimental health and environmental impacts and changing public attitudes, has led to a demand for sustainable alternatives to herbicides. However, as we have seen, non-chemical practices tend to be labour intensive and, therefore, expensive, as well as less effective than herbicides.

Biological control may offer a means of controlling mikania sustainably and without adverse effects. Barbora (2001) argues that because the overall impact of weeds on tea yield is high, chemical control has become necessary, but in consideration of the resulting environmental pollution and herbicide residues in soil and tea, other methods such as biological control need be developed. He suggests augmentative biocontrol, as it is specific and can be discontinued if necessary; this would ideally be a bioherbicide that could be used in conjunction with chemical herbicides. An alternative that has been developed in recent years is classical biological control, as introduced in the next section and covered in depth in Chapter 10, this volume.

Biological control

Arthropods and pathogens

Plantations were the sites of some of the first attempts at biological control of mikania, but when the introduced insect agent failed to establish, interest waned (Cock *et al.*, 2000).

As part of this first initiative, Teoh *et al.* (1985) recorded over 75 arthropod species in association with mikania during a short duration survey in Malaysia. Although these included a number of potentially useful natural enemies, most were either polyphagous or pests of economically important crops, and hence not appropriate for augmentative or classical biocontrol. In Kerala, Abraham *et al.* (2002b) identified 19 insect and one mite species associated with mikania, though all proved to be generalist feeders (some on crops) and so unsuitable as biological control agents. In addition, Abraham *et al.* (2002c) identified four pathogens infecting the weed, but again, none of these was host specific.

A concerted effort at developing a new biological control strategy was initiated when Indian scientists raised the need for a sustainable control measure for mikania in the Western Ghats of Kerala and the tea gardens in Assam. The resulting project looked at pathogens infecting mikania both in India (Sankaran *et al.*, 2001) and over the native range of the plant in the New World (Barreto and Evans, 1995). Even though nine pathogens were identified as causing various diseases on mikania in India, none was found to have the potential for use in biocontrol (Sreenivasan and Sankaran, 2001). However, a rust pathogen, viz. *Puccinia spegazzinii*, commonly found infecting the plant throughout its native range, was selected and assessed as a classical biological control agent (Ellison *et al.*, 2008). This rust was found to be specific to the genus *Mikania*, and is able to infect only a few species in the genus. It is highly damaging (with leaf, petiole and stem infections leading to cankering and death) and has a broad environmental tolerance. The pathogen was imported to India in August 2004, and after completing the required host-specificity tests, it was released in Assam and Kerala in 2005–2006. The release sites in Kerala included an agricultural system with mixed cropping of coconut and areca nut, and natural moist deciduous forests heavily invaded by mikania. The releases were successful in the sense that the rust had spread to the native population of mikania at all sites within a week of release, although it did not persist on the field population of mikania beyond a period of 3–4 months – especially when the environmental conditions at the release sites became unsuitable (high temperature and low humidity) for disease spread. Low inoculum load and inappropriate time of release are considered to be the main reasons for failure in survival of the rust in the field (Sankaran and Suresh, 2013b; see Chapter 10, this volume for further details).

Parasitic plants

The parasitism of *Cuscuta campestris* on mikania was first demonstrated at Shenzhen Xianhu Botanic Garden in China in 2003. It led to a decrease in various growth parameters of mikania by the end of the first month, and after 2 months it had adversely affected a range of parameters that contribute to the invasiveness of the weed, including its photosynthetic rate. The parasitic plant spread rapidly, extending over an area of 20 m^2 and up to 5 m from the initial point of infestation within 2 months (Deng *et al.*, 2003). Later studies by Shen *et al.* (2011) demonstrated the immense potential of *C. campestris* as a biological control agent against mikania in China, though its effect in controlling the weed infestation also depends on the intensity of its parasitism.

Chiu and Shen (2004) observed two *Cuscuta* species (tentatively identified as *C. campestris* and *C. reflexa*) parasitizing and killing mikania and another weed, *Asystasia intrusa*, in oil palm estates in Sumatra and West Kalimantan in Indonesia. *C. campestris* is the most widespread species in this genus, but it is a generalist parasite with a wide host range and inflicts substantial losses on crops (Mishra, 2009); *C. reflexa* too is known to parasitize a number of economically important plants (Baruah *et al.*, 2003). Chiu *et al.* (2002) found that another parasitic plant, *Cassytha filiformis*, parasitizes mikania and reduce infestations significantly, but this is also parasitic on certain ornamental plants. An argument put forward to support the use of parasitic plants against mikania is that once they have controlled the weed, they are themselves far easier to control with herbicides than mikania (Zhang *et al.*, 2004). Despite this being so, there is no question but that the use as a biological control agent of an insect or a pathogen with such a wide host range as these parasitic plants would be rejected out of hand, with the reasoning that it is 'a lesser evil' dismissed. There is no logic for being more compromising because the putative biocontrol agent is a plant.

Grazing

A different type of biological control involves grazing weeds using sheep. This method was successful in rubber plantations in Malaysia (see above). In one study, sheep preferentially grazed mikania and other weeds such as *Asystasia coromandeliana, Ottochloa nodosa, Brachiaria mutica* and *Paspalum* spp., leaving the leguminous cover crop intact. The spread of the legumes controlled any further spread of the weeds (Arope *et al.*, 1985). Nevertheless, not all weeds are palatable to the sheep, and this is a major impediment to this method (Stöber, 1993). In addition, grazing cannot be used as an option in all situations.

Conclusions

Mikania micrantha has serious impacts on the productivity and profitability of plantations across the Asia–Pacific region, which the current management efforts are either unable to contain or, if they work, can only do so at a cost that is unsustainable. Economic assessments have proved that mikania invasion in plantations causes cost escalation or income reduction, or both. Managers frequently fail to decide how to make the trade-off.

The export of plantation crops is crucial for earning foreign exchange in many a country in the Asia–Pacific region, and mikania invasion and the ensuing low crop productivity can affect a country's economy.

Further, excessive herbicide residues affect the marketability of the crops and products, especially their export, and the impact of herbicides on the environment and on human and animal health cannot be overemphasized.

At least as importantly, weed invasion frustrates the sustainability of cropping and impinges heavily on the livelihood of farmers and plantation owners. Regular monitoring of sites, with early detection and control of mikania before it can spread far and wide is the most appropriate strategy to avoid damage. However, wherever an invasion is unmanageable, a judicious mix of methods, such as physical (uprooting before seed setting and burning), cultural (e.g. legume cover cropping or grazing wherever possible) and chemical (application on to cut stems after slashing) needs be employed to alleviate the problem in the short term. In this case, caution is warranted to avoid soil disturbance, ensure the non-toxicity of fodder and use lower dosages of comparatively safer herbicides. This strategy may be followed until biological control methods are attempted (Ellison *et al.*, 2014) and established in the field.

To synchronize with this, of late, plantation managers have started to attach priority to weed control in their plantation prescriptions, and this is well assisted by the improved methods developed through many years of research.

References

Abraham, M. and Abraham, C.T. (2000) Weed flora of rubber plantations in Kerala. *Indian Journal of Natural Rubber Research* 13, 86–91.

Abraham, M. and Abraham, C.T. (2002) Pre-emergence herbicides for the control of *Mikania micrantha* H.B.K. *Journal of Tropical Agriculture* 40, 81–82.

Abraham, M., Abraham, C.T. and George, M. (2002a) Competition of mikania with common crops of Kerala. *Indian Journal of Weed Science* 34, 96–99.

Abraham, M., Abraham, C.T. and Joy, P.J. (2002b) Natural enemies on *Mikania micrantha* H.B.K. in Kerala. *Journal of Tropical Agriculture* 40, 39–41.

Abraham, M., Abraham, C.T. and Varma, S.A. (2002c) Native pathogens of *Mikania micrantha* H.B.K. – an introduced weed in India. *Indian Journal of Weed Science* 34, 152–153.

Ahmad Faiz, M.A. (1992) Comparison of three weeding methods in rubber cultivation. *Planters' Bulletin of the Rubber Research Institute of Malaysia* 212/213, 99–101.

Alif, A.F.M. [also cited as Anwar Ismail, A.] (2001) Impact and management of selected alien and invasive weeds in Malaysia with some action plans instituted for biological diversity. In: *Proceedings of the 3rd International Weed Science Congress, Foz do Iguassu, Brazil, 6–11 June 2000.* CD-ROM, MS No. 446. International Weed Science Society, Oxford, Mississippi.

Arope, A.B., Ismail, T.B. and Chong, D.T. (1985) Sheep rearing under rubber. *Planter* 61, 70–77.

Barbora, A.C. (2001) Weed control in tea plantations: current scenario in northeast India. In: Sankaran, K.V., Murphy, S.T. and Evans, H.C. (eds) *Alien Weeds in Moist Tropical Zones: Banes and Benefits. Proceedings of a Workshop, Kerala Forest Research Institute, Peechi, India, 2–4 November 1999.* Kerala Forest Research Institute, Peechi, India and CABI Bioscience, UK Centre (Ascot), Ascot, UK, pp. 107–111.

Barooah, A.K. (2011) Present status of use of agrochemicals in tea industry of eastern India and future directions. *Science and Culture* 77, 385–390.

Barreto, R.W. and Evans, H.C. (1995) The mycobiota of the weed *Mikania micrantha* in southern Brazil with particular reference to fungal pathogens for biological control. *Mycological Research* 99, 343–352.

Baruah, I.C. and Gogoi, A.K. (1995) Phytosociological studies of sugarcane fields in Assam. *World Weeds* 2, 107–115.

Baruah, I.C., Rajkhowa, D.J., Deka, N.C. and Kandali, R. (2003) Host range study of *Cuscuta reflexa* Roxb in Assam. *Indian Journal of Forestry* 26, 414–417.

Brooks, S.J. and Setter, S.D. (2014) Increased options for controlling mikania vine (*Mikania micrantha*) with foliar herbicides. In: Baker, M. (ed.) *Proceedings of the 19th Australasian Weed Conference, Hobart, Australia, 1–4 September 2014.* Tasmanian Weed Society, Hobart, Australia, pp. 409–412.

Brooks, S.J., Panetta, F.D. and Galway, K.E. (2008) Progress towards the eradication of mikania vine (*Mikania micrantha*) and limnocharis (*Limnocharis flava*) in northern Australia. *Invasive Plant Science and Management* 1, 296–303.

CABI (2005) *Tectona grandis.* CABI Forestry Compendium datasheet. CAB International, Wallingford, UK. Available at: http://www.cabi.org/fc/datasheet/52899 [restricted access] (accessed 28 October 2014).

Chee, Y.K., Alif, A.F.[M.] and Chung, G.F. (1992) Management of weeds in plantation crops in 2000. In: Kadir, A.A.S.A. and Barlow, H.S. (eds) *Pest Management and the Environment in 2000.* CAB International, Wallingford, UK, pp. 270–280.

Chiu, S.B. and Shen, H. (2004) Growth studies of *Cuscuta* spp. (dodder parasitic plant) on *Mikania micrantha* and *Asystasia intrusa. Planter* 80, 31–36.

Chiu, S.B., Chan, S.M. and Siow, A. (2002) Biological control of *Mikania micrantha* – a preliminary finding. *Planter* 78, 715–718.

Chung, C.F. (2010) Weed and cover crop management in oil palm. *Planter* 86, 857–871.

Chung, G.F., Balasubramaniam, R. and Cheah, S.S. (2000) Recent development in spray equipment for effective control of pests and weeds. *Planter* 76, 65–84.

Cock, M.J.W., Ellison, C.A., Evans, H.C. and Ooi, P.A.C. (2000) Can failure be turned into success for biological control of mile-a-minute weed (*Mikania micrantha*)? In: Spencer, N.R. (ed.) *Proceedings of the X International Symposium on Biological Control of Weeds, Bozeman, Montana, 4–14 July 1999.* USDA Forest Service, Forest Health Technology Enterprise Team, Morgantown, West Virginia, pp. 155–167.

Day, M.D., Kawi, A., Tunabuna, A., Fidelis, J., Swamy, B., Ratutuni, J., Saul-Maora, J., Dewhurst, C.F. and Orapa, W. (2011) The distribution and socio-economic impacts of *Mikania micrantha* (Asteraceae) in Papua New Guinea and Fiji and prospects for its biocontrol. In: *23rd Asian-Pacific Weed Science Society Conference, The Sebel Cairns, Australia, 26–29 September 2011: Weed Management in a Changing World. Conference Proceedings, Volume 1.* Asian-Pacific Weed Science Society, pp. 146–153. http://apwss.org/apwss-publications.htm (accessed 14 July 2017).

Day, M.D., Kawi, A., Kurika, K., Dewhurst, C.F., Waisale, S.-M.J., Saul-Maora, J, Fidelis, J., Bokosou, J., Moxon, J., Orapa, W. and Senaratne, K.A.D. (2012) *Mikania micrantha* Kunth (Asteraceae) (mile-a-minute): its distribution and physical and socio economic impacts in Papua New Guinea. *Pacific Science* 66, 213–223.

Deng, X., Feng, H.-L., Ye, W.-H., Yang, Q.-H., Xu, K.-Y., Cao, H.-L. and Fu, G. (2003) A study on the control of the exotic weed *Mikania micrantha* by using the parasitic *Cuscuta campestris. Journal of Tropical and Subtropical Botany* 11, 117–122. [In Chinese, English abstract.]

Derksen, A. and Dixon, W. (2009) *Interim Report: Florida Cooperative Agricultural Pest Survey for Mile-a-minute,* Mikania micrantha *Kunth in Miami–Dade County.* Florida Department of Agriculture and Consumer Services, Miami, Florida.

Ellison, C.A. (2004) Biological control of weeds using fungal natural enemies: a new technology for weed management in tea? *International Journal of Tea Science* 3, 4–20.

Ellison, C.A., Evans, H.C., Djeddour, D.H. and Thomas, S.E. (2008) Biology and host range of the rust fungus *Puccinia spegazzinii*: a new classical biological control agent for the invasive, alien weed *Mikania micrantha* in Asia. *Biological Control* 45, 133–145.

Ellison, C.A., Day, M.D. and Witt, A. (2014) Overcoming barriers to the successful implementation of a classical biological control strategy for the exotic invasive weed *Mikania micrantha* in the Asia–Pacific region. In: Impson, F.A.C., Kleinjan, C.A. and Hoffmann, J.H. (eds) (2014) *Proceedings of the XIV International Symposium on Biological Control of Weeds, Kruger National Park, South Africa, 2–7 March 2014.* University of Cape Town, Rondebosch, South Africa, pp. 135–141.

FAOSTAT (2011) Exports/commodities by country, Papua New Guinea, 2011. Available at: http://faostat3.fao.org/browse/rankings/commodities_by_country_exports/E (accessed 23 February 2017).

Gogoi, A.K. (2001) Status of mikania infestation in north-eastern India: management options and future research thrust. In: Sankaran, K.V., Murphy, S.T. and Evans, H.C. (eds) *Alien Weeds in Moist Tropical Zones: Banes and Benefits. Proceedings of a Workshop, Kerala Forest Research Institute, Peechi, India, 2–4 November 1999.* Kerala Forest Research Institute, Peechi, India and CABI Bioscience, UK Centre (Ascot), Ascot, UK, pp. 77–79.

Gopakumar, B. and Motwani, B. (2013) Factors restraining the natural regeneration of reed bamboo *Ochlandra travancorica* and *O. wightii* in Western Ghats, India. *Journal of Tropical Forest Science* 25, 250–258.

Gray, B.S. and Hew, C.K. (1968) Cover crops experiments in oil palm on the west coast of Malaysia. In: Turner, P.D. (ed.) *Oil Palm Developments in Malaysia.* Incorporated Society of Planters, Kuala Lumpur, pp. 138–151.

Gurusubramanian, G., Rahman, A., Sarmah, M., Ray, S. and Bora, S. (2008) Pesticide usage pattern in [the] tea ecosystem, their retrospects and alternative measures. *Journal of Environmental Biology* 29, 813–826.

Holm L.G., Plucknett, D.L., Pancho, J.V. and Herberger, J.P. (1991) *The World's Worst Weeds: Distribution and Biology.* Kreiger, Malabar, Florida.

Hutauruk, C., Lubis, Y.R. and Lubis, R.A. (1982) Field evaluation of several herbicides for controlling *Mikania* sp. (Compositae) in immature oil palm. *Bulletin Pusat Penelitian Marihat [Bulletin, Marihat Research Station, Indonesia]* 1, 30–36.

Jones, P. (2012) Growing biosecurity. In: *Partners in Research and Development, Special Report, Papua New Guinea.* Australian Centre for International Agricultural Research (ACIAR), Canberra, pp. 18–19.

KFD (2013) Forest statistics 2012–2013. Kerala Forest Department. Available at: www.forest.kerala.gov.in (accessed 28 October 2016).

Kuo, Y.-L., Chen, T.-Y. and Lin, C.-C. (2002) Using a consecutive-cutting method and allelopathy to control the invasive vine, *Mikania micrantha* H.B.K. *Taiwan Journal of Forest Science* 17, 171–181.[In Chinese, English abstract.]

Li, W.-H., Zhang, C.-B., Jiang, H.-B., Xin, G.-R. and Yang, Z.-Y. (2006) Changes in soil microbial community associated with invasion of the exotic weed *Mikania micrantha* H.B.K. *Plant and Soil* 281, 309–324.

Macanawai, A.R., Day, M.D., Tumaneng-Diete, T. and Adkins, S.W. (2011) Impact of *Mikania micrantha* on crop production systems in Viti Levu, Fiji. In: *23rd Asian-Pacific Weed Science Society Conference, The Sebel Cairns, Australia, 26–29 September 2011: Weed Management in a Changing World. Conference Proceedings, Volume 1.* Asian-Pacific Weed Science Society, pp. 304–312. Available at: http://apwss.org/apwss-publications.htm (accessed 14 July 2017).

Manabendra, D. and Rudrapal, M. (2011) Tea mosquito bug *Helopeltis theivora* Waterhouse: a threat for tea plantation in north-east India. *Asian Journal of Biochemical and Pharmaceutical Research* 1, 70–73.

Mangoensoekarjo, S. (1978) Mile-a-minute (*Mikania micrantha* H.B.K.) control in immature oil palm. In: Amin, L.L. (ed.) *Proceedings, Plant Protection Conference, Malaysia, 1978.* Rubber Research Institute of Malaysia, Kuala Lumpur, pp. 381–397.

Mangoensoekarjo, S. (1979) Glyphosate trials on plantation crops in north Sumatra. In: *Simposium Herbisida Roundup 3, Medan, Indonesia, 1979.* [No publisher details available.]

Mangoensoekarjo, S. and Soewadji, R.M. (1973) The influence of cover crops on rubber. II. The growth of rubber. *Bulletin Balai Penelitian Perkebunan Medan* 4, 127–134.

Matthews, S. (2004) *Tropical Asia Invaded: The Growing Danger of Invasive Alien Species.* Global Invasive Species Programme, Cape Town. Available at: www.issg.org/pdf/publications/GISP/Resources/TropicalAsiaInvaded.pdf (accessed 28 October 2014).

Mishra, J.S. (2009) Biology and management of *Cuscuta* species. *Indian Journal of Weed Science* 41, 1–11.

Muraleedharan, P.K. and Anitha, V. (2000) The economic impact of *Mikania micrantha* on teak plantations in Kerala. *Indian Journal of Forestry* 23, 248–251.

Murdiati, T. and Stoltz, D.R. (1987) Investigation of suspected plant poisoning of north Sumatran cattle. *Penyakit Hewan* 19, 101–105.

Nair, V.K.B. (1968) *Mikania cordata* (Burm. f.) B.L. Robinson – an alien new to South India. *Rubber Board Bulletin* 9, 28–29.

Nambiar, E.K.S. and Brown, A.G. (1997) *Management of Soil, Nutrients and Water in Tropical Plantation Forests.* ACIAR Monograph No. 43. Australian Centre for International Agricultural Research (ACIAR), Canberra.

Puzari, K.C., Bhuyan, R.P., Dutta, P. and Nath, H.K.D. (2010) Distribution of *Mikania* and its economic impact on tea ecosystem of Assam. *Indian Journal of Forestry* 33, 71–76.

Rajkhowa, D.J., Barua, I.C., Bhuyan, R.P. and Yaduraj, N.T. (2005) *Weed Management in Tea.* Technical Bulletin, National Research Centre for Weed Science, Jabalpur, India.

RRIM (1965) Effect of fertilizers and cover plants on early yield of young rubber. *Planters' Bulletin of the Rubber Research Institute of Malaysia* 77, 56–64.

Sankaran, K.V. and Pandalai, R.C. (2004) Field trials for controlling mikania infestation in forest plantations and natural forests in Kerala. KFRI Research Report No. 265, Kerala Forest Research Institute, Peechi, India.

Sankaran, K.V. and Sreenivasan, M.A. (2001) Status of *Mikania* infestation in the Western Ghats. In: Sankaran, K.V., Murphy, S.T. and Evans, H.C. (eds) *Alien Weeds in Moist Tropical Zones: Banes and Benefits. Proceedings of a Workshop, Kerala Forest Research Institute, Peechi, India, 2–4 November 1999.* Kerala Forest Research Institute, Peechi, India and CABI Bioscience, UK Centre (Ascot), Ascot, UK, pp. 67–76.

Sankaran, K.V. and Suresh, T.A. (2013a) *Invasive Alien Plants in the Forests of Asia and the Pacific.* Food and Agriculture Organization of the United Nations, Regional Office for Asia and the Pacific, Bangkok, Thailand.

Sankaran, K.V. and Suresh, T.A. (2013b) Evaluation of classical biological control of *Mikania micrantha* with *Puccina spegazzinii.* KFRI Research Report No. 472, Kerala Forest Research Institute, Peechi, India.

Sankaran, K.V., Muraleedharan, P.K. and Anitha, V. (2001) Integrated management of the alien invasive weed *Mikania micrantha* in the Western Ghats. KFRI Research Report No. 202, Kerala Forest Research Institute, Peechi, India.

Seller, B.A., Lancaster, S.R. and Langeland, K.A. (2014) Herbicides for post-emergence control of mile-a-minute (*Mikania micrantha*). *Invasive Plant Science and Management* 7, 303–309.

Shen, H., Hong, L., Chen, H., Ye, W.-H., Cao, H.-L. and Wang, Z.-M. (2011) The response of the invasive weed *Mikania micrantha* to infection density of the obligate parasite *Cuscuta campestris* and its implications for biological control of *M. micrantha. Botanical Studies* 52, 89–97.

Shen, S.-C., Xu, G.-F.., Zhang, F.-D., Jin, G.-M., Liu, S.-F., Liu, M.-Y., Chen, A.-D. and Zhang, Y.-H. (2013) Harmful effects and chemical control study of *Mikania micrantha* H.B.K in Yunnan, south-west China. *African Journal of Agricultural Research* 8, 5554–5561.

Soedarsan, A., Kuntohartono, T. and Mangoensoekarjo, S. (1977) Weed control in plantation crops in Indonesia. *Menara Perkebunan* 45, 183–188. [In Indonesian.]

Sreenivasan, M.A. and Sankaran, K.V. (2001) Management of *Mikania micrantha* in Kerala – potential of biological and chemical methods. In: Sankaran, K.V., Murphy, S.T. and Evans, H.C. (eds) *Alien Weeds in Moist Tropical Zones: Banes and Benefits. Proceedings of a Workshop, Kerala Forest Research Institute, Peechi, India, 2–4 November 1999.* Kerala Forest Research Institute, Peechi, India and CABI Bioscience, UK Centre (Ascot), Ascot, UK, pp. 122–130.

Stöber, S. (1993) Weed control by integration of sheep in permanent tree crops in west Malaysia. In: *Proceedings of the 4th International IFOAM Conference on Non-chemical Weed Control, 5–9 July 1993, Dijon, France.* Association Colleque IFOAM, Dijon, France, pp. 213–218.

Suharti, M. and Sudjud, D.A. (1978) Experiment on *Mikania micrantha* control with herbicides. *Lembaga Penelitian Hutan Laporan* 281, 1–30.

Sukasman, M. (1979) A short note on weed control experiments with Roundup in tea plantations. In: *Simposium Herbisida Roundup 3, Medan, Indonesia, 1979.* [In Indonesian. No publisher details available.]

Sutedjo, K. and Lubis, J.R. (1971) Some results of experiments and practical use of herbicides in tea estates in north Sumatra. *Proceedings of the Third Conference of the Asian-Pacific Weed Science Society, Federal Hotel, Kuala Lumpur, Malaysia, June 7th to 12th, 1971, Volume 1.* Asian-Pacific Weed Science, Society, pp. 60–69.

Tajuddin, I. (1986) Integration of animals in rubber plantations. *Agroforestry Systems* 4, 55–66.

Teng, Y.T. and Teh, K.H. (1990) Wallop (glyphosate+dicamba): a translocative broad spectrum herbicide for effective general weed control in young and mature oil palm. *BIOTROP Special Publication* 38, 165–174.

Teoh, C.H., Chung, G.F., Liau, S.S., Ghani Ibrahim, A., Tan, A.M., Lee, S.A. and Mariati, M. (1985) Prospects for biological control of *Mikania micrantha* H.B.K in Malaysia. *Planter* 61, 515–530.

Tewari, D.N. (1992) *A Monograph on Teak (*Tectona grandis *Linn. f.).* International Book Distributors, Dehradun, India.

Ti, T.C., Pee, T.Y. and Pushparaj, E. (1971) Economic analysis of cover policies and fertilizer used in rubber cultivation. In: Ng, S.K. and Rajarao, J.C. (eds) *Proceedings of the Rubber Research Institute of Malaysia Planters' Conference, Kuala Lumpur, Malaysia, 1971.* RRIM, Kuala Lumpur.

Tjitrosemito, S. (2005) Efficacy tests of herbicide mixture Eskamin to control weeds in young oil palm plantation. Internal report, BIOTROP, Bogor, Indonesia.

Wang, T., Su, Y.-J. and Chen, G.-P. (2008) Population genetic variation and structure of the invasive weed *Mikania micrantha* in southern China: consequences of rapid range expansion. *Journal of Heredity* 99, 22–33.

Waterhouse, D.F. and Norris. K.R. (1987) *Biological Control: Pacific Prospects.* Inkata Press, Melbourne, Australia.

Watson, G.A., Wong, P.W and Narayanan, R. (1964) Effects of cover plants on soil nutrient status and on growth of *Hevea.* IV. Leguminous creepers compared with grasses, *Mikania cordata* and mixed indigenous covers on four soil types. *Journal of the Rubber Research Institute of Malaysia* 18, 123–145.

Widjaja, E.A. and Tjitrosoedirdjo, S.S. (1991) The development of weeds under *Dendrocalamus asper* (Schultz. F.) Back. ex Heyne plantation in Lampung, Sumatra, Indonesia. In: *Proceedings of the 13th Asian-Pacific Weed Science Society Conference, Vol. 1, Jakarta, Indonesia, 15–18 October 1991.* Asian-Pacific Weed Science Society, Taipei, Taiwan, pp. 225–227.

Widyatmoko and Riyanto, H. (1986) Weed management in newly opened area. Experience from Gunung Madu sugarcane plantations, Lampung. In: Madkar, O.R., Soedarsan, A. and Sastroutomo, S.S. (eds) *Proceedings of the Eighth Indonesian Weed Science Conference, Bandung, Indonesia, 24–26 March 1986*. Himpunan Ilmu Gulma Indonesia (HIGI), Bandung, Indonesia, pp. 143–151

Wirjahardja, S. (1975) Autecological study of *Mikania* spp. *BIOTROP Newsletter* 11. Also available as: Wirjahardja, S. (1976) Autecological study of *Mikania* spp. In: *Proceedings of the Fifth Asian-Pacific Weed Science Society Conference, Tokyo, Japan, October 5th to 11th, 1975*. Asian-Pacific Weed Science Society, pp. 70–73. Available at: http://apwss.org/apwss-publications.htm (accessed 22 August 2017).

Zhang, L.Y., Ye, W.H., Cao, H.L. and Feng, H.L. (2004) *Mikania micrantha* H.B.K. in China – an overview. *Weed Research* 44, 42–49.

5 *Mikania micrantha*: its Status and Impact on People and Wildlife in Nepal

Hem Sagar Baral[1,2*] and Bhaskar Adhikari[3]

[1]Zoological Society of London, Nepal Office, Kathmandu, Nepal; [2]School of Environmental Sciences, Albury-Wodonga Campus, Charles Sturt University, Albury, New South Wales, Australia; [3]Royal Botanic Garden Edinburgh, Edinburgh, Scotland, UK

Introduction

Nepal is primarily a mountainous country and is landlocked between the two most populous nations of the world, India and China. The country is roughly rectangular in shape and has an area of 147,181 km^2. It lies between latitudes 26°22′N and 30°27′N, and longitudes 80°04′E and 88°12′E, occupying a central position in the Himalayas where the flora and fauna of the western and eastern Himalayas converge. The country is nearly 800 km long and less than 200 km wide and covers one third of the entire Himalayan range. The altitude ranges from just 60 m above sea level (asl) to Mount Everest (Sagarmatha) which, at 8848 m asl, is the world's highest point. This variation in altitude within such a short distance (less than 200 km) has resulted in a wide variety of habitats. In addition, Nepal lies on several biogeographical boundaries: the Eastern Asiatic and Irano–Turanian elements of the Holarctic kingdom and the Indomalayan elements of the Palaeotropical kingdom meet in Nepal to create a unique and rich floristic and faunal diversity; in all, six floristic provinces converge here.

Because of Nepal's wide range of climatic and altitudinal variation, it is endowed with many different types of ecosystems and habitats (see Box 5.1 and Fig. 5.1 within the box).

To protect such diverse natural resources, the Government of Nepal (GoN), has formulated several acts, policies and guidelines. The most important of these is the National Parks and Wildlife Conservation Act 2029 (1973), by which the Government has created a network of protected areas throughout the country. These protected areas cover more than 23% of the country's land mass (see Fig. 5.2).

There has been intentional and unintentional introduction of many alien species into Nepal in recent decades. The country's varied ecosystems and habitats can support a wide variety of non-native plants, and over 100 introduced plants are now so well established that they have become weeds (HMGN/MFSC, 2002). Recently *Mimosa diplotricha* has been also noted in the only lowland protected area of east Nepal, Koshi Tappu Wildlife Reserve. These invasive plant species are threatening and destroying Nepal's natural environment and impinging upon its agriculture (Tiwari *et al.*, 2005). Species such as *Amaranthus viridis*, *A. spinosus*, *Argemone mexicana*, *Bidens pilosa*, *Cassia tora*, *Ageratina adenophora* (= *Eupatorium adenophorum*), *Parthenium hysterophorus* and *Lantana camara*, as well as *Mikania micrantha*

Corresponding author. E-mail: hem.baral@zsl.org or hem.baral@gmail.com

Box 5.1. Nepal's contrasting landscapes.

1. Although Nepal is a small country, almost all types of vegetation are squeezed into the diverse topography, with tropical, subtropical, temperate, subalpine and alpine zones represented in five distinct physiographic zones (Inskipp, 1989).

2. The Terai (also spelt Tarai), which contains mostly tropical vegetation, is a narrow strip of land lying north of the Indian border and south of the foothills of the Siwaliks. The altitudinal range is from 60 to 300 m asl. Nearly 50% of Nepal's growing population lives in the Terai area, though it occupies only 12% of the country's total area. It is a continuation of the Indo-Gangetic alluvial plain and contains the most fertile agricultural land in Nepal. The Terai also includes the dry Bhabar region on its northern edge, which consists of gravelly soil and stone washed down from the hills. The Bhabar is not suitable for agriculture because of its dry soil and porous nature.

3. The Siwaliks, which contain tropical and subtropical vegetation, are the first of the Himalayan foothills. They rise steeply from the Terai, north of the Bhabar region. They occupy 13% of the total area of the country. The soil is shallow and erodible, and has little potential for cultivation, but in many parts of Nepal, these hills have been denuded because of timber extraction and fodder collection. In central and west Nepal, the Siwaliks are separated from the next range of mountains to the north – the Mahabharat range – by broad, gently sloping valleys known as duns. Rapti Dun, home to Chitwan National Park, is a fine example of a dun valley.

4. The mid mountains region (containing subtropical and lower temperate vegetation) is a broad complex of mountains and valleys, including the Mahabharat range, which runs the length of Nepal and beyond. Before the malaria eradication programme in lowland Nepal, which began in 1954, this used to be the most populated region of the country. Most inhabited valleys, for example the Kathmandu Valley and the Pokhara Valley, lie within this region, the former with one tenth of the country's current population. Previously, this region was heavily degraded and deforestation was rampant, but now, with efforts from local communities, several small patches of forests remain and are well managed under community forestry programmes.

5. The high mountains (containing lower temperate, upper temperate and subalpine vegetation) are less cultivated than the mid mountains and lowland Nepal because the land is less fertile. The forests are in a better condition here than in other zones and include one of the least disturbed forests. Some of the forests in the trans-Himalayan zone also fall within this zone.

6. The fifth zone is the High Himal, which includes alpine or high Himalayan flora. Most of this area is under permanent snow and has very little cultivation as the area is only suitable for grazing. Ten of the world's highest peaks, above 8000 m, lie within this zone, and attract tourists from all over the world.

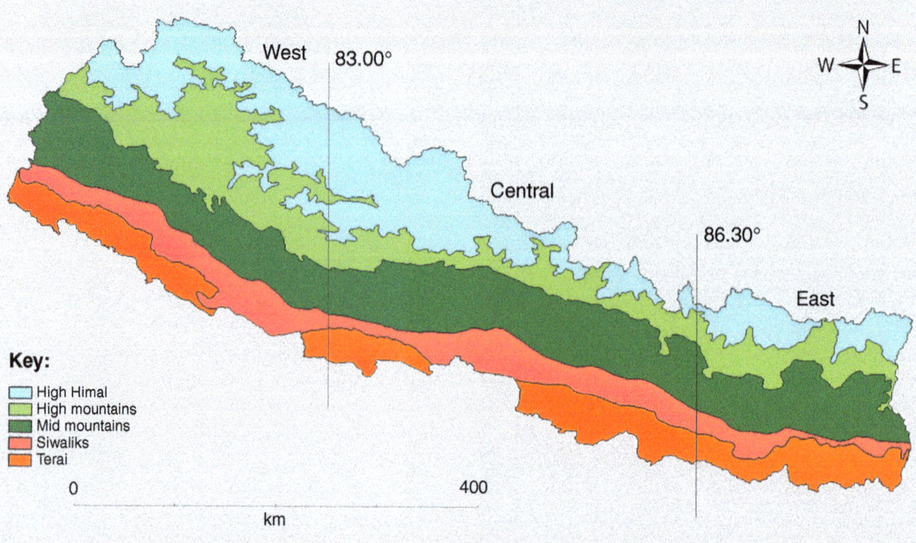

Fig. 5.1. The physiographic zones of Nepal.

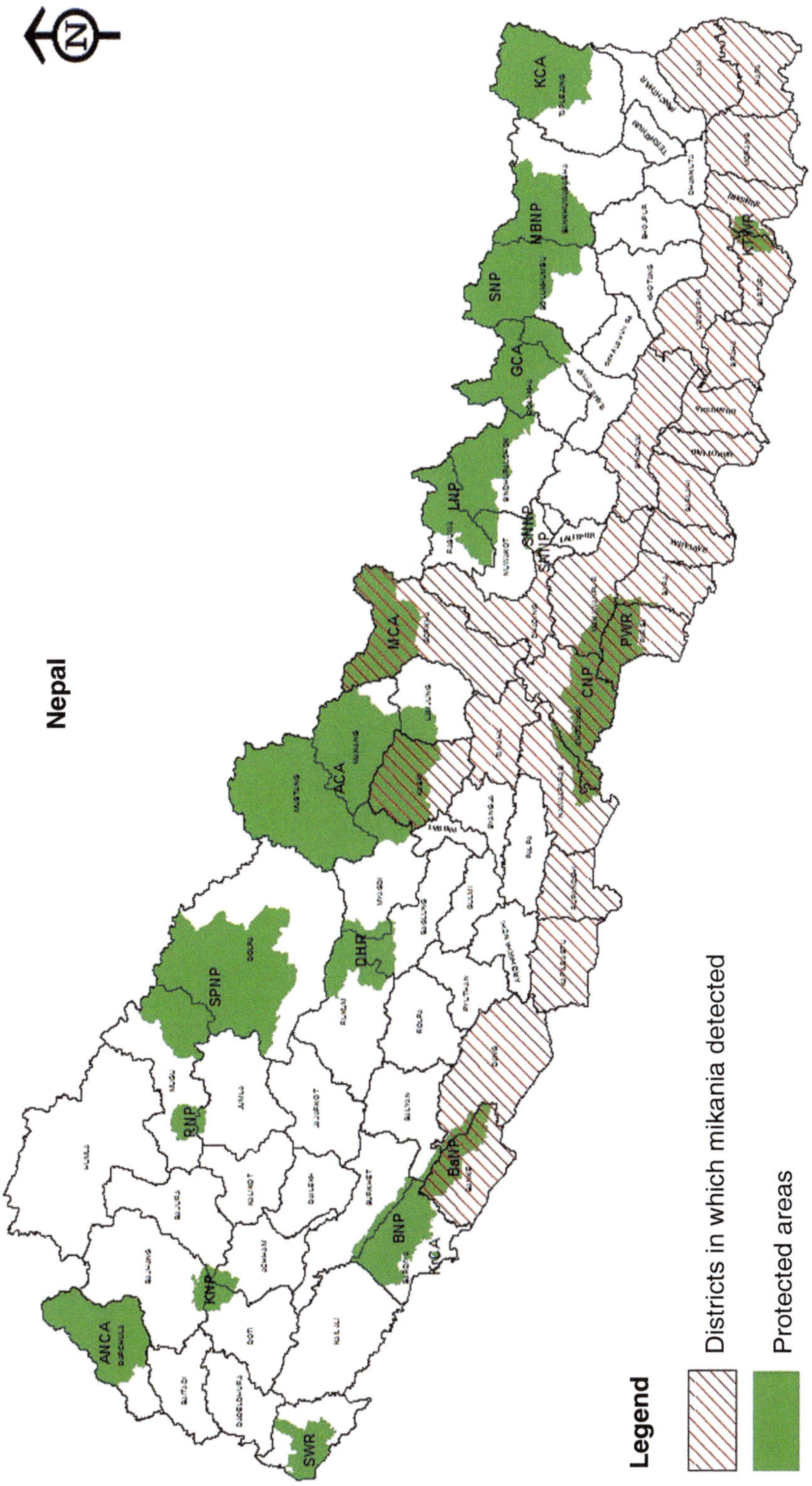

Fig. 5.2. Nepal's mikania weed infestation in relation to districts and protected areas (which are designated by their initials).

(mikania weed) are so common in agricultural land that they have changed the species composition of fallow and cultivated fields. Exotic aquatic weeds such as *Alternanthera philoxeroides*, *Eichhornia crassipes* (water hyacinth), *Ipomoea* spp., *Leersia hexandra*, *Myriophyllum aquaticum* and *Pistia stratiotes* are invading Nepal's wetlands (HMGN/MFSC, 2002; Poudel *et al.*, 2005; Sikawoti, 2007). *E. crassipes* and *M. micrantha* weed have been identified as specific threats to birds, with devastating impacts in some national parks and reserves (Inskipp *et al.*, 2013). Of the non-native species that have begun to show invasive characteristics, *M. micrantha* has emerged as the most serious weed in the tropical and subtropical areas of Nepal (Tiwari *et al.*, 2005; Rai *et al.*, 2012; Murphy *et al.*, 2013).

Arrival and spread of mikania weed

Records in the National Herbarium and Plant Laboratories (KATH), Kathmandu, show that *M. micrantha* was first recorded in the country in 1963 from eastern Nepal (see Box 5.2). This first record is from the major

Box 5.2. *Mikania micrantha* in Nepal: a fact sheet.

Local names:	Lahare banmara, Bahudale jhar, Bahramase, Bire lahara, Tite lahara, Bakhre lahara, Pyangri lahara, Pani lahara
First report in Nepal:	Kitamura, S. (1966) in *The Flora of Eastern Himalaya*
First herbarium record:	Eastern Nepal: Ilam-Jogmai, 08.12.1963 (H. Hara, H. Kanai, S. Kurosawa, G. Muratta, M. Togashi and T. Tuyama 6306311 (KATH, TI)
Uses:	Fodder for goats and compost material in eastern Nepal
Distribution:	In Nepal, it has been reported from the lowlands up to 1300 m above sea level (asl), east from Jhapa and Ilam districts, west to Banke District in west Nepal

Fig. 5.3. *Mikania micrantha* showing (a) branch with flowers, (b) close up of single flower, and (c) mature seed. Photos courtesy B. Adhikari.

tea-growing district of Ilam. It is likely that the species was introduced to Nepal as a contaminant of tea saplings or seed from the north-eastern Indian state of Assam, which is the main source of Ilam's tea planting material (Siwakoti, 2007). However, according to Rai and Rai (2013) there are different opinions on the introduction of *Mikania* into eastern Nepal; many people think that flooding in the Koshi River is the most probable cause of introduction, while a few think that birds and wild animals are the main carriers. Although mikania was introduced more than 50 years ago, it did not emerge as a serious weed until two decades ago. Its present distribution shows that it has spread westwards from its area of introduction, through the Terai region and into the Siwaliks of eastern and central Nepal (Fig. 5.2), and is poised to invade western Nepal. Its potential to spread to higher elevations in Nepal is not known, but at present it is found at altitudes of up to 1300 m (Poudel *et al.*, 2005); in its native range it is found up to 2000 m asl.

Within the last two decades mikania weed has not only travelled nearly 800 km westwards, but in doing so has created havoc for farmers and wildlife conservationists. It has emerged as a problem weed in three protected areas: the Koshi Tappu Wildlife Reserve (KTWR), Parsa Wildlife Reserve (PWR) and Chitwan National Park (CNP) (Fig. 5.2). All of these sites are Important Bird Areas (IBAs); the first is also a Ramsar Site and the last is a World Heritage Site (Baral and Inskipp, 2005). Banke National Park (BaNP), a newly established park adjacent to Bardia National Park (BNP), is the westernmost locality where the weed has been reported. In addition, the weed is currently found in at least eight other IBAs that are not protected: Mai Valley, Urlabari forest groves, Dharan forests, Barandabhar forests, Nawalparasi forests, Lumbini farmlands, Jagadishpur Reservoir and Dang Deukhuri Foothill forests. Outside protected areas and forests, the spread of mikania weed has been exacerbated by the close interaction between people and nature. For example, villagers living near Koshi Tappu inadvertently helped to spread the weed by collecting grass contaminated with *M. micrantha* seed for thatching their houses (Yadav, 2010) and, as described in Box 5.3, human-mediated spread has been identified as the major factor in its spread in and around Chitwan National Park (Murphy *et al.*, 2013).

The fauna and flora of the tropical and subtropical regions of the Terai and the Siwaliks are relatively poorly known because they have been surveyed less extensively than other parts of the country. While the need to compile an inventory of biodiversity has been acknowledged (HMGN/MFSC, 2002), unless the invasive threat is addressed, many biological resources, including those with biochemical and genetic potential, may be lost before they are even discovered.

Impact of *Mikania micrantha*

Ecological impact

Although the introduction of an alien species may increase the number of species in a particular site in the short term, it will lead to a decrease in species diversity if native species are reduced or eventually displaced from the habitat or region (Begon *et al.*, 1990).

Many of the forests of the tropical region of eastern and central Nepal are now covered by a blanket of *M. micrantha*. The scale of the problem appears to be escalating, with undisturbed riverine forest especially prone to invasion (Poudel *et al.*, 2005; Murphy *et al.*, 2013). The plant invades the forest to such an extent that it smothers the ground flora, which prevents seedlings of other species from growing and thereby decreases the species diversity of the forest (authors' observations). The regeneration of the dominant tree species, for example *Shorea robusta* (sal), *Dalbergia sissoo* (local names: sisham or sissau) and *D. latifolia* (satisal), is completely blocked in highly infested areas (Adhikari, 2004). These trees are the most valuable timber trees of Nepal (see below) and *Dalbergia* spp. are also nitrogen-fixing trees and improve soil fertility – a vital function in the so often nutrient-poor soils of tropical forest ecosystems (Orwa *et al.*, 2009).

A study that was conducted in a core area and community forest buffer zone of Chitwan National Park found that riparian sites, forest edges, grassland with sparse tree and shrub cover, and the lower canopy areas of both natural and planted forests were seriously invaded by *M. micrantha* (Sapkota, 2007). Most small trees, shrubs and herbaceous plants were completely smothered. Of the major tree species, *D. sissoo* in planted forest and small, i.e. less than 17 cm dbh (diameter at breast height) *Bombax ceiba* were the worst affected, with trees failing to regenerate. Murphy *et al.* (2013) (see Box 5.3) recorded *M. micrantha* across 44% of habitats sampled in Chitwan National Park, with 15% having more than 50% coverage by the weed; the highest densities were recorded in riverine forest, tall grass and wetland habitats (see Fig. 5.4a).

Though there are no quantitative data on the changes, many plant species considered as important for non-timber forest products in Nepal are reported to be seriously threatened by the rampant expansion of mikania weed (Bhola Bhattarai, Federation of Community Forestry Users Nepal (FECOFUN), 2005, personal communication). In addition, according to the people of eastern Nepal, many medicinal plants are disappearing at an alarming rate (see Economic and social impacts section below).

The impact of mikania weed on Nepal's wetland ecosystems is most obvious in the Koshi Tappu Wildlife Reserve, one of the most important Ramsar sites in Nepal and an IBA (Baral and Inskipp, 2005). The Reserve lies on the flood plains of the Sapta-Koshi River in the Terai of eastern Nepal at about 78 m asl. It is home to the last surviving population of the endangered wild buffalo or arna (*Bubalus arnee*), as well as the globally threatened Ganges river dolphin (*Platanista gangetica*), nilgai (*Boselaphus tragocamelus*), smooth-coated otter (*Lutrogale perspicillata*), gharial (*Gavialis gangeticus*) and red-crowned roof turtle (*Kachuga kachuga*) (Thapa and Dahal, 2009). Over 500 species of birds have been recorded in the Reserve, including 19 globally threatened species, e.g. the swamp francolin (*Francolinus gularis*), the lesser adjutant (*Leptoptilos javanicus*) and the white-rumped and slender-billed vultures (*Gyps bengalensis* and *G. tenuirostris*), together with several near-threatened species (Inskipp *et al.*, 2013; Baral, 2016). The Reserve is also an important stopover for migratory birds (Inskipp and Inskipp, 1991).

Mikania has spread to cover the entire Koshi Tappu Reserve, even open water (Poudel *et al.*, 2005). Although the weed does not grow in water, it climbs over other wetland species, using them as a support, and in this manner can cover the normally exposed water surfaces of the wetland. This prevents sunlight and oxygen from reaching the water column and submerged plants, which has both direct (oxygen deprivation) and indirect (failure of food supply) effects on consumers, i.e. aquatic animals. The submerged plants that die as a result of lack of sunlight and oxygen begin to decay, and consume more oxygen in the process. The decrease in soluble oxygen disturbs the ecological balance of the aquatic ecosystem and the entire aquatic flora and fauna is affected.

The weed has equally serious consequences on wildlife populations. Birds are known to be directly affected by its spread (Baral, 2002). In Koshi Tappu, populations of several species of birds have plummeted; terrestrial feeding species, such as thrushes, pipits, and some babblers that require open forest floors, are all affected (Inskipp *et al.*, 2013). The weed can quickly cover the forest floor as well as trees and shrubs, and initial observations on the impact on birds that feed on the forest floor indicate a significant decline of two species of *Zoothera* thrushes, orange-headed thrush (*Z. citrina*) and scaly thrush (*Z. dauma*), and Tickell's thrush (*Turdus unicolor*) within the last 8 years (authors' observations). The impact of mikania on the swamp francolin is not clearly understood, but the Reserve's population of swamp francolin is of international significance and a study is therefore underway.

Similarly, in areas with high infestations *M. micrantha* has reduced the available food for mammalian herbivores; including the great one-horned rhinoceros (*Rhinoceros*

Fig. 5.4. *Mikania micrantha* in Chitwan National Park, Nepal. (a) The invasive alien plant *M. micrantha* smothering native vegetation in Chitwan National Park. Photo courtesy of S.T. Murphy. (b) areas worst affected by *M. micrantha* in the Park are home to the endangered one-horned rhinoceros (*Rhinoceros unicornis*), whose browse has been overwhelmed by infestations of the invasive plants *M. micrantha* and *Chromolaena odorata*. Photo courtesy of S.T. Murphy. (c) A rhinoceros census using elephants to reach inaccessible areas allowed the distribution of *M. micrantha*, to be mapped in the Park. Photo courtesy of the Zoological Society of London/National Trust for Nature Conservation, Nepal.

unicornis) in the Chitwan National Park (DNPWC, 2009; Lamichhane *et al.*, 2014) (see Box 5.3). This Park in southern central Nepal, bounded by the Mahabharat range to the north and the Siwalik hills and Terai forests to the south (Sapkota, 2007), is one of the few remaining undisturbed vestiges of the Terai (UNESCO, 2014). The vegetation is predominantly sal moist deciduous climax forest. The Park has a particularly rich flora and fauna and, in addition to being home to one of the last populations of the great one-horned rhinoceros, it is one of the last refuges of the royal Bengal tiger (*Panthera tigris*) (Mishra and Jefferies, 1991). A study by Murphy *et al.* (2013), which was part of a wider census of the one-horned Indian rhinoceros in the Park, found that the habitats most seriously affected by *M. micrantha* were those that were also home to the highest numbers of rhinoceros (Fig. 5.4b). The scarcity of normal browse in the Park has prompted rhinoceroses to raid villagers' crops, which is causing an increasing incidence of direct conflict between the wildlife and people living on the periphery of the Park. The reduction of grazing habitat is likely to have a severe impact on the overall herbivore population of Chitwan and other protected areas of Nepal, ultimately

Box 5.3. Invasive weeds, communities and the one-horned rhinoceros.

The Zoological Society of London, with CABI and the National Trust for Nature Conservation in Nepal, undertook a 3-year project based in Chitwan and Bardia National Parks and Shuklaphanta Wildlife Reserve. Funded by the Defra (UK Department for Environment, Food and Rural Affairs) Darwin Initiative, the team implementing the 'Crisis to biological management – rhinoceros, grassland and public engagement – Nepal' project carried out a number of initiatives to try to conserve the one-horned rhinoceros and other species in this Park and protect their Terai grassland habitat. Tackling invasive weeds was a key component.

The major problem species, including *Mikania micrantha*, *Chromolaena odorata* and *Lantana camara*, are smothering Chitwan's grasslands and forests and blocking pathways for the free movement of wild animals (see Fig. 5.4a). In addition, native grasses and other plants that provide essential food for the rhinoceros are being outcompeted and may be on the decline. *M. micrantha* had been identified as a particular threat to the remaining rhinoceros habitat in Chitwan (Murphy *et al.*, 2013) (Fig. 5.4b). It has been estimated that the risk of rhinoceroses being poached is up to ten times more likely when they forage outside the Park, and mikania was thought likely to be a major driver for animals to leave the protected area.

To establish the extent of the mikania problem (as part of a census of rhinoceros numbers), surveys of the Park were carried out on the backs of elephants, thus allowing access to areas inaccessible via vehicle or on foot (Fig. 5.4c). The surveys allowed an invasive plant map of the area to be drawn, so that the species can be monitored. It was found that mikania has the highest incidence in the wetter areas, but additional casual observations made during the survey suggest that *C. odorata* is more common in the drier zones of Chitwan.

While short-term and longer-term plans are needed in order to manage and eventually prevent the spread of these invasive weeds, it was first necessary to understand the ecology of the weeds and the habitat types that exist within the parks, as well as the socio-economics of park usage by local and visiting communities.

Local community dependence on natural resources in the core area of the Park was found to be high, with two factors exacerbating the spread of *M. micrantha* (Murphy *et al.*, 2013). First, the range and volume of resources (e.g. fodder) collected and the distances travelled all contribute to the spread of the weed. Secondly, and of even greater significance, is the annual burning of the grasslands in the Park by local communities, which is estimated to affect 25–50% of the total area, and favours the emergence of *M. micrantha* infestations (see Chapters 2 and 7, this volume).

The ultimate aim of the project was to contribute to the development of a management plan to control the invasive weeds for the benefit of the rhinoceros, the environment and the local communities. Core elements of a management plan for *M. micrantha* need to incorporate actions to control grassland burning, reduce the spread of the weed and raise awareness of the best practices for local resource management by local communities (Murphy *et al.*, 2013).

affecting umbrella species like the royal Bengal tiger, which depend on the ungulates as their main prey item. The effects of mikania on the apex predator of the lowland Nepal is worrying, and further steps to control and manage this invasive weed need to be a high priority for conservationists in Nepal. The impact of mikania on other vertebrates – amphibians, reptiles and fish – remains poorly understood and needs to be studied. Chapter 10, this volume, provides a summary of the biological control of the weed using a rust fungus that has been successfully implemented in a number of countries in Asia; its introduction into Nepal is being considered (Poudel *et al.*, 2005).

Economic and social impacts

Forests have a vital role in maintaining ecological balance and in economic development; timber, energy sources, animal fodder and other non-wood products all originate from them. Pristine forest is also a major tourist attraction. The sal-dominated forests of the Terai that have been invaded by mikania are a valuable source of timber; sal itself produces high-grade durable timber that is used in construction, and the rosewood timber of *Dalbergia* spp. has outstanding properties and is used in a wide variety of products, from veneer and furniture to musical instruments and boat keels. Timber from these forests is used for the elaborate and extensive wood-carving seen in the temples of Nepal. Non-wood products include sal seed, a human and animal food resource, while *Dalbergia* spp. provide fodder and fuelwood, and parts of these trees are valued for pesticidal and medicinal properties (Orwa *et al.*, 2009); additionally, resin and sal leaves are used in special Hindu and Buddhist rituals.

All forests in Nepal that are not privately owned are National Forests. However, the GoN has handed over the management of some of these forests to various types of groups with different management objectives. Community forest user groups (CFUGs) undertake the development, conservation and utilization of community forests for the collective benefit of the community (HMGN/MFSC, 2002). This collective benefit includes the production of forest products: timber, fuel, fodder and non-wood forest products, such as grass for paper production, sal seed for oil and, in addition, medicinal plants.

Nepal is rich in medicinal plant resources, which provide an emerging local industry, and together with the growth of the country's main export market in India, present opportunities for improving rural livelihoods (Bhattarai and Karki, 2004). The authorities are trying to put measures in place to prevent increased harvesting of forest resources from threatening the medicinal plant resource base, but the impact of invasive species could prove overwhelming.

Although fewer medicinal plants are found in the Terai than in the higher altitude regions of the country (Sah and Dutta, 1996), they are directly linked with the livelihoods of the local people. For example, *Pogostemon benghalensis* (locally called rudilo), which is reported to be on the decline as a result of the mikania invasion, is harvested and the oil extracted and used as a stimulant and to stop bleeding, while the fresh leaves are used to clean wounds and promote healing; a patchouli-like oil is also distilled from the leaves (Shrestha *et al.*, 2013).

CFUGs in some districts in the Terai started to cultivate medicinal plants to boost their economies, but this proved impossible to sustain following the introduction and spread of mikania (Tiwari *et al.*, 2005), as the costs associated with the management and control of mikania far outweigh the financial benefits from the cultivation of medicinal plants. In the Humse Dumse Community Forest of Jhapa District, East Nepal, the estimated annual clearance cost for mikania was (Nepalese) Rs. 70,000 (US$1000) in 2002, but the area invaded is expanding rapidly, so the figure may have risen substantially since then.

Even though agriculture is the main occupation of the Nepalese population, production is insufficient to meet annual demand. Consequently, people encroach on forestland for cultivation. Because of the proximity of cultivated land to the forested areas, mikania can find its way from the forest into farmers' fields, and this increases the weeding costs year on year (Tiwari *et al.*, 2005). Moreover, in agricultural situations, mikania may be one of several invasive weeds the farmers has to contend with. For example, in the 4 years after the arrival of *A. philoxeroides* at the National Agricultural Research Centre in Khumaltar, it spread through the entire farm, with rice fields, paths, canals and embankments (bunds) all now covered with the weed (Jagat Devi Ranjit, National Agricultural Research Council, Khumaltar, 2015, personal communication).

The results of the assessment of the invasive alien plant species of Nepal undertaken by IUCN-Nepal show that there are six high-risk invasive alien plants threatening its natural ecosystems, namely *A. adenophora*, *C. odorata*, *E. crassipes*, *I. carnea*, *L. camara* and *M. micrantha* (Tiwari *et al.*, 2005). Addressing the mikania problem may provide mechanisms through which the other weeds can be tackled (Poudel *et al.*, 2005).

The response of people living around the Chitwan National Park to invasive plants, investigated by Rai *et al.* (2012), was found to be complex and changed over time, but the aggressive growth of *M. micrantha* meant that it was perceived as more deleterious than other, longer established species. Given external support, people will try to control the weed, but in the absence of such support, their reaction is to 'make the best out of the worst situation', e.g. by using *M. micrantha* for forage and grazing (despite its harmful effects on livestock; Siwakoti, 2007), though Rai *et al.* (2012) noted that *M. micrantha* was of secondary preference, not being an adequate substitute for any of the traditional forest products used in everyday life in Nepal.

Current Management Practices in Nepal

While no management and control frameworks have yet been implemented at a national level to control the spread of mikania, attempts have been made at a local level in some parts of eastern Nepal (e.g. Box 5.4). Local people and CFUGs are encouraged to plant medicinal plants and tree saplings after clearing mikania manually.

Box 5.4. Community engagement and mikania management.

The role of local communities and limitations of manual mikania management in wildlife areas are indicated by a study conducted in the Koshi Tappu Wildlife Reserve core area, where the infestations remain severe, and in the buffer zones of the Reserve (Siwakoti, 2007).

A collaborative effort between the community and the Reserve authority to protect both the biodiversity of the Reserve and the crops from animals focused on the most seriously invaded part of the buffer zones, the eastern embankment, which was handed over to the community to manage.

The key first step was to explain to people that wildlife depredation of crops had intensified because mikania had overgrown the grasses that formed the natural fodder of wild animals. Community forest user groups (CFUGs) were created, each with responsibility for an area of community forest of 2–10 ha. These groups cleared mikania from their areas using manual labour, planted saplings of forest trees and also constructed fences to protect the crops from wildlife incursions.

The forest has begun to recover, and people are able to harvest grass (for fodder and thatching), firewood, etc., in the community forest areas (and some have also planted medicinal herbs), while animal impacts on the crops have lessened. All of this has helped to reduce conflicts between the local people and the Reserve authorities.

Nevertheless, the core area remains heavily infested by mikania and is a source of reinfestation. Although there is community interest in tackling this, a sustainable method is needed; manual measures are impractical on such a large area and the use of herbicides is inappropriate in a protected area.

This initiative was financed partly by user groups and partly by conservation groups, but Siwakoti (2007) emphasized that communities must see visible benefits if their efforts are to be sustained.

Farmers simply dig out the plants that encroach on their agricultural land. Koshi Camp, a private wildlife lodge in the Koshi Tappu Wildlife Reserve, spends more than US$500 p.a. on the manual control of the weed in a privately managed habitat with an area of 4 ha. People in the northern part of Koshi Tappu have reported that a native climber, *Trichosanthes* sp. (local name indrayani), can suppress mikania on a local scale, though this has not yet been investigated (Poudel *et al.*, 2005). Sapkota (2007) reported the results of experiments that suggested mulching with manually cleared mikania could suppress its growth, and that the parasitic plant *Cuscuta reflexa* had potential as a control agent, but these approaches need further evaluation.

There are also attempts at the management of mikania through utilization. In Jhapa District in eastern Nepal, the Nepal Organic Agricultural Society in collaboration with Damak Municipality has started to make compost using mikania collected from highly infested forest areas. In the Koshi Tappu Wildlife Reserve, managers have encouraged local people to collect mikania as fodder for livestock. However, such management practices could do more harm than good, given that mikania can germinate from its plentiful small seeds and from mere fragments of stem of the weed when it is collected from highly infested forest areas; these could then provide a mechanism for the spread of mikania to new areas. Hence, such exercises could increase damage to native plants that have already been badly affected by mikania weed, as well as ultimately aiding the growth and spread of the weed (Poudel *et al.*, 2005). Furthermore, a suggestion for opening forest areas for cattle to graze mikania needs careful evaluation as anecdotal reports suggest that cattle may prefer other vegetation. Although manually cleared mikania has been used to some extent as cattle and goat fodder by communities bordering the Koshi Tappu Wildlife Reserve, it was mixed it with grass because, they reported that the mikania caused decreased milk yields and abdominal disorders in the animals (Siwakoti, 2007).

The Way Forward

The *First National Workshop for Stakeholders* – Mikania micrantha *Weed Invasion in Nepal*, organized by Himalayan Nature and sponsored by IUCN-Nepal and CABI, was held in November 2004 in Kathmandu (Poudel *et al.*, 2005). The aims were to assess the state of knowledge of mikania invasion in Nepal and consider management options. The meeting provided basic information to researchers, managers and conservationists from concerned agencies in the environment, forestry and agricultural sectors in order to facilitate and synergize efforts to control mikania. Although manual, cultural (utilization) and chemical measures were considered at the workshop, they were discounted as practical or sustainable wide-scale solutions. The limitations of manual control and utilization have been described above, while herbicide use in Nepal is very low, and is not considered appropriate for forest areas because of its adverse impacts on native biodiversity. The participants heard about work in India and China on the classical biological control of mikania using a rust fungus (Ellison, 2001; Ellison *et al.*, 2003), and agreed that using co-evolved natural enemies was the priority approach; the need to build awareness of the mikania problem and consult all stakeholders on its solution was highlighted (see below). There are also lessons to be learnt from other countries (Cock *et al.*, 2000; Chapter 10, this volume).

Perhaps most significantly, this mikania initiative led to the importance of the biological control approach for tackling the problems imposed by invasive alien plant species in Nepal to be understood for the first time, and the value of including this management approach in the Nepal Biodiversity Strategy (NBS) (HMGN/MFSC, 2002).

Following the NBS, the National Biodiversity Strategy and Action Plan (NBSAP) (GoN/MoFSC, 2014) recognizes *M. micrantha* as a serious threat to native ecosystems, but beyond stating that there are more than 100 such plants that have become invasive weeds in Nepal, it proposes little to deal with them apart from a statement declaring that they should be eradicated. An underlying major problem is that information on the status, ecology and distribution of invasive alien species (IAS) in Nepal is rather inadequate, and what is known has yet to be compiled into a single information source, such as a book. This lack of accumulated information has hampered initiatives to control mikania at both the governmental and non-governmental level. To address the more general problem of invasive weed management, a dossier has been published by IUCN-Nepal detailing the impacts of all the major weeds of Nepal (Tiwari *et al*., 2005).

There is a need for national IAS experts to coordinate with other stakeholders of society who are directly and indirectly affected by mikania. This exercise will help to assess the real threat from the weed and estimate more accurately its effects on people and wildlife; the *First National Workshop for Stakeholders* held in 2005 was a good beginning on which to build. It recommended that IAS should be flagged and set as a priority on a cross-sectoral basis, with awareness building at all levels, and in the case of mikania, a proper investigation of the spread of the weed should be made. The potential for community forestry to contribute at the grassroots level was highlighted, as these groups have their own mechanisms for creating awareness.

The need for a legal framework to coordinate the response to the invasive species threat is clear from other chapters of this book, and was also raised at the *First National Workshop for Stakeholders*. Nepal already has in place protected area management strategies (GoN/MoFSC, 2014) that are well placed to handle invasive species issues. Indeed, some protected areas have been piloting control operations for mikania, for example Chitwan National Park, and these lessons could be shared. Equally, people could be made aware of the serious threat this weed poses through existing institutions such as CFUGs and FECOFUN. In addition, there is a need for mechanisms to enable stakeholders in forestry, agricultural and environmental agencies (governmental and non-governmental) to work together. Moreover, given the interest in the biological control of mikania, the appropriate national organizations need to be identified, and their functions in the process of regulating and implementing biological control clarified (Murphy, 2014). There is now a good example of what can be achieved from the progress that has been made with the biological control of *M. micrantha* in Papua New Guinea, Fiji and Taiwan (Ellison *et al*., 2008; Day *et al*., 2013).

Given the growing problem of invasive species in Nepal and more globally, an international workshop on invasive species was held at Chitwan National Park by the National Trust for Nature Conservation in March 2014 (Thapa *et al*., 2014). This workshop was significant as it brought all major experts to the same platform for sharing knowledge and for formulating an agenda for the future. Although the mikania problem was the eye-opener and perhaps led to this international workshop, various other invasive alien species were also discussed as well as mikania. However, now the time has come for Nepal to decide on how best to control mikania, and this workshop will provide a good experience on which to build and address other species. If we do not act soon, it may be too late and this weed and other invasive species might spread into more areas than currently known.

References

Adhikari, B. (2004) *Inventory and Assessment of Invasive Alien Plant Species of Nepal*. Report for IUCN (The World Conservation Union), Kathmandu.

Baral, H.S. (2002) Invasive weed threatens protected area. *Danphe* 11, 10–11.

Baral, H.S. (2016) *Birds of Sapta Koshi Floodplains*. Himalayan Nature, Kathmandu.

Baral, H.S. and Inskipp, C. (2005) *Important Bird Areas in Nepal: Key Sites for Conservation*. Bird Conservation Nepal, Kathmandu and BirdLife International, Cambridge, UK.

Begon, M., Harper, J.L. and Townsend, C.R. (1990) *Ecology: Individuals, Populations and Communities*, 2nd edn. Blackwell, Oxford, UK.

Bhattarai, N.K. and Karki, M.B. (2004) Conservation and management of Himalayan medicinal plants in Nepal. In: Donoghue, E.M., Benson, G.L., and Chamberlain, J.L. (eds) *Sustainable Production of Wood and Non-Wood Products. Proceedings of the IUFRO Division 5 Research Groups 5.11 and 5.12, Rotorua, New Zealand, 11–15 March 2003.* General Technical Report PNW-GTR-604, USDA Forest Service, Pacific Northwest Research Station, Portland, Oregon, pp. 45–50.

Cock, M.J.W., Ellison, C.A., Evans, H.C. and Ooi, P.A.C. (2000) Can failure be turned into success for biological control of mile-a-minute weed (*Mikania micrantha*)? In: Spencer, N.R. (ed.) *Proceedings of the X International Symposium on Biological Control of Weeds, Bozeman, Montana, 4–14 July 1999.* USDA Forest Service, Forest Health Technology Enterprise Team, Morgantown, West Virginia. pp. 155–167.

Day, M.D., Kawi, A., Fidelis, J., Tunabuna, A., Orapa, W., Swamy, B., Ratutini, J., Saul-Maora, J. and Dewhurst, C.F. (2013) Biology, field release and monitoring of the rust *Puccinia spegazzinii* de Toni (Pucciniales: Pucciniaceae), a biocontrol agent of *Mikania micrantha* Kunth (Asteraceae) in Papua New Guinea and Fiji. In: Wu, Y., Johnson, T., Sing, S., Rhagu, R., Wheeler, G., Pratt, P., Warner, K., Center, T., Goolsby J. and Reardon, R. (eds) *Proceedings of the XIII International Symposium on Biological Control of Weeds, Waikoloa, Hawaii, 11–16 September 2011.* USDA Forest Service, Forest Health Technology Enterprise Team, Morgantown, West Virginia, pp. 211–217.

DNPWC (2009) *The Status and Distribution of the Greater One-Horned Rhino in Nepal.* Department of National Parks and Wildlife Conservation, Ministry of Forests and Soil Conservation, Government of Nepal, Kathmandu.

Ellison, C.A. (2001) Classical biological control of *Mikania micrantha.* In: Sankaran, K.V., Murphy, S.T. and Evans, H.C. (eds) *Alien Weeds in Moist Tropical Zones: Banes and Benefits. Proceedings of a Workshop, Kerala Forest Research Institute, Peechi, India, 2–4 November 1999.* Kerala Forest Research Institute, Peechi, India and CABI Bioscience UK Centre (Ascot), Ascot, UK, pp. 131–138.

Ellison, C.A., Evans, H.C. and Ineson, J. (2003) The significance of intraspecies pathogenicity in the selection of a rust pathotype for the classical biological control of *Mikania micrantha* (mile-a-minute weed) in Southeast Asia. In: Cullen, J.M., Briese, D.T., Kriticos, D.J., Lonsdale, W.M., Morin, L. and Scott, J.K. (eds) *Proceedings of the XI International Symposium on Biological Control of Weeds, Canberra, Australia, 27 April–2 May 2003.* CSIRO (Commonwealth Scientific and Industrial Research Organisation) Entomology, Canberra, pp. 102–107.

Ellison, C.A., Evans, H.C., Djeddour, D.H. and Thomas, S.E. (2008) Biology and host range of the rust fungus *Puccinia spegazzinii*: a new classical biological control agent for the invasive, alien weed *Mikania micrantha* in Asia. *Biological Control* 45, 133–145.

GoN/MoFSC (2014) *Nepal National Biodiversity Strategy and Action Plan 2014–2020.* Government of Nepal, Ministry of Forests and Soil Conservation, Kathmandu.

HMGN/MFSC (2002) *Nepal Biodiversity Strategy.* His Majesty's Government of Nepal, Ministry of Forests and Soil Conservation, Kathmandu.

Inskipp, C. (1989) *Nepal's Forest Birds: Their Status and Conservation.* ICBP Monograph No. 4, BirdLife International [formerly International Council for Bird Preservation], Cambridge, UK.

Inskipp, C. and Inskipp, T. (1991) *A Guide to the Birds of Nepal,* 2nd edn. Christopher Helm, London.

Inskipp, C., Baral, H.S., Inskipp, T. and Stattersfield, A. (2013) The state of Nepal birds 2010. *Journal of Threatened Taxa* 5, 3473–3503.

Kitamura, S. (1966) Compositae. In Hara. H. (ed.) *The Flora of Eastern Himalaya: Results of the Botanical Expedition to Eastern Himalaya Organized by the University of Tokyo 1960 and 1963.* University of Tokyo, Tokyo.

Lamichhane, B.R., Subedi, N., Chapagain, N.R., Dhakal, M., Pokheral, C.P., Murphy, S.T. and Amin, R. (2014) Status of *Mikania micrantha* invasion in the rhino habitat of Chitwan National Park, Nepal. In: Thapa, G.J., Subedi, N., Pandey, M.R., Thapa, S.K., Chapagain, N.R. and Rana, A. (eds) *Proceedings of the International Conference on Invasive Alien Species Management, Chitwan, Nepal, 25–27 March 2014.* National Trust for Nature Conservation, Kathmandu, pp. 52–60.

Mishra, H.R. and Jefferies, M. (1991) *Royal Chitwan National Park: Wildlife Heritage of Nepal.* King Mahendra Trust for Nature Conservation and The Mountaineers, Kathmandu.

Murphy, S.T. (2014) Galvanizing action for the management of invasive alien species. In: Thapa, G.J., Subedi, N., Pandey, M.R., Thapa, S.K., Chapagain, N.R. and Rana, A. (eds) *Proceedings of the International Conference on Invasive Alien Species Management, Chitwan, Nepal, 25–27 March 2014.* National Trust for Nature Conservation, Kathmandu, pp. 1–6.

Murphy, S.T., Subedi, N., Jnawali, S.R., Lamichhane, B.R., Upadhyay, G.P., Kock, R.

and Amin, R. (2013) Invasive mikania in Chitwan National Park, Nepal: the threat to the greater one-horned rhinoceros *Rhinoceros unicornis* and factors driving the invasion. *Oryx* 47, 361–368.

Orwa, C., Mutua, A., Kindt, R., Jamnadass, R. and Simons, A. (2009) Agroforestree database: a tree reference and selection guide, version 4.0. Available at: www.worldagroforestry.org/treedb/ (accessed 5 May 2016).

Poudel, A., Baral, H.S., Ellison, C., Subedi, K., Thomas, S. and Murphy, S. (2005) Mikania micrantha *Weed Invasion in Nepal: a Summary Report of the First National Workshop for Stakeholders, Kathmandu, Nepal, 25 November 2004.* Himalayan Nature and IUCN, Kathmandu and CAB International, Wallingford, UK.

Rai, R.K. and Rai, R. (2013) Assessing the temporal variation in the perceived effects of invasive plant species on rural livelihoods: a case of *Mikania micrantha* invasion in Nepal. *Conservation Science* 1, 13–18.

Rai, R.K., Scarborough, H., Subedi, N. and Lamichhane, B. (2012) Invasive plants – do they devastate or diversify rural livelihoods? Rural farmers' perception of three invasive plants in Nepal. *Journal for Nature Conservation* 20, 170–176.

Sah, S.P. and Dutta, I.C. (1996) Inventory and future management strategies of multipurpose tree and herb species for non-timber forest products in Nepal. In: Leakey, R.R.B., Temu, A.B., Melney, K. and Vantomme, P. (eds) *Domestication and Commercialization of Non-timber Forest Products in Agroforestry Systems. Proceedings of an International Conference, Nairobi, Kenya 19–23 February 1996.* Non-Wood Forest Products No. 9, Forestry Department, Food and Agriculture Organization of the United Nations, Rome.

Sapkota, L. (2007) Ecology and management issues of *Mikania micrantha* in Chitwan National Park, Nepal. *Banko Janakari* 17, 27–39.

Shrestha, T.K., Tamang, B. and Khadgi, A. (2013) *A Study on Control of* Mikania micrantha *in Lowland of Protected Areas in Nepal.* Prepared by Lumbini Environmental Services Pvt. Ltd, Kathmandu, for the Ministry of Forests and Conservation, Department of National Parks and Wildlife Conservation, Kathmandu. Available at: http://dx.doi.org/10.13140/RG.2.1.1171.9765.

Siwakoti, M. (2007) Mikania weed: a challenge for conservationists. *Our Nature* 5, 70–74.

Thapa, I. and Dahal, B.R. (2009) Sustainable wetland management for wildlife and people at Koshi Tappu Wildlife Reserve. *Banko Janakari* 19 (special issue), 36–39.

Thapa, G.J., Subedi, N., Pandey, M.R., Thapa, S.K., Chapagain, N.R. and Rana, A. (eds) (2014) *Proceedings of the International Conference on Invasive Alien Species Management, Chitwan, Nepal, 25–27 March 2014.* National Trust for Nature Conservation, Kathmandu.

Tiwari, S., Adhikari, B., Siwakoti, M. and Subedi, K. (2005) *An Inventory and Assessment of Invasive Alien Plant Species of Nepal.* IUCN (The World Conservation Union), Kathmandu.

UNESCO (2014) World Heritage Site: Chitwan National Park. Available at: http://whc.unesco.org/en/list/284 (accessed 1 January 2014).

Yadav, N.P. (2010) How *Mikania micrantha* (lahare banmara) destroy the forest and agricultural crops in residential area of Nepal. Forestry Nepal website, Dr N.P. Yadav's Blog, 27 January 2010. Previously available at: www.forestrynepal.org/article/1549/4534 (accessed 12 June 2014).

6 Impact and Management of Invasive Alien Plants in Pacific Island Communities

Warea Orapa*

(Formerly, Plant Protection Advisor, Secretariat of the Pacific Community), National Agriculture Quarantine and Inspection Authority, Port Moresby, Papua New Guinea

Introduction

Invasive alien plants are among the most important groups of invasive species affecting both the socio-economic well-being of Pacific island people and the island ecosystems, which have a unique biological diversity. Invasive species in general have an impact on food security, human access to natural resources, personal and national incomes, and human health. The impacts of alien plants can be positive (e.g. as food, building materials) or negative (as weeds). Although global concern about the invasiveness of alien plants has been promoted by mostly environmental interests, the responses to reducing or preventing their impacts do not always receive the necessary level of attention by developing Pacific island governments or agencies until the socio-economic sectors have been visibly implicated. However, efforts to address and minimize existing or new biosecurity issues associated with all invasive species are making some headway in the Pacific, even in the presence of obstacles such as chronic deficiencies in national and regional institutional capacities. This chapter focuses on the issues concerning invasive alien plant management or prevention and the negative environmental and socio-economic impacts of selected invasive alien plants in some of the developing Pacific Island States and Territories (PICTS).

Geography and spread of alien plants

The 22 developing PICTs (Fig. 6.1) include some of the smallest and most isolated countries in the world. The total combined population of the region is estimated to be 7.2 million people, who are spread over 30 million km^2 of predominantly water. Most of the countries are very small in land and population size. Papua New Guinea is the largest and it alone contains 80% of the combined land mass of the PICTS and 78% of the total Pacific population. The Pacific region is characterized by social and cultural diversity, and most PICTs have been subject to foreign colonization. Invasive alien plant species may be quietly threatening the socio-economic status of each island country, but this increasingly important problem needs the awareness and recognition that has already been attained by other environmental and socio-economic issues facing the PICTs.

In the past, the majority of Pacific islands have benefited from being isolated from the major land masses and population centres. Located in a vast expanse of ocean, very few plant introductions occurred

E-mail: warea.orapa@gmail.com or worapa@naqia.gov.pg

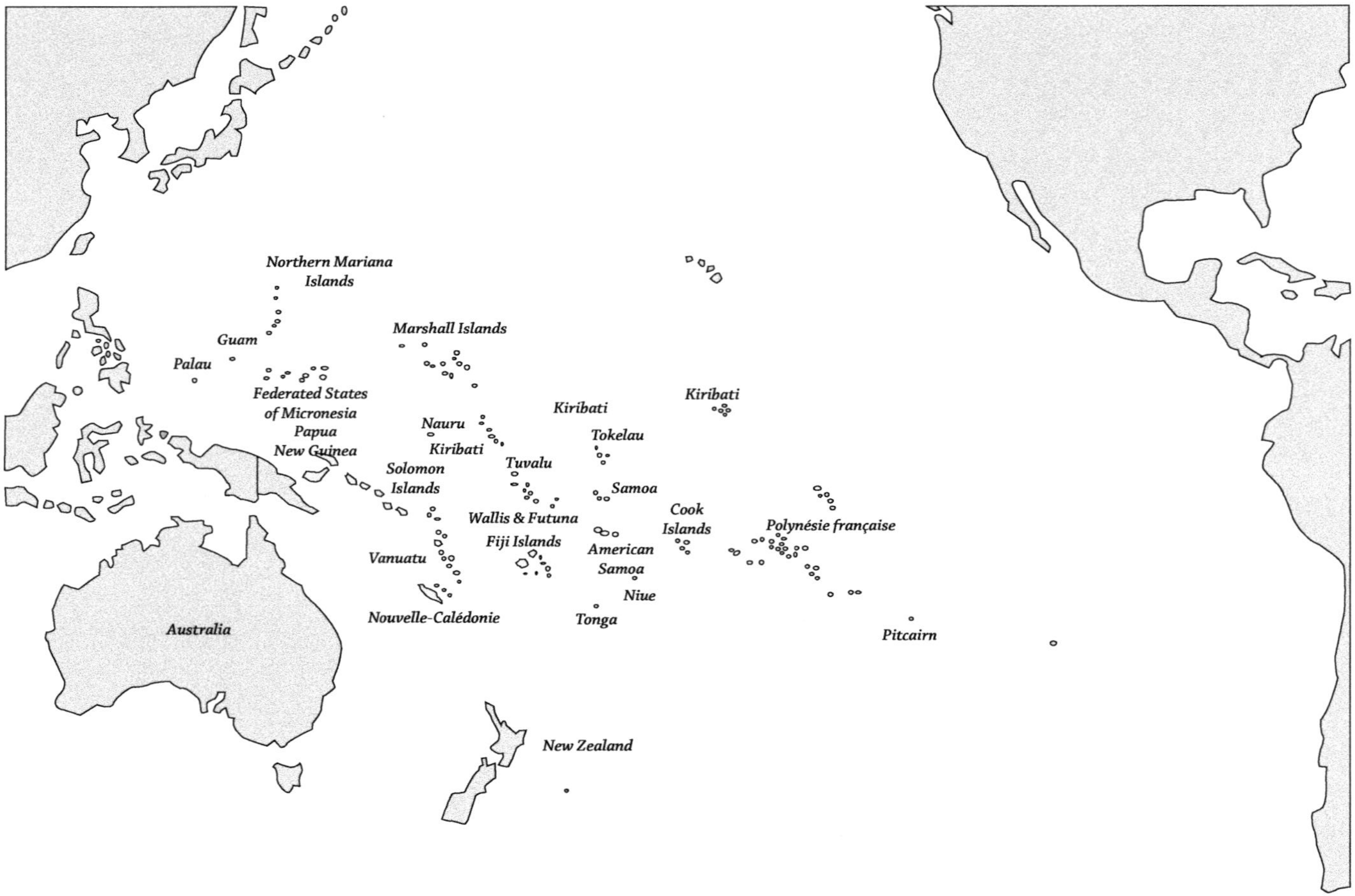

Fig. 6.1. Map of the Pacific Community and its 22 member developing countries and territories. Courtesy Secretariat of the Pacific Community (SPC).

during the 3000–4000 years of human presence in the Pacific, until the arrival of colonizers and the rise of global trade and travel during the last two centuries. As Mack and Lonsdale (2001) and Barnard and Waage (2004) have correctly pointed out, we humans have been the main culprits in the dispersal of vascular plants worldwide. The distribution patterns of many invasive plant species have historical linkages, particularly with European powers and their spheres of influence. In recent times though, introductions and inter-island spread have been accelerated by trade and travel within and outside the region. There has been an increasing trend to introduce plants for aesthetic reasons (as ornamentals), as well as for soil and pasture improvement and forestry, and in some cases for food. Many such plants have escaped to become weedy. The kinds of traits that have made escaped ornamentals highly desirable to gardeners, landscapers and the nursery industry also tend to make them invasive: they reproduce and establish easily, and grow rapidly under varied conditions (Li *et al.*, 2004). None the less, while many invasive alien plants have been moved around the world intentionally, many invasive plant problems in the Pacific have been caused by accidentally introduced plants associated with human activities such as trade and travel.

Impact on biodiversity

Invasive plants can change landscapes and whole ecosystems and contribute to reductions in biological diversity. The globally significant biodiversity areas of the Pacific region, such as those in New Guinea (the second highest biodiversity centre in the world) and New Caledonia (Mittermeier and Mittermeier, 1997; Mittermeier *et al.*, 2005; Papineau and Blanfort, 2008) are threatened. Invasive species threaten the critical ecosystems (in terms of species extinctions) of New Caledonia, where as many as 74% of all plant species, 67% of mammals, 32% of birds, 89% of reptiles and 100% of amphibians are endemic. Similarly, islands in eastern Melanesia and the Polynesia–Micronesia region are also classified as critical areas (Mittermeier *et al.*, 1999, 2005) that are threatened by habitat destruction and invasive alien species (IAS). The island archipelagos of eastern Polynesia contain a diverse range of habitats, including montane forests with a highly endemic flora (e.g. 50% of vascular plants on Nuku Hiva and 45% in Tahiti), which are threatened more by invasive species than by human-mediated habitat destruction (Meyer, 2004). Oceanic islands are regarded as 'extinction hotspots' because species are more likely to be endemic and consist of only small populations, the ecosystems are simple and island species often lack defence mechanisms against IAS. A few invasive alien plant species could easily account for as much as 55% of future species extinctions (Liz Dovey, Secretariat of the Pacific Regional Environment Programme (SPREP), 2005, personal communication). Any loss of biological diversity on islands will have serious socio-economic implications for the livelihoods of indigenous people.

Socio-economic impacts

It is impossible to discuss the social and economic impacts of invasive plants separately because very close linkages exist between these two sectors in the Pacific islands. Economic gains at individual or community levels can be reduced by alien plant invasions, which can then contribute significantly to social problems such as health and poverty.

The negative impacts of invasive plants can be in the form of losses suffered as a consequence of the presence of a particular species or the costs (energy, time and money) of minimizing its impacts through surveillance, prevention or control programmes. In addition, indirect losses such as lost opportunities resulting from weed invasion can be significant. The financial losses from alien plant invaders can be substantial, but are difficult to quantify in any Pacific island situation. In most cases, it is difficult to differentiate the various impacts that may arise at the same time or to estimate their monetary value, individually or in total, in the absence of good research data. Conversely, it has long been recognized by traditional subsistence food farmers that crop yields fare poorly in the presence of competing weeds. Poor crop yields, whether due to weeds or other factors, nearly always determine the socio-economic status of individuals, families and even communities. The impacts of alien invasive plants would, therefore, definitely include reduced income from primary production (agriculture, horticulture, grazing, forestry and fisheries) and in some service sectors (e.g. loss of tourism earnings) through reduced crop yields or through increased production costs, infrastructure maintenance or direct impacts on human welfare, such as health.

This chapter outlines the many challenges that the Pacific islands face in mitigating the impacts of invasive plant species, and discusses the species involved and the socio-economic impacts of selected alien plant invaders for which published or anecdotal information exists. It concludes by reviewing the management of plant invasions in the Islands, including some case studies and future threats.

Obstacles to Addressing Invasive Plant Issues

Awareness and regional collaboration

While global and regional agencies and their various foci of expertise and country representatives have recognized IAS as one of the threats to environmental sustainability (Reaser *et al.*, 2003; Barnard and Waage, 2004; SPREP, 2014), the general awareness of IAS issues in the Pacific has been low.

The *Seventh Pacific Islands Conference on Nature Conservation and Protected Areas* (Seventh PICNCPA) held in 2002 recognized the threats to island biodiversity by invasive species and developed 5 year targets from 2003 to 2007 in an Action Strategy (SPREP, 2004). As a result of these collective concerns, regional actions were undertaken such as the setting up of the Pacific Invasives Learning Network (PILN) for the exchange of information on invasive species, and the identification and development of selected demonstration projects to address general invasive species issues. The PILN is a network of national or local island-based multisectoral groups working to identify, develop and implement selected invasive species prevention or management projects on islands.

One of the main topics discussed during the Eighth PICNCPA, held in 2007, was partnership (SPREP, 2009), and this led on to the establishment of the Pacific Invasives Partnership (PIP) in 2010. The PIP was established to support PILN by promoting coordinated planning and assistance from regional and international agencies in the Pacific on invasive species management; it brought together all of the agencies working on invasive species, including the PILN, under one umbrella. The PIP is housed at the SPREP based at Apia in Samoa (see http://www.sprep.org/Pacific-Invasives-Partnership/invasive-partnerships). The primary focus of most agencies involved in the PIP is biodiversity conservation, especially that of threatened or critically endangered species. However, there is a significant interest in addressing the impacts of IAS on the economic and social sectors of the islands, and the Secretariat of the Pacific Community (SPC, until 1998 the South Pacific Commission; also designated plain 'Pacific Community') (www.spc.int/) has been at the forefront of that regional mandate, although resourcing implications have been an ongoing issue.

During the decade from 2002, the SPC has continued to increase awareness of invasive plant pests and diseases, and to incorporate these into national sectoral development plans and regionally harmonized draft biosecurity legislation for countries to adopt. At the medium-term policy level, the Pacific Islands Regional Millennium Development Goals (SPC, 2004) recognize the problem of IAS among the many other island developmental issues, and member countries are expected to take the necessary actions.

In 2009, the SPREP and SPC collaborated on the support of the fledgling PIP to obtain approval from the member countries for regional guidelines on invasive species management (Tye, 2009). These guidelines became useful between 2010 and 2014 (David Moverley, Invasive Species Advisor, SPREP, February 2014, personal communication) for the planning and implementation of the Global Environment Facility (GEF)-funded PIP to be implemented in nine of the ten GEF-eligible countries (www.sprep.org/table/ias/). Recently, in annual meetings of the 15 Pacific Islands Forum Leaders in 2013 and 2014, lobbying efforts by the PIP have raised the profile of invasive species among governments.

The Ninth PICNCPA, held in 2013, produced and adopted the *Framework for Nature Conservation and Protected Areas in the Pacific Islands Region 2014–2020* (SPREP, 2014). This important document resulted from extensive regional consultation, and to some extent consolidated all of the concepts discussed at the PICNCPA meetings series since its inception in 1975. Invasive species management remains an important component under Objective 4, 'Protect and recover threatened species and preserve biodiversity, focusing on species and genetic diversity of ecological, cultural and economic significance'; and Objective 5, 'Manage threats to biodiversity, especially climate change, invasive species, over-exploitation and habitat loss and degradation'. Invasive species control in these two Objectives are covered by Target 9, which states that: 'By 2020, invasive alien species and pathways are identified and prioritized, priority species are controlled or eradicated, and measures are in place to manage pathways to prevent their introduction and establishment'.

The SPC has been working on an ongoing basis with individual countries and territories to increase their levels of awareness of invasive plant pests and diseases, and to help countries develop their national biosecurity awareness, legislative frameworks and capacity to improve the prevention of new introductions while managing existing threats. At the medium-term policy level, the Pacific Islands Regional Millennium Development Goals (SPC, 2004) recognize the problem among the many other developmental issues, and expected member countries to take necessary actions. Yet despite this awareness, as well as strategic actions on the regional scene, individually the PICTs are faced with many other significant developmental issues so the level of attention that governments give to invasive species issues in general can be described as low priority.

Despite these efforts on the regional scene, individually, the PICTs are still faced with many other significant developmental challenges so the level of attention that governments allocate to addressing invasive species issues in general will continue to remain low or negligible. The recognition of the wider issue of invasive species is still a long way from being recognized at the community and national levels in many countries because of other competing priorities. However, all is not gloomy, as some countries are already taking actions – from preventing the arrival of new invasive species to tackling individual invasive plants (although more often, only those in limited, anthropogenic situations). Officials are increasingly recognizing the wider issues of invasive species and their consequences, for example because of the consistent inclusion of the rules of the World Trade Organization (WTO) Sanitary and Phytosanitary (SPS) Agreement in regional or bilateral trade agreements such as the Pacific Agreement of Closer Economic Relations or the Melanesian SpearHead Group's Trade Agreement. At the community level, some examples of efforts against invasive plants are covered in the later part of this chapter.

Information and national capacity

The most important challenge facing invasive alien plant management at the country level is the general lack of information on the impacts of existing alien plant invaders and new threats. This results in inadequate or lack of attention paid to addressing invasive plant issues by governments. Traditional sources of information such as books and periodicals, if available, only cover topics such as the usefulness of plants or the management of weeds in large cropping systems.

A major reason for the absence of information on invasive plants is that there is often no core group of nationally based invasive plant scientists and managers at middle decision-making levels to persuade governments to include invasive species issues on their national development agendas. Most agriculture department personnel traditionally concentrate on work on arthropod pests and plant diseases rather than on weeds. In some small PICTS, such as Tokelau, Nauru, Kiribati, Tuvalu, Niue, and Wallis and Futuna, serious lack of capacity may mean that only a few individuals undertake all plant and animal health-related activities. Lack of skilled personnel also means that very few or no quantitative data exist. The absence of both qualitative and quantitative data on the impacts of invasive plants means that it can be difficult to convince governments to pay attention or fund programmes to address invasive plant issues. As a result, resources are used elsewhere. Where actions have been taken by island governments, they have been mostly one-off, knee-jerk measures based on anecdotal or perceived concerns, or the result of a few concerned individuals who have been able to convince their governments, instead of a systematic policy-based effort against invasive plants and other IAS. This affects the development and sustainability of long-term invasive plant management strategies in most PICTs.

Historically, government environment departments, now mandated to address IAS issues, did not develop capacity and experience in this field. The new responsibility accorded to these departments following the signing of the Convention on Biological Diversity (CBD) in 1992 by some PICTs has, to date, led more to talk than to tangible actions on the ground, even after two decades of global recognition of the impacts that IAS have on biological diversity.

On a positive note, though, the Internet is quickly making information available to those who can afford it, thus providing the impetus for some progressive island communities to start invasive species management activities. Some workers are also linked to web-based discussion fora such as *Pestnet*, a public email discussion service (pestnet@yahoogroups.com) for pests and diseases of crops in the Pacific that has been extended to cover other issues and has a global subscribership; in addition they may have access to a number of web-based databases on invasive species. Hence, groups with interests in biodiversity conservation have developed active invasive species management programmes in Palau, the Federated States of Micronesia, Fiji, French Polynesia, New Caledonia and American Samoa.

To facilitate trade among PICTs and with their trading partners, the SPC initiated the development of a regional Pacific Islands Pest List Database (PLD) during the mid-2000s. The database contains simple pest lists and occurrence data as well as some interception data for each member PICT. This information on pest occurrences is available on the Web (www.spc.int/pld), but more detailed information for the PICTs can be found in the CABI Invasive Species Compendium (www.cabi.org/isc/). When the SPC developed the PLD, each PICT was to keep its own national database, containing both new pest reports and data on all pest interceptions at its international borders. Unfortunately, updating and utilizing this database has continued to be a major obstacle, largely attributed to lack of expertise in the surveying, identification and recording of authenticated data. For invasive plants, an additional and more useful resource about species occurring on, or threatening, some small island PICTs was the database maintained by the United States Department of Agriculture (USDA) Forest Service's Institute of Pacific Islands Forestry Pacific Island Ecosystems at Risk (PIER) project (www.hear.org/pier). Unfortunately, this excellent database is no longer being updated, but it is being maintained.

Conflicts, perceptions and border security

Invasive plant problems in the Pacific islands are increasingly the result of 'escaped' cultivated plants, although there are significant proportions of accidentally introduced plants too. Nearly all intentional introductions of plant species have been facilitated by horticulturalists, pastoralists, agronomists, foresters, botanists, floriculturists and ornamental gardeners. These groups are the most difficult among which to spread the message that non-native plants can become invasive. They nearly always consider only the benefits of introducing new species and give little consideration to the potential weediness of a species. This has been the case since the arrival of the early European explorers in the region, and is still continuing even in the face of new import restrictions.

Most PICTs lack quarantine personnel with the appropriate training in risk assessment. Many officers involved in border biosecurity have no formal training at all and have learnt only from their peers. This inadequacy has often contributed to lax import screening and permit approval processes, thereby allowing the introduction of potential plant invaders. Even in today's climate of increasing awareness and risk assessment protocols, it is rare for an import permit application by another division of the responsible ministry to be refused on the basis that the plant in question might become invasive. Plant import permits are nearly always issued based on the plant's perceived socio-economic benefits. Given the lack of appropriate training, most quarantine personnel in the Pacific are unfamiliar with (or pay little attention to) proper import risk assessment procedures and are not normally aware of the potential threats from plant imports.

Sometimes, importers aggressively emphasize the benefits of the plant species they wish to import to the extent that quarantine personnel with little training can be intimidated into issuing import permits. Aggressive legumes (e.g. *Leucaena leucocephala*, *Centrosema pubescens*, *Calypogonum mucunoides*, *Mucuna pruriens*) and grasses (e.g. *Pennisetum purpureum*, *Panicum maximum*, *Brachiaria mutica*) have been introduced to many island countries, mainly as animal fodder or for soil improvement. Sadly, many of these have jumped the fence and are now weeds. It is also not uncommon for quarantine officers in the close-knit societies in the Pacific to allow plants to be imported by relatives or influential people for fear of personal repercussions if they do not. In the past, the screening practice of quarantine divisions involved releasing a live plant import, after a prescribed period of detainment, if it harboured no visible arthropod pest or disease (i.e. no insects or symptoms were noted). However, one of the historical failures of this screening method was that no consideration was given to the weed potential of the imported plant itself.

The smuggling of plants for various reasons is a serious threat to Pacific island communities as it can be difficult to detect, even in the presence of stringent border protection systems. Smuggling probably continues despite increasing levels of awareness among Pacific islanders that they should not return home with living biological material. Plant smuggling is difficult to prevent in the first instance because many plants are seen as necessary by a cross section of communities. Too often, blame is placed on Pacific island women for returning home from other islands or the continental USA, New Zealand and Australia with ornamental seeds, or even cuttings in their bags, though such blame may be unfair as the claims are difficult to prove. Male islanders could be equally guilty but in island societies, where women are always associated with flowers, it is they that tend to attract the blame.

Trading or movement of used equipment, particularly that of road-building, military and logging machinery, but personal effects as well, could serve as pathways for new invasive species introductions. The accidental spread of many other plants continues unabated because seeds are frequently too small to detect. Plants that have been spread in this way include most of the invasive members of the Asteraceae, as discussed later in this chapter.

Even with accidentally introduced plant invaders, there are often conflicts of interest because people have adopted some of the species for various local uses in food, herbal medicine or building materials, or for aesthetic purposes. One such plant is *Mikania micrantha* (mikania weed). Despite being widely considered a troublesome weed, in Fiji and Samoa some people consider it to be useful for treating fresh cuts. Some farming communities prefer mikania to other tougher weeds because it suppresses those weeds and is then easier to control manually. These differing interests among various groups of people can make invasive weed management in the Pacific rather difficult, but also interesting in many ways. Efforts aimed at managing some alien plants will always bring their share of opposition, particularly at a local level, with most arguments based on some limited instances of utilization of the invasive plant concerned.

The implications of IAS for the socio-economic status of communities or islands are still a vague concept for many islanders. Many colleagues in the production sector have often posed questions like 'What is it about invasive species?' or 'Why should I be concerned about invasive species?'. Many workers in the production sectors in the Pacific tend to associate invasive species with concerns only in the environmental sector, associating themselves more with the traditional terms 'pests', 'diseases' and 'weeds'. Similarly, environmental interests (e.g. Meyer, 2000) tend to leave out 'weeds' or 'ruderals' as being of no concern to natural environments or habitats. In reality, such divisions are artificial as there are many alien plant invaders of concern to both of these sectors in the Pacific (see Tables 6.1 and 6.2). There is frequently a failure by these interest groups to recognize that in the Pacific islands, the environmental and the socio-economic sectors form a continuum. The indigenous Pacific island peoples are very close to their environment and have depended on the associated biological diversity for their survival. Therefore, the negative impacts of alien plants in the islands should be of equal concern to everyone.

A related contentious debate concerns where lines should be drawn when it comes to categorizing introduced plants as 'invasive' species; this is because of the real or perceived benefits of some invasive plants by some sections of society.

For example, there are genuine arguments in Fiji and in French Polynesia about whether the introduced African tulip tree, *Spathodea campanulata*, should be considered to be a positive addition to the flora. It is often argued that this species quickly revegetates disturbed forests after fire in the absence of a native ecotonal successional community. In some of the high islands of eastern Polynesia, such as Tahiti and Rarotonga, invasion of hills by the African tulip tree is considered to be a serious biodiversity issue (Meyer, 2004), but some local people have told this author that they prefer tree cover to fern cover over the rugged topography. On the one hand, those advocating the usefulness of invasive plants such as the African tulip tree argue that adverse perceptions of the impacts of invasive species are 'anthropomorphically driven' – advanced by agriculturalists because these species grow where crops should be growing or by conservationists because they grow where native plants should be growing. On the other hand, other workers who are familiar with the issues surrounding the African tulip tree, such as Auld and Senililo-Nagatalevu (2003) and Meyer (2004), contend that it has serious negative impacts on the biodiversity of the Fijian islands as well as in agriculture and forestry, even in areas where fire disturbance is absent. The aggressiveness of the African tulip tree (see Fig. 6.5c) in Fiji, Eastern Polynesia and some islands of the Bismarck Archipelago in Papua New Guinea makes it a serious biodiversity threat because its ability to grow aggressively and fill forest gaps through suckering and seed production allows it to replace native hardwood species that are useful for building houses, canoes and other traditional wooden items. Additionally, African tulip trees are easily felled by storms so represent a danger to homes, road traffic, power lines and people in cyclone prone areas.

Invasive Plants in the Pacific

A whole range of introduced plants occur and have an impact on life on the Pacific islands. To examine these plants and their impacts would require a whole book by itself, so here we focus on the impacts of the Asteraceae as a representative group. First though, we consider the lists of important invasive alien plants that have been identified by workers in two different sectors: agriculture and the environment. These sector-related invasive plant lists that have been developed for the Pacific islands (Tables 6.1 and 6.2) are interesting in that some species are of common interest to both sectors, while others are of relevance to one or the other.

Plant lists

Compiling an accurate list of alien plants to include those already exhibiting invasiveness and potential or 'sleeper' weeds for every single island group in the Pacific is an enormous and difficult task. The relative lack of published information and trained manpower necessary for compiling accurate lists, and the isolation of many islands from major travel routes, contribute to the difficulty. Where information is available, it varies from country to country, with a few countries having more up-to-date data than others. However, estimates are that over 250 invasive alien plants may be of importance to the Pacific islands region. Ranked lists containing only the plants identified by individual PICTs as their ten most troublesome weeds are given in Tables 6.1 and 6.2.

Table 6.1 contains plants listed and ranked by agricultural workers during SPC regional plant protection meetings in 2002 and 2004 (SPC, unpublished internal reports). For some island countries, more detailed field survey-based data on weeds and likely threats are available from surveys by particular workers (Swarbrick, 1997; author's unpublished data). For some PICTs, more detailed lists of weed species are available in the regional PLD.

The list in Table 6.2 was developed from information provided during a Pacific regional invasive species workshop organized by the SPREP in 1999 (Meyer, 2000). Contributions to this list were thus mostly from the environmental sector. Based on this, Meyer (2000) provided a preliminary list of over 30 plant invaders occurring in a number of Pacific island countries, and these are also incorporated into Table 6.2.

Although Tables 6.1. and 6.2 contain some differences in plant rankings due to the differences in the interests of participants who communicated the information at the meetings concerned, a good number of invasive alien plants appear in both tables, indicating their significance to both the socio-economic (Table 6.1) and environmental (Table 6.2) sectors. The position of a species regarded as troublesome on one island may differ in an adjacent island or country depending on the physical features of the island, the island's land-use practices and the impacts of the species on natural resources utilization, and the conservation values attached to the natural biota of the islands. It is important to note that these lists are crude and may not truly represent the reality on the ground as they were based on information from country participants attending regional meetings, and many participants normally have little to do with invasive plants in their countries.

Some troublesome Asteraceae

Many of the most common and important invasive plants in the Pacific islands belong to the family Asteraceae. This fact, together with the existence of active weed management programmes to control some species of that family, make it appropriate to discuss the weedy Asteraceae of the Pacific here. The majority of troublesome species in the family have been introduced into the region accidentally, but some have been intentionally introduced (e.g. *Sphagneticola trilobata* and *Tithonia diversifolia* for ornamental reasons).

Table 6.1. An agricultural sector list of invasive plants of the Pacific Island Countries and Territories (PICTs) as ranked by the number of countries considering them among their 'top ten' weeds at the Secretariat of the Pacific Community (SPC) meetings in 2002 and 2004, and as ranked by Swarbrick (1997). The order given is based on the 2002 rankings.

Plant species (family)	No. PICTs ranking plant in their 'top 10 weeds' list		As ranked by Swarbrick (1997)[a]
	2002	2004	
Mikania micrantha (Asteraceae)	12	10	3
Cyperus rotundus (Cyperaceae)	10	6	1
Merremia peltata (Convolvulaceae)	10	11	23
Mimosa diplotricha (Fabaceae)	8	8	2
Sida rhombifolia (Malvaceae)	7	6	NR
Mimosa pudica (Fabaceae)	7	1	5
Sphagneticola trilobata (Asteraceae)	5	8	NR
Spathodea campanulata (Bignoniaceae)	5	7	NR
Solanum torvum (Solanaceae)	5	3	8
Bidens pilosa (Asteraceae)	4	1	6
Antigonon leptopus (Polygonaceae)	4	4	NR
Chromolaena odorata (Asteraceae)	4	4	21
Lantana camara (Verbenaceae)	3	5	4
Stachytarpheta urticifolia (Verbenaceae)	3	2	20
Parthenium hysterophorus (Asteraceae)	3	3	NR
Clidemia hirta (Melastomataceae)	3	2	NR
Kyllingia polyphylla (Cyperaceae)	3	2	26
Eichhornia crassipes (Pontederiaceae)	3	1	10
Clerodendrum chinense (Verbenacaeae)	2	5	9
Clerodendrum quadriloculare (Verbenaceae)	2	4	NR
Coccinia grandis (Curcubitaceae)	2	4	NR
Imperata cylindrica (Poaceae)	2	2	NR
Piper aduncum (Piperaceae)	2	2	NR
Ageratum conyzoides (Asteraceae)	2	1	NR
Senna tora (Fabaceae)	2	1	NR
Piper auritum (Piperaceae)	1	3	NR
Acacia farnesiana (Fabaceae)	1	3	NR
Miconia calvescens (Melastomataceae)	1	1	28
Costus speciosus (Zingiberaceae)	1	1	NR
Cardiospermum grandiflorum (Sapindaceae)	1	1	NR
Sorghum halepense (Poaceae)	1	1	NR
Rottboellia cochinchinensis (Poaceae)	1	1	NR
Broussonnetia papyrifera (Moraceae)	1	1	NR
Albizia chinensis (Fabaceae)	NR	2	NR
Euphorbia hirta (Euphorbiaceae)	NR	2	NR
Cassytha filiformis (Cassythaceae)	NR	2	NR
Commelina benghalensis (Commelinaceae)	NR	2	NR
Xanthium strumarium (Asteraceae)	NR	2	NR
Ocimum gratissimum (Lamiaceae)	NR	2	NR
Sida acuta (Malvaceae)	NR	1	17

[a]NR, not reported.

The family is characterized by highly adapted floral and seed forms with various mechanisms for seed dispersal which, together with the grasses (Poaceae), make them the commonest members of weed flora everywhere. The most aggressive Asteraceae

Table 6.2. An environmental sector list of the 33 most significant invasive plant taxa ranked by order of the number of Pacific Island Countries and Territories (PICTs) where the plant is considered to be dominant (D), followed by the number of PICTs where it is considered to be of moderate (M) occurrence and the sum of these (D + M) (Meyer, 2000). Information in this table excludes Papua New Guinea, the Solomon Islands and New Zealand, but includes Hawaii. The order given is based on the D rankings.

Plant name and family	D	M	D+M
Lantana camara (Verbenaceae)	14	1	15
Leucaena leucocephala (Fabaceae)	13	3	16
Pennisetum spp. (*P. clandestinum, P. polystachyon, P. purpureum, P. setaceum*) (Poaceae)	11	2	13
Psidium spp. (*P. guajava* + *P. cattleianum*) (Myrtaceae)	6+4	5+1	16
Mikania micrantha (Asteraceae)	8	0	8
Paspalum spp. (*P. conjugatum, P. distichum, P. urvillei*) (Poaceae)	7	6	13
Mimosa diplotricha (Fabaceae)	7	2	9
Merremia peltata (Convolvulaceae)	7	0	7
Adenanthera pavonina (Fabaceae)	5	2	7
Clerodendrum spp. (*C. chinensis, C. japonica, C. paniculatum, C. quadriloculare*) (Verbenaceae)	5	2	7
Passiflora spp. (*P. foetida, P. laurifolia, P. ligularis, P. tripartita, P. quadrangularis, P. rubra*) (Passifloraceae)	4	10	14
Rubus spp. (*R. argutus, R. ellipticus, R. glaucus, R. moluccanus, R. nivalis, R. rosifolius*) (Rosaceae)	4	6	10
Syzygium spp. (*S. cumini, S. floribundum, S. jambos*) (Myrtaceae)	4	4	8
Panicum spp. (*P. maximum* + *P. repens*) (Poaceae)	3+1	3+0	7
Eichhornia crassipes (Pontederiaceae)	4	3	7
Falcataria moluccana [= *Paraserianthes falcataria, Albizia falcataria*] (Fabaceae)	4	2	6
Clidemia hirta (Melastomataceae)	4	0	4
Acacia spp. (*A. confusa, A. farnesiana, A. mearnsii, A. melanoxylon, A. spirorbis*) (Fabaceae)	3	5	8
Spathodea campanulata (Bignoniaceae)	3	5	8
Hedychium spp. (*H. coronarium, H. flavescens, H. gardnerianum*) (Zingiberaceae)	3	4	7
Sphagneticola trilobata (Asteraceae)	3	4	7
Melinis minutiflora (Poaceae)	3	4	7
Sorghum spp. (*S. halepense* + *S. sudanense*) (Poaceae)	2+1	1+1	5
Chromolaena odorata (Asteraceae)	3	1	4
Ardisia elliptica (Myrsinaceae)	3	0	3
Ischaemum spp. (*I. polystachyum* var. *chordatum, I. timorense*) (Poaceae)	3	0	3
Albizia spp. (*A. chinensis, A. lebbeck, A. saman* = *Samanea saman*) (Fabaceae)	2	6	8
Cestrum spp. (*C. diurnum* + *C. nocturnum*) (Solanaceae)	2+0	2+1	5
Cecropia spp. (*C. obtusifolia, C. peltata*) (Cecropiaceae)	2	1	3
Coccinia grandis (Curcubitaceae)	2	1	3
Imperata cylindrica (Poaceae)	2	0	2
Tecoma stans (Bignoniaceae)	1	4	5
Stachytarpheta spp. (*S. urticifolia* + *S. jamaicensis*) (Verbenaceae)	1+0	7+1	9

in the region and across habitats are chromolaena (*Chromolaena odorata*), which is restricted to the north-western Pacific and Papua New Guinea, and mikania, which has been ranked among the top three weeds in the region (Table 6.1).

Mikania

Mikania is among the most prominent of the many invasive plants common in the vast Pacific islands region as it is present in many of the countries within it (see Fig. 6.2a). It appears that mikania spread around the region after World War II, and that with increasing movement of people and trade, and it may still be spreading. In Papua New Guinea, it was first recorded in the 1960s (Henty and Pritchard, 1973), but may have been present earlier. In Fiji, mikania was recorded as early as 1906, but here too it may have arrived much earlier (Smith and Carr, 1991). It was reported from Tonga in the 1980s but has not been recorded in subsequent surveys conducted by this author. On some islands, where warm and moist conditions prevail, mikania is one of the commonest weeds. Its importance has been recorded by PICT, and it was listed among the top ten weeds in eight countries in 2000, ten in 2002 and 11 in 2004 (Tables 6.1 and 6.2).

In Samoa, mikania is obviously the most common weed, and it is present in almost all situations, including cropping areas (as illustrated in Fig. 6.2), but despite it being recognized as invasive by the environmental sector, many farmers are clearly unaware of its hidden negative costs and regard it as a useful plant to have around. Nearly all of the farmers interviewed said mikania was appealing to them because it had certain positive qualities, such as promoting the retention of soil moisture and nutrients, and for the ease with which they could control it manually. However, when the same farmers were made aware of the financial costs of their efforts, they were all surprised and their perceptions of the plant changed (M.D. Day and author's personal observations). In addition, despite the purported benefits of mikania leaves for the treatment of cuts in Samoa and Fiji, the author has yet to see people resorting to harvesting mikania leaves for this purpose. Furthermore, studies have implicated mikania as a major host for the cucumber mosaic virus (Davis *et al.*, 2005), a virus that causes a serious disease of kava (*Piper methysticum*), a culturally important drink and export revenue earner for Fiji and Vanuatu. The continued existence and spread of mikania will, therefore, mean ongoing losses for kava growers.

In Papua New Guinea, mikania is a serious weed under cocoa, oil palm, coconuts, bananas, sweet potato, vanilla, and a host of other traditional food crops (see Fig. 6.2). It invaded the New Guinea islands regions first, but is now reported from all 15 lowland provinces (Day *et al.*, 2012). Under cocoa (see Fig. 6.2b), yields are reported to be higher when mikania is absent than when it is present (J. Konam, Papua New Guinea Cocoa and Coconut Institute, 2004, personal communication).

The biological control of mikania weed was first attempted in the Pacific region in the 1980s with the introduction of *Liothrips mikaniae* from Trinidad to the Solomon Islands, although the thrips failed to establish (Cock *et al.*, 2000). An ambitious regional biological control project commenced in Papua New Guinea and Fiji in 2006, using the rust fungus *Puccinia spegazzinii*. Initially, the introduction of one or both of the butterflies *Actinote anteas* and *A. thalia pyrrha* had also been considered under this project as classical biological control agents. The two *Actinote* spp. were tested and used for the control of the closely related invasive weed chromolaena in Indonesia, but both have been reported to attack and survive on mikania equally well (de Chenon *et al.*, 2002). Later, in 2006, the two butterfly species were imported into Fiji, but they died out before host-specificity testing could be undertaken (Day *et al.*, 2013a). The rust fungus has, however, established widely in Papua New Guinea and Fiji, and is giving good control of the weed, particularly in areas with regular rainfall, such as the Gazelle Peninsula of New Britain Island in Papua New Guinea and eastern Viti Levu in Fiji. More recently, it was introduced into Vanuatu (M.D. Day, Department of Agriculture and Fisheries, Australia 2016, personal communication), where it is now established and spreading; there are also plans to release it into the Cook Islands, Guam (Q. Paynter, Landcare Research, New Zealand, 2014,

Fig. 6.2. Impacts of *Mikania micrantha* (mikania weed) in the Asia–Pacific region. (a) *M. micrantha* in flower. Photo courtesy the author; (b) Mikania weed invading a cocoa plantation in Samoa. Photo courtesy M.D. Day; (c) Mikania weed reducing productivity in a young oil palm plantation in Malaysia. Photo courtesy Chung Gait Fee; (d) Mikania weed competing easily with coffee in Samoa. Photo courtesy the author; (e) Mikania weed is common in sugarcane plantations (here, in Assam, India). Photo courtesy C.A. Ellison; (f) Mikania weed competing with a sweet potato crop in Samoa. Photo courtesy the author.

personal communication; Paynter, 2014a) and Palau (M.D. Day, Department of Agriculture and Fisheries, Australia, 2016, personal communication) (see Chapter 10, this volume).

Chromolaena

Chromolaena (shown in flower in Fig. 6.3a) arrived in South-east Asia in about 1930 (McFadyen, 2002). Its spread into the

Fig. 6.3. Impacts of *Chromolaena odorata* on Pacific island communities: (a) *C. odorata* in flower; (b) impacts of the weed on the daily lives of villagers; (c) invasion of a coconut plantation; (d) invasion of food gardens, reducing yields of crops; (e) large thickets growing on grazing land; (f) dry *C. odorata*, which burns easily and can become a threat to fencing and villages. Photos courtesy the author.

western Pacific is believed to have occurred as a result of accidental introductions during and after World War II on contaminated military and other machinery used in road works, agriculture or logging. Chromolaena is present in and spreading from several islands in the western Pacific, particularly Micronesia and Papua New Guinea. It has been reported as absent from the central and south-west Pacific region, east of Bougainville Island in Papua New Guinea to eastern Polynesia (Orapa, 2004). The occurrence of chromolaena on Bougainville threatens the neighbouring Solomon Islands and it may not be long before it arrives there because there is no international quarantine border control between the two countries.

In Papua New Guinea, chromolaena is present in several lowland provinces and islands from sea level up to 1000 m above (mean) sea level (asl) and is of major concern (Bofeng *et al.*, 2004; Orapa, 2004; Orapa *et al.*, 2004; Day and Bofeng, 2007). Local spread is most likely due to movement of seeds on machinery, vehicles and planting materials, but spread via clothing, wind and water dispersal are also locally important. The population of chromolaena at Lae on mainland Papua New Guinea appears to have originated from the Gazelle Peninsula of New Britain Island; E.E. Henty, a botanist working with the Papua New Guinea Department of Forests brought plants from there to Lae in about 1970 with the intention of growing them through to flowering stage for a book project. The botanist admitted planting the chromolaena plants in his garden at Awilunga, 9 km west of Lae, and did not destroy them after drawing the fertile parts (B.M. Waterhouse, Northern Australia Quarantine Strategy, 2001, personal communication). This appears to have resulted in the current chromolaena infestations in the agriculturally important Markham and Gudsup Valleys.

Chromolaena tends to invade and thrive best in the disturbed areas of Papua New Guinea, and these are often areas important to agriculture or for other socio-economic activities. It prevents access to food and forest resources in seasonally dry as well as disturbed forest areas because it forms dense stands. It invades young eucalypt plantation forests and plantations of oil palm, coconut, vanilla, bananas, abaca (*Musa textilis*) and cocoa in New Britain, and is a major invader of subsistence food gardens (Figs 6.3b–d). Clearing and uprooting can be very difficult and time-consuming for families who depend on the cultivation and harvesting of crops including bananas, sweet potatoes, yams and taro. In addition, subsistence farmers, particularly on Misima Island in the south-east of Papua New Guinea, have anecdotal concerns that tubers of yams grown on land cleared of chromolaena are smaller than those grown on land that was covered by other, native, vegetation. This may be due to allelopathic properties of chromolaena.

Probably the most significant impacts on production have been in grazing land where smallholder cattle farmers cannot afford to keep their natural pastures free of chromolaena. The leaves are toxic to grazing animals and are avoided by them in most cases, thus allowing large thickets to establish on properties (Fig. 6.3e). Farmers in the south of New Ireland and parts of the lower Markham Valley in Papua New Guinea have had little choice but to reduce or sell off their stock. In the Gudsup Valley, adjoining and located west of the Markham Valley, it was estimated in 2003 that the following 10 years would see chromolaena infesting 30–40% of grazing properties if left uncontrolled (L.S. Kuniata, Ramu Sugar Limited, 2003, personal communication). Wooden fence posts are often burnt down when chromolaena dies back and burns following flowering in August–September (Fig. 6.3f). Chromolaena bush fires are hotter than normal grass fires and threaten even trees adapted to fire. In addition, such fires can damage infrastructure and lead to loss of lives and property. In Timor Leste and parts of Papua New Guinea, such as the seasonally dry Markham Valley and south-east New Ireland, traditional villages are constantly under threat from chromolaena-fuelled bush fires. These can also kill fruit trees such as mango and reduce native plant and animal species, including those adapted to normal grassland fires.

Other social impacts of chromolaena are significant too. In Timor Leste, it was reported to make up over 60% of all vegetation, which has led to the loss of the useful kunai grass (*Imperata cylindrica*) grasslands, which is a useful resource for thatching in rural homes, although it is invasive elsewhere (PIER, 2013), as well as providing natural pastures for livestock (McWilliam, 2000).

Biological control programmes against chromolaena have been undertaken in Papua New Guinea (Orapa *et al.*, 2002; Bofeng *et al.*, 2004; Day and Bofeng, 2007),

Timor Leste (Day *et al.*, 2013b), Guam, the Federated States of Micronesia, the Commonwealth of the Northern Mariana Islands and Palau (Esguerra, 1998; Muniappan *et al.*, 2004, 2007), Indonesia (de Chenon *et al.*, 2002; Wilson and Widayanto, 2002) and the Philippines (Aterrado and Bachiller, 2002). Physical and chemical control efforts were made to eradicate this alien shrub from the Marshall Islands (Muniappan *et al.*, 2007), but these have not been reported as successful.

The implementation of a classical biological control project of chromolaena in Papua New Guinea (Bofeng *et al.*, 2004) had cost nearly AU$600,000 by the time the project ended in 2007. A socio-economic assessment of this project was published by Day *et al.* (2013a), and reported that chromolaena is considered to be under substantial/significant control in nine provinces in Papua New Guinea. Farmers have reported a 50% increase in income from the sale of agricultural products since the weed has been brought under control. The most effective biological agent was found to be the gall fly *Cecidochares connexa*.

Wedelia

Wedelia, or Singapore daisy (*Sphagneticola trilobata*), is probably the next most noteworthy invasive plant species (Fig. 6.4). It is an aggressive carpet-forming weed which can grow in both open and semi-shady conditions, and it is already a serious weed in Fiji, French Polynesia, Samoa and Tonga. In 1988, when J. L. Swarbrick conducted weed surveys in 16 PICTs for the SPC, wedelia was not reported as an important weed (Swarbrick, 1997). However, over a decade later, in 2002, five PICTs considered wedelia to be among their top ten weeds, and by 2004 this number had increased to eight PICTs (Table 6.1). It has become particularly invasive in the wetter parts of the islands and has been recorded as showing aggressive behaviour even at 1600 m altitude in the central highlands of Papua New Guinea. In Fiji and French Polynesia, wedelia is a particularly serious weed of fruit orchards, paddocks, village greens, wastelands and riverbanks.

Wedelia has been largely spread for ornamental reasons, but in several countries it is also grown on newly cut road edges to prevent erosion.

Other invasive Asteraceae

Other notable common and aggressive Asteraceae of economic importance to the Pacific islands because of their negative impacts include Mexican sunflower (*Tithonia diversifolia*), which is very aggressive and competes with crops in cooler areas such as the highlands of Papua New Guinea and the Austral Group of islands in French Polynesia, common dandelion (*Taraxacum officinale*), synedrella weed (*Synedrella nodiflora*), vernonia weed (*Vernonia cinerea*), thickhead (*Crassocephalum crepidioides*), cobbler's pegs

Fig. 6.4. As shown in these two photos, wedelia (*Sphagneticola trilobata*) is quickly becoming an important invasive plant with implications for agricultural and livestock production. Photos courtesy the author.

(*Bidens pilosa*, *B. alba*), *Eleutheranthera ruderalis*, tobacco weed (*Elephantopus mollis*), *Pseudoelephantopus spicatus*, *Pluchea carolinensis* and *P. indica*. *P. carolinensis* has aggressively invaded the Loyalty Islands in New Caledonia, and other islands in Kiribati, Nauru, the Marshall Islands, Espirito Santo (Vanuatu) and the Society Islands (French Polynesia) (PIER, 2013). *P. indica* commonly occurs along the runway of Funafuti Airport in Tuvalu and requires regular slashing (author's personal observation). The impacts of these plants vary from island to island, with their severity depending on the agro-ecological or anthropomorphological characteristics of the islands where they occur.

A potentially troublesome species likely to become important to human health in the region is *Parthenium hysterophorus* (parthenium weed) because of its allergenic pollen – although it is not the only species that can contribute to health problems. Parthenium is of increasing concern as a common weed in farmland in New Caledonia and some islands of Vanuatu. It was found growing in two small areas in Lae, Papua New Guinea, in 2000. Efforts to eradicate the weed from these two outbreak areas commenced in 2002 and repeated treatments with herbicides, as well as monitoring are continuing (Day, 2014). A small outbreak that is occurring on Tubuai Island in French Polynesia may spread to other larger islands if it is not controlled early on.

Known noxious species absent from the region include *Ageratina adenophora* and *Austroeupatorium inulifolium*, among others. Many of these are known to occur in neighbouring countries (New Zealand, Australia and Indonesia). While the PICTs already have quarantine services in place, national capacities to detect potentially detrimental alien plants species are low. Some island countries are served only by direct shipping or air routes from within the region, so the risks to those are lower. None the less, ongoing surveillance work by the SPC and a few other initiatives in the region are of an irregular nature and are unlikely to assist in the detection of new alien plant threats. It would be ideal to have technical capacity in the PICTs that works full-time on managing these threats.

Socio-economic Impacts

It is extremely difficult to give monetary values to the impact of invasive plants in the Pacific because of the fragmented nature of the region, the prevalence of the subsistence (though resilient) nature of Pacific island economies, and the simple lack of data related to invasive plants. Yield losses in traditional subsistence production systems are normally caused by a range of abiotic or biotic factors, including rapidly growing native plants. Alien plant invasions can be said to be an increasing problem in production systems, including both traditional subsistence farming and commercial crop production, particularly in areas that have become linked by roads or ports. This has meant increasing costs in terms of monetary or labour inputs. Additional costs may simply reflect the number of extra man-days spent weeding, as in a cocoa and yam production system in Vanuatu (McGregor, 1999). Unfortunately, even farmers who produce commercial crops such as cocoa, coffee and coconuts do not understand the links between the costs of production, yields and the presence of troublesome weeds. Too often, farmers regard the costs of weed management as a normal part of farming and may not factor them into their production costs. This perception may be true in most cases, because weeds of various kinds will always interfere with production, but for some invasive plants like mikania, which require weekly attention, the effort and costs are not warranted even in small production systems.

In contrast to the Pacific islands, the Australian Weeds Cooperative Research Centre was able to estimate the total impact of weeds in that country to be of the order of AU$4 billion over the 5 year period from 1997/98 to 2001/02 as measured in terms of the loss of economic surplus (Sinden *et al.*, 2004); in New Zealand, the national flora increased tenfold in 150 years as a result of plant introductions, and alien weeds cost the national economy NZ$100 million p.a. even when less tangible costs such as loss of parks or biological diversity are excluded (Williams and Lee, 2001). Similar statistics cannot be derived for the developing Pacific islands owing to the lack of quantitative data. Even where available, the data are normally specific to the arthropod pests and diseases of a few agricultural crops and then only in the countries where some research capacity exists, such as Guam, Fiji, New Caledonia, Papua New Guinea and Tonga.

The impact of invasive plants on the socio-economic sector in the Pacific is best illustrated by looking at a few selected invasive plants. Much of the discussion that follows is, therefore, based on anecdotal evidence or limited published information from a few countries.

Common invaders: their costs

The most significant socio-economic impacts on communities are caused by those species that have a direct impact on production, and interfere with land and natural resource utilization and other human interests. Common invasive plants with impacts on agriculture and the environment are listed in Tables 6.1 and 6.2. The more common ones include mikania, wedelia and chromolaena, whose impacts have already been discussed, and also nutsedge (*Cyperus rotundus*), Navua sedge (*Kyllingia polyphylla*), broomweeds (*Sida* spp.; Fig. 6.5a), devil's fig (*Solanum torvum*), ratstail (*Starchytarpheta* spp.), sensitive plants (*Mimosa* spp.), elephant grass (*Pennisetum purpureum*; Fig. 6.54b), Guinea grass (*Panicum maximum*), para grass (*Brachiaria mutica*) and Johnson grass (*Sorghum halepense*). In a number of countries, these invade fallow land, making subsequent cultivation difficult or costly, and can cause crop losses through weed competition. The crops affected usually include banana, taro, yam, coffee, cocoa, coconut, vanilla, sweet potato and numerous other tree fruits. The invasive weeds contribute to loss of income by depressing yields and threaten food security for many rural families.

Increasingly, very aggressive invasive alien plants, like the African tulip tree (Fig. 6.5c), merremia (*Merremia peltata*; Fig. 6.5d), spiked pepper (*Piper aduncum*; Fig. 6.5e) and chromolaena (Fig. 6.3) are having an impact on rural incomes and food security in island countries by direct competition with crop plants and indirect competition through a reduction in the value of land and forage resources for smallholder farming (e.g. Hartemink and O'Sullivan, 2001; Orapa, 2001; Auld and Seniloli-Nagatalevu, 2003). The losses or costs in monetary terms of such invaders are difficult to estimate. In Papua New Guinea, weeds such as chromolaena and mikania are silently but rapidly encroaching on rural subsistence communities, commercial plantation operations and grazing, and agroforestry areas (Orapa, 2001). Small-scale semi-subsistence agriculture is the mainstay of the Papua New Guinean economy, accounting for 30% of foreign exchange earnings, and is the lifeline of 85% of the population. Any widespread invasion by alien plant invaders causes serious and immeasurable negative impacts on this sector. Similar comments could be made about invasive species in other countries, particularly the Solomon Islands, Vanuatu and Fiji.

Many invasive plants can cause damage to infrastructure, particularly to roads. Thick roadside stands of elephant grass reduce water flow in roadside drainage, increase sedimentation and cause storm water to flow on to and damage roads and even bridges during rains. This is particularly common in Papua New Guinea, and it increases the costs of road maintenance for the government. In addition, the resulting high frequency of potholes and the encroachment of weeds, mainly elephant grass, on to roads can result in high vehicle operating costs and increased fatal accidents for road users. As already described above, the fire hazard created by invasive species like chromolaena affects infrastructure as well as agriculture and biodiversity.

In the western parts of New Caledonia, a study by Blanfort *et al.* (2008) reported serious impacts of invasive plants on pastures. These authors found that 30% of the cattle farms they studied had more than

Fig. 6.5. Socio-economic impacts of different invasive plants on Pacific islanders: (a) a cattle property invaded by broomweed (*Sida acuta)* in Vanuatu; (b) encroachment of roadsides by elephant grass (*Pennisetum purpureum*), requiring constant road maintenance and causes traffic accidents in Papua New Guinea (PNG); (c) African tulip tree (*Spathodea campanulata*) invading forest on Viti Levu, Fiji; (d) land invasion by the aggressive merremia (*Merremia peltata*) is a biodiversity and food security issue in many small islands, such as Samoa; (e) invasions by spiked pepper (*Piper aduncum*) are having an increasing impact on forest regeneration and subsistence agriculture in Fiji and PNG; (f) water hyacinth (*Eichhornia crassipes*) in the Sepik River, PNG, has caused serious hardships, including the death from fatigue of the man paddling his canoe in this picture. Photos courtesy the author.

40% of their land overrun with weeds; 50% of the farms were considered to have deteriorated because more than 25% of their land was invaded by weeds (both alien and indigenous), with serious consequences for pasture growth, nutritional levels, carrying capacity and eventual productivity. The estimated annual loss in profit was as high as US$150/ha. However, the economic costs of weed control can reach US$180/ha/year, which the authors described as a significant amount for the low-input farming systems characteristic of New Caledonia.

Upland areas

In Papua New Guinea, an unspecified number of weed species (many of which are likely to be aliens) in coffee were reported to cause significant yield losses of between 30 and 80% when weed cover in plantations exceeded 30% (Manzan, 2000). These losses were attributed to the presence of mostly low-growing broadleaved weeds and the grasses *Setaria glauca* and *Paspalum paniculatum*, but no reference is made in such yield-loss studies, conducted at research station level, to the enormous combined losses suffered by thousands of smallholder coffee farmers in rural villages. In the highlands of Papua New Guinea. over 2 million people depend on small-scale coffee production, which makes up nearly 80% of the country's coffee exports. Individually, smallholder farmers have very little money but spend a lot of time and energy controlling weeds such as elephant grass, Mexican sunflower, molasses grass (*Melinis minutiflora*) and the legume *Desmodium intortum* (introduced for soil and pasture improvement).

Lowland areas

In the moist lowlands of Papua New Guinea, Vanuatu, Samoa, Fiji and the Solomon Islands, the production of lowland food and other high-value crops has meant a constant battle for farmers with weeds such as mikania, nutsedge, itch grass (*Rottboellia cochinchinensis*), Johnson grass, chromolaena and merremia, among many others. High-value crops, such as oil palm, cocoa, vanilla, taro and kava, bring healthy returns for farmers so the energy and financial costs of weed control are justified. For some crops, such as coconuts or lowland coffee, the efforts required are not worth the income, but many families have little choice. The biological control programmes being planned or implemented in various countries against mikania and chromolaena (see above), if successful, would help to alleviate both the impacts and the management costs in all sectors.

Case studies

Two case studies are presented here, the first on invasive *Sida* spp. in pasture in Papua New Guinea (Box 6.1) and the second on the impact of weeds on squash cultivation in Tonga (Box 6.2).

Impacts on traditional resources

In the Pacific, indigenous people have been and continue to be sustained by subsistence agriculture and fishing. Most land and marine areas are communally owned, including the land for farming and sourcing building materials, and the sea for food and inter-island trade. Social structures such as the chieftaincies in many indigenous island societies are centred on natural resource ownership and utilization. The loss of indigenous biological resources as a result of invasive species could, therefore, lead not just to lost opportunities in economic gains but also to the loss of social status or values. For instance, one day modern synthetic bungee-jumping equipment will become common among Pentecost islanders in Vanuatu if the traditionally preferred native vines used for this famous ages-old practice are replaced in the country's flora by weak alien vines such as merremia. Similarly, there are already concerns that the uninhibited spread of the fast-growing and invasive false kava (*Piper auritum*) may spell disaster for the production and consumption of true kava or yagona (*P. methysticum*). The false kava plant lacks the potent kava lactones found in true kava, which is mixed and consumed as a cultural drink in many of the island countries. It is feared that kava growers may resort to mixing the drink using the roots of the false kava plant, so reducing its quality, and thus the growers' income (due to consumers refusing to buy poor quality kava) and the

Box 6.1. Case study: *Sida* spp. in pasture in Papua New Guinea.

A useful case study of the economic costs of an invasive alien plant in animal production occurred when the El Niño-related climatic phenomenon in the Pacific region caused a drought and triggered an invasion by three *Sida* spp. (see Fig. 6.5a) in Papua New Guinea in 1998. Ramu Sugar Ltd (RSL), a Papua New Guinean company that primarily produces sugar but also some prime beef in the Ramu and Markham valleys, made some calculations on the cost of controlling these sidas. Before 1997 (pre-drought) the company was spending 60–80,000 Kina (PGK) (1 PGK=US$0.72 in September 1997, so this equates to US$43,200–57,600) on herbicidal sprays for all weed control on 15,000 ha of land annually. When the severe drought began in early 1997, most of the valuable pasture grasses died out owing to stress or were burnt in fires that were common for weeks throughout Papua New Guinea, as reported by Allen and Bourke (2001). At the end of the drought in mid-1998, significant tracts of pasture were rendered useless by the invasion of *S. acuta* and, to a lesser extent, *S. cordifolia* and *S. rhombifolia*. Efforts such as slashing and ploughing the sida using tractors proved expensive and difficult and did not solve the problem. RSL had three tractors dedicated to slashing sida all year round, and their costs for applying 2,4-D (the preferred herbicide for sida control) to just 2000 ha twice a year were US$116,000 (1997), US$136,000 (1998), US$98,000 (1999) and US$21,000 (2000) (L.S. Kuniata, RSL, 2003, personal communication).

In smallholder cattle and cropped (particularly coffee) areas infested by *Sida* spp. or other alien plant invasions, amelioration costs are often too high to attempt control measures (Orapa, 2001).

In 1999, RSL and the Papua New Guinea National Agricultural Research Institute (NARI) explored the options available for sida control. It was noted that successful control had been achieved in northern Australia by using, among other agents, the leaf-feeding sida beetle, *Calligrapha pantherina*. In late 1999, Dr L.S. Kuniata of RSL travelled to Darwin, Australia, and returned to Papua New Guinea with a consignment of the beetles. Following successful quarantine screening at the NARI laboratories during December 1999, the *Calligrapha* beetle was field released as a classical biological control agent at Gudsup in January 2000. The use of this agent significantly reduced both the costs of control (i.e. herbicides, tractor fuel, health risks and time) and the weed problem in the Markham and Gudsup valleys and elsewhere (Kuniata and Korowi, 2004). Within a few months, the beetle spread rapidly and reduced the sida stands at six initial release sites; and within 12 months, the beetle had spread through more than 150 km of the valley and reduced large areas of the weeds. The agent has since spread naturally into the highland areas; it has also spread on to New Britain and New Ireland in the Bismarck Archipelago as a result of single releases there. The significant reductions in the cost of sida control, as well as the yield increases experienced by RSL, appear to be repeated on a smaller scale in smallholder paddocks in Papua New Guinea. The cost to RSL of undertaking the sida biocontrol work was PGK25,000 (US$8000) and the expected benefits from just the cattle farms in the Ramu and Markham valleys was estimated at PGK2 million (US$700,000) (L.S. Kuniata, RSL, 2003, personal communication). Following the successful control of *Sida* spp. in Papua New Guinea, the biocontrol agent was released in Fiji (from 2002) and Vanuatu (from 2004).

special cultural status of kava. Sadly, however, popular concern about the impacts of invasive plants and action to counter these are uncommon in most of the developing Pacific islands, and those impacts are often not recognized through lack of awareness until serious damage or social hardships occur, as in aquatic weed invasions (see Aquatic plant invasions: case studies of successful containment below).

The loss of many native species of plants traditionally used for food, medicines or building materials as a result of invasion by exotic plants is an area of concern. The replacement of useful grasses such as kunai grass (itself invasive in some countries), wild sugarcane (*Saccharum robustum*) and arrowshaft grass (*Miscanthus floridulus*) by more aggressive and less useful grasses such as elephant grass and molasses grass in the central highland valleys of Papua New Guinea is having significant impacts on rural life. Village communities that previously relied on the native grasses for roof thatching and constructing walls in traditional buildings now have to resort to using expensive corrugated iron sheets and timber. The traditional grasses that were once harvested

Box 6.2. Case study: squash in Tonga.

The weed flora associated with squash farming in Tonga is dominated by three introduced herbaceous species: *Commelina benghalensis*, *Coronopus dydymus* and *Cyperus rotundus*. These plants are not really capable of invading and having an impact on natural areas as they survive best in cultivated systems, but they are worth discussing because the cost of controlling them is a constant preoccupation for many farmers in Tonga.

Squash is the most important export crop for this small South Pacific kingdom, contributing up to 55% of gross domestic product (GDP) and representing 96% of the total exports from Tonga. The squash industry is, therefore, a major source of employment, foreign exchange and food security, and earned the kingdom 10.8 million Pa'anga (1 Pa'anga = US$0.50, so this equates to US$5.4 million) in 2002 (Government of Tonga, 2004). Every year, during the squash-growing season from August to October, Tongan farmers spend 25% of their time weeding or, if they can afford it, rely on very expensive paid labour. Many squash growers also use large quantities of paraquat (Gramoxone®) or glyphosate to suppress these weeds before planting, and follow this up with weekly sprays of glyphosate until there is full ground coverage by the crop canopy. In July 2003, the rates for hired labour costs for weeding stood at US$2/h, and this increased to US$3/h in 2006, which eroded the farmers' earnings. The average return on labour input was estimated at US$12/labour–hour, but this would have been higher if the competing weeds and crop diseases had been controlled effectively (S. Halavatau, SPC, 2006, personal communication).

locally at no cost by villagers have become a saleable commodity, and villagers from areas where there are still some natural stands of kunai or arrowshaft grass, predominantly those in the Southern Highlands region, are now selling to those wanting these resources (author's personal observation). It has been recounted above how chromolaena has had similar impacts in Timor Leste.

The invasion of secondary forest areas by alien plants that form monospecific stands can result in the loss of plant species used in traditional medicine. Theoretically, the loss of a useful species would also mean the loss of traditional knowledge of its use in subsequent generations of people because the knowledge is passed on orally and through the active use of the medicinal plant. The loss of such plants and of the traditional knowledge associated with them as a result of invasion by alien plants can mean a loss for modern medicine, as the cures for many of humankind's illnesses may yet be found in little-known plants.

Managing Plant Invasions

In the Pacific islands, the dilemma is how to find effective solutions for invasive species with the meagre human and financial resources that are available to the island states. Too often, responses to pest problems have involved knee-jerk reactions to outbreaks. Even then, the emphasis has been focused on a few species that are perceived to have impacts on primary production. From a broad perspective, little attention has been paid to invasive plant management as a whole, although plant invaders may be the most widespread pests and have the most significant impacts on natural resource management, food security and other forms of human welfare. None the less, the region has achieved success where coordinated action has been taken either nationally or regionally (see Tables 6.3 and 6.4). The most significant successes have come from the biological control of a limited number of plant invaders in both small and large island situations (Table 6.3). Containment or eradication attempts are listed in Table 6.4 and are further discussed below in the section entitled 'Institutional responses to plant invasions').

Weed biological control has been implemented regularly for the extensive subsistence farming systems that are found in many Pacific island countries. This has depended on feasibility and the availability of biological control agents and relied on

Table 6.3. Invasive plants for which biological control attempts have been made in the Pacific Islands Countries and Territories (PICTs). Based on Winston *et al.* (2014), with additional information from Brinon (2008) and the author's unpublished data.

Weed	PICTs where biocontrol agents were released[a]	Agents released	Status
Acanthocereus tetragonus	New Caledonia	*Hypogeococcus festerianus*	Agent lost in quarantine
Chromolaena odorata	Guam, CNMI, PNG, Palau, FSM, Timor Leste	*Pareuchaetes pseudoinsulata*, *Ceccidochares connexa*	Good control in Guam; results promising in PNG and elsewhere
Clidemia hirta	Fiji, Solomon Islands	*Liothrips urichi*	Good control in Fiji and possibly in some areas of the Solomon Islands
Coccinia grandis	CNMI	*Acythopeus burkhartorum*?	Results promising
Cyperus rotundus	Tonga, Fiji, Cook Islands	*Bactra venosana*	Failed; ineffective
Eichhornia crassipes	PNG, Fiji, Solomon Islands, Vanuatu	*Neochetina eichhorniae*, *N. bruchi*	Excellent to good control
Elephantopus mollis	Fiji, Vanuatu, Tonga, Niue	*Tetraenaresta obscuriventris*	Established but impact not significant
Lantana camara	Vanuatu, Fiji, PNG, New Caledonia, Cook Islands, Samoa, Tonga, Niue, Solomon Islands, Palau, FSM, CNMI	*Teleonemia scrupulosa*, *Uroplata girardi*, *Octotoma scabripennis*, *Calycomysa lantanae*, *Ophiomyia lantanae*	Effective agents, others have given varying levels of control
Miconia calvescens	Tahiti	*Colletotrichum gloeosporioides* f. sp. *miconiae*	Variable impact, significant mortality of seedlings; impact best at higher elevations with cool temperatures and high humidity
Mikania micrantha	Solomon Islands, Fiji, PNG	*Liothrips mikaniae*	Failed to establish in quarantine in PNG; released on Solomon Islands, but failed to establish
	Fiji, PNG, Vanuatu	*Puccinia spegazzinii*	Established and spreading; having impact, particularly in PNG
Mimosa diplotricha	PNG, Fiji	*Heteropsylla spinulosa*	Excellent to good control
	Samoa	*Heteropsylla spinulosa*	Agent is established but has not given control
Mimosa pigra	PNG	*Acanthocelides* spp.	Introduced from Australia and releases in two provinces continuing; establishment yet to be confirmed
Opuntia stricta	New Caledonia	*Dactylopius opuntiae*	Successful
Pistia stratiotes	PNG, Vanuatu	*Neohydronomus affinis*	Excellent control
Salvinia molesta	PNG, Fiji	*Cyrtobagous salviniae*	Excellent control
Sida spp. (*S. acuta*, *S. retusa*, *S. rhombifolia*)	PNG, Fiji, Vanuatu	*Calligrapha pantherina*	Good control, particularly of *S. acuta* and *S. rhombifolia*; up to 99% weed control in some areas (SPC, 2008)
Tribulus cistoides	PNG	*Microlarinus lypriformis*	Excellent control
Xanthium strumarium	Fiji, PNG	*Epiblema strenuana*	Failed to establish in the field
	Fiji	*Euaresta aequalis*, *Nupserha vexator*	Failed to establish in the field

[a]CNMI, Commonwealth of the Northern Mariana Islands; FSM, Federated States of Micronesia; PNG, Papua New Guinea.

Table 6.4. Invasive alien plants for which management (eradication or containment) efforts have been made in the Pacific Islands Countries and Territories (PICTs) by either the Secretariat of the Pacific Community (SPC) or other PICT agencies.

Species	Location[a]	Status	Year eradication implemented[b]
Acacia concinna	Viti Levu, Vanua Levu, Fiji	Eradication was attempted from five known sites on both Viti Levu and Vanua Levu; species eradicated from two but still occurs on the other sites; follow-up visits ongoing (SPC, 2008)	2004
Albizia chinensis	American Samoa	Eradication efforts started in 2002, but this plant was very widespread even then and still is; eradication is unlikely to be successful (Space and Flyn, 2002)	2002
Antigonon leptopus	Niue	Inadequate resourcing stopped efforts to eradicate in 2003, but Space *et al.* (2004) reported that eradication work is still ongoing	2002–2003, 2004
	FSM, Palau	Eradication efforts started 2002; the weed is still present on some of the FSM islands (Pohnpei) and Palau, but is uncommon and may be considered contained (Space *et al.*, 2003)	2002
Chromolaena odorata	Marshall Islands	Eradication efforts continued until at least 2008 (SPC, 2008), but failed; biological control implementation recommended (Day and Winston, 2016)	2002
Imperata cylindrica	FSM, Palau	Eradication efforts started in 2002 (Space *et al.*, 2003); success unknown	2002
Merremia tuberosa	Niue	Eradication attempts started in 2002 (SPC, 2008), but have been unsuccessful; the weed was reported as especially aggressive on Niue in 2005 (www.iucngisd.org/gisd/species.php?sc=1279)	2002
Mikania micrantha	FSM	Eradication efforts started in 2002, but failed; the weed is considered widespread on Kosrae and biocontrol is recommended (Pacific Invasives Initiative, 2012)	2002
Mimosa diplotricha	Niue, Wallis and Futuna	Effort discontinued owing to inadequate resources in both territories	2002–2003
Mimosa pigra	Madang, PNG	Single effort at Madang not sufficient to destroy mimosa stand, but efforts ongoing (SPC, 2008); eradication failed, biological control implemented (Winston *et al.*, 2014)	1999
Parthenium hysterophorus	Lae, PNG	Original stand destroyed; herbicidal treatments of regrowth continued in 2009; declared eradicated (Shabbir, 2012); monitoring continuing (Day, 2014)	2002–2012
	Espirito Santo, Vanuatu	Destroyed at sole outbreak site at Lugainville airfield by 2004	2003-2004
	Tubuai Island, French Polynesia	Advice given for eradication but no information available if this was carried out	2003
Piper auritum	FSM	Eradication efforts started in 2002 and were still underway in 2004 on Pohnpei, where the plant is currently considered contained (www.hear.org/pier/species/piper_auritum.htm)	2002–2004?
Spathodea campanulata	FSM	Eradication efforts started in 2002 but have not been successful; still considered an important weed on some islands (CABI, 2013)	2002
Sphagneticola trilobata	Niue	Inadequate resourcing stopped efforts to eradicate in 2002, second attempt in 2008 but unsuccessful (Space *et al.*, 2004; SPC, 2008)	2002, 2008

[a]FSM, Federated States of Micronesia; PNG, Papua New Guinea.

[b]And year ended (successfully or unsuccessfully) when information available.

international collaboration. A workshop was held in New Zealand in 2009 with the goal of developing a regional strategic plan for undertaking biological control of widespread invasive species in the Pacific islands on a more cooperative and collaborative basis. The outcomes of this workshop formed the basis of an excellent assessment of the potential of classical biological control in the Pacific islands (Dodd *et al.*, 2011). A prioritization exercise of Pacific weed targets for biological control was also included in the document based on Paynter (2010).

Most other weed management methods are frequently not practical for managing widespread weed infestations in non-commercial crop areas, pastures, forests and aquatic systems. As we have seen, several species have been targeted for classical biological control with some success to date, and a number of others are currently being targeted or projects against them planned (Table 6.3). Most recently, a 5 year project to develop weed biocontrol for the Cook Islands has been funded by New Zealand's Ministry of Foreign Affairs and Trade through its International Development Fund (Paynter, 2014a,b).

Although biological control has by no means solved all the region's weed problems, there are two famous success stories that we have not yet touched on that illustrate both the range and severity of invasive plant impacts, and the extraordinary effect that successful classical biological control can have. These further demonstrate that the battle against invasives is never won and that continued vigilance will always be needed against new threats; they are described in the aquatic plant case studies that follow in the next section.

Aquatic plant invasions: case studies of successful containment

The most spectacular impacts of invasive plant species anywhere in the Pacific seem to be those associated with aquatic systems, especially in Papua New Guinea, where large river systems depended upon daily by the rural people are constantly under threat from exotic plant invasions. The three most noxious water weeds in the region have been the floating fern salvinia (*Salvinia molesta*; see Box 6.3), the dangerous water hyacinth (*Eichhornia crassipes*; see Box 6.4) and, to some extent, the floating aroid water lettuce (*Pistia stratiotes*). Water lettuce was of temporary concern before being brought under biological control in Papua New Guinea in the 1980s (Laup, 1986). Small infestations in Vanuatu were brought under successful biological control with the introduction of the host specific agent *Neohydronomus affinis* from Papua New Guinea in 2004. Unfortunately, new infestations became common as a result of horticultural spread of the plant in Fiji around 2010 (author's personal observation), but its impacts are not discussed here.

Future threats

The successes against salvinia and water hyacinth should not prompt complacency. The occurrence and spread of the giant sensitive plant (*Mimosa pigra*) and the aquatic plant *Limnocharis flava*, both aliens with an affinity for wetland ecosystems, are worrying. The presence of the very prickly *M. pigra* in Papua New Guinea (Orapa and Julien, 1996) and the neighbouring Papua Province of Indonesia (Sulistyawan and Hartono, 2004) could be very dangerous, not just to humans, but also to biodiversity given the behaviour of this species in northern Australia (Miller *et al.*, 1981), Thailand (Napompeth, 1983) and Cambodia (Samouth, 2004). *L. flava*, an aquatic herb that is used as a vegetable, and is known to be invasive and weedy in rice paddies in Asia, has been found at two locations in Papua New Guinea, apparently spread from Papua Province, Indonesia, by people crossing the international land border (B.M. Waterhouse, Northern Australia Quarantine Strategy, 2004, personal communication). If they spread further, these two alien plants could cause unpredictable environmental as

Box 6.3. Case study: salvinia in Papua New Guinea.

The invasion of the vast waterways of the Sepik River system in Papua New Guinea by salvinia (*Salvinia molesta*) was the first non-flooding-related natural disaster experienced by the Sepik people. This small floating alien fern from Brazil invaded the Sepik during the mid-to-late 1970s. Salvinia forms thick mats on the surface of still or slow-moving freshwater bodies. In very good conditions, i.e. at high nutrient (particularly NPK) levels and in warmer water temperatures, the weed can grow very rapidly (Room and Thomas, 1986a,b). As the salvinia invasion of the Sepik took hold, the weed's rapid growth clogged and rendered useless numerous lakes and canals for canoe or small boat traffic. Salvinia covered an estimated 31.7 km^2 in 1977, which increased to 79.4 km^2 in 1979, with 47.1 km^2 of this in the shallow Chambri Lakes alone (Mitchell, 1979a; Richards, 1979). At the height of the salvinia infestations in 1982, more than 500 km^2 of the Sepik River and its associated lakes system were affected (Laup, 1985).

During the salvinia crisis in the Sepik, production in the 'solpis' (salt fish) industry, which is based on the introduced tilapia fish (*Sarotheroden mossambicus*), was adversely affected. The annual catch of 170 t of fresh tilapia in 1978, which yielded 22.6 t of dried salted fillets, was estimated to be 30% less than it would have been in the absence of salvinia (Mitchell, 1979b; Mitchell *et al.*, 1980). The estimated monetary losses for the year May 1998–May 1999 in five lower Sepik River oxbow lagoons with stable sudds (floating mats) of salvinia covering 80–90% of the water surface were Kina 1300–9700 (≈US$400–3000) (Mitchell 1979a; Richards, 1979). By the mid-1980s, the solpis industry had totally collapsed. This not only affected cash income in the area but also decreased the availability of cheap protein to potential consumers who thenceforth had inadequate protein in their diets. In addition, the weed invasion depressed tourism and boat movement was affected. Of greater importance was that salvinia affected the lives of thousands of resource-poor villagers living along the length and breadth of the Sepik River floodplains by preventing them from moving freely. Every aspect of life in the Sepik is influenced by the river. The saying among the Sepik that 'children learn to paddle first before walking' sums up how important open waterways are in the lives of the people in that region. The problem was so severe that some affected villages originally located around oxbow lakes were relocated to near the main river.

Fortunately, the salvinia problem ended when, under a United Nations Development Programme (UNDP)-supported classical biological control project, the host-specific curculionid weevil *Cyrtobagous salviniae*, found in Brazil and introduced into Australia by the Commonwealth Scientific and Industrial Research Organisation (CSIRO), was introduced into the Sepik in 1982. The weevils were reared in tanks and in an oxbow lake near Angoram in the Lower Sepik River and redistributed over a 2 year period into 130 other lakes and lagoons (Thomas and Room, 1986). The control of salvinia in the Sepik, and also elsewhere in Australia and Sri Lanka, represents one of the best textbook examples of how the dramatic impacts of invasive plants on rural communities can be reversed with suitable natural enemies backed by good science.

However, the Sepik River people were not then free of invasive plants, as 3 years after the end of the salvinia problem, a new crisis was looming: invasion by another South American aquatic weed, the water hyacinth. This problem is described in Box 6.4.

well as social and economic problems in the vast wetlands of mainland Papua New Guinea and other larger islands in the southwest Pacific.

Unfortunately, these are just two of an increasing onslaught of invasive plants threatening new areas of the Pacific. In Viti Levu, Fiji, Keppel and Watling (2011) reported on four plant species – *Pinanga coronata*, *Schefflera actinophylla*, *Swietenia macrophylla* and the African tulip tree (see above) – which have recently been observed invading primary lowland tropical rainforest, having been transported from other islands where they are already problematic. Another species, *Antigonon leptopus*, has attractive pink flowers and is being introduced in many areas as an ornamental; however, the invasive nature of the plants has been recognized and it is recommended that introductions should be closely monitored to prevent spread. In Guam, *A. leptopus* is becoming so widespread that it already threatens local diversity (Burke and DiTommaso, 2011).

Box 6.4. Case study: water hyacinth in Papua New Guinea.

The water hyacinth (*Eichhornia crassipes*) is probably the most troublesome aquatic weed throughout the tropics and subtropics, having being recorded between latitudes 40°N and 45°S (Gopal, 1987). Interestingly, the early arrival and spread of this noxious invader of freshwater bodies in Papua New Guinea is well documented. It was first reported on 9 June 1962, from a gold-dredging pond near Bulolo, 60 km south-west of Lae in Morobe Province. Efforts to eradicate water hyacinth from three out of 55 gold-dredging ponds found to be infested began immediately. In April 1963, the ponds were treated with a combination of manual labour to remove the weed, as well as spraying with 0.2% 2,4-D ester and burning uprooted plants with diesel fuel. Nevertheless, regrowth occurred in treated ponds and follow-up treatments were necessary. In October 1962, water hyacinth was gazetted as a 'notifiable noxious weed', making it obligatory for landholders to notify an Inspector of Plants of its presence and making it illegal to take it out of the Wau-Bulolo area (Mitchell, 1979b; Papua New Guinea Department of Primary Industry unpublished data). Despite these preventive measures, the weed persisted and eventually spread to most other parts of the country (Julien and Orapa, 2001). In 1986, water hyacinth was carelessly introduced into a village within the Sepik River floodplains by a villager returning from the eastern town of Madang. This incident occurred just as the salvinia problem was receding (see Box 6.3), and was quickly identified as a potential threat to the natural ecology and socio-economic status of the recovering region (Laup, 1985; Julien and Orapa, 2001).

In 1989, within 3 years of introduction into the Sepik, water hyacinth was already increasing in the Lower Sepik River around Angoram and was threatening to invade all of the available waterways upstream and cause problems more severe than those that salvinia had caused – because of its larger size. By November 2001, water hyacinth cover of the lakes and canals was estimated at 3 km^2 (M. Julien and author's unpublished data). Over the next few years, the invasion intensified and began to cause social and economic hardships for nearly 100,000 people living on the Sepik River and threatened the entire 1500 km of the river system. The spread of water hyacinth upstream was mostly facilitated by boat movement: plants caught under the engines were cleaned out by operators and left in areas that had previously been free of the weed. There were also several unsubstantiated reports that people from villages in areas already affected by water hyacinth intentionally introduced the weed to unaffected areas owned by rival villages upstream out of jealousy, because the latter were seen to be enjoying fishing and open waterways to their sago forests – the staple food of people in these wetland areas.

The invasion of water hyacinth had an impact on the food security, health and daily lives of people dependent on the open waterways in the Sepik River and its wetlands system, with fishing and access to other natural resources becoming extremely difficult. It reduced access to backwater villages by tourists and government workers, as all travel in the region is heavily reliant on boats or motorized canoes. This disruption to all boat traffic had an impact on income earnings by villagers too, because access to local fresh produce markets for the sale of fish, artefacts, vegetables, sago, crocodile skins and rubber were severely affected. While the impact on the income levels of villages dependent on the river is difficult to estimate, costs as a direct result of the aquatic weed invasions were certainly increased, especially for fuel. There was anecdotal evidence that the costs of servicing outboard engines was placing strains on the rural economy in water hyacinth-infested areas, as boat owners had to charge more for ferrying people and cargo owing to increased operating costs.

In extreme cases, reports suggest there may have been deaths of villagers as a result of water hyacinth choking waterways. In 1992, three people died at a village on the Keram River, a tributary of the Sepik, simply because they could not get past water hyacinth-infested waterways to the hospital at Angoram in time for the treatment of their ailments, which included a case of snakebite. The worst affected village was Tambali, 20 km south-west of Angoram (see Fig. 6.6a). This village is located 5 km away from the main river on the bend of a 9 km long oxbow lake. In 1994, it was totally isolated by water hyacinth that choked up to 90% of the lake area. One villager died from fatigue after struggling for hours when his canoe got caught in the middle of the lake among the weeds; 2 weeks before he met his eventual fate, this author and co-workers rescued the same person and his daughter from a similar situation (see Fig. 6.5f). The long and heavy dugout canoes used in the Sepik River are normally powered by outboard motors, but engines are useless among thick mats of water hyacinth. The weed also prevented access to schools, health centres and markets, all located in Angoram, an hour downstream of the main river by motorized canoe. Water-related skin diseases and other ailments

Fig. 6.6. Case studies of infestation by water hyacinth (*Eichhornia crassipes*) and its biological control in Papua New Guinea: (a) infestation of Tambali Lagoon; (b) weed cover after control by release of water hyacinth weevils (*Neochetina bruchi* and *N. eichhorniae*) as control agents; (c) water hyacinth weevils; (d–f) infestation of Waigani Lake before, during and after the release of the release of the moths *Xubida infusellus* and *Niphograpta albiguttalis* as control agents. Photos courtesy Commonwealth Scientific and Industrial Research Organisation (CSIRO).

increased at Tambali Village as a result of humans sharing their source of water for washing, cooking and drinking with pigs (author's personal observation). Eventually, the district health authorities and police had to go to the village and shoot all the pigs to reduce the waterborne diseases. Had eradication of this dangerous weed been successful in the 1960s when it was attempted in the Wau–Bulolo area, such human costs three decades later would have been avoided.

In other parts of Papua New Guinea, water hyacinth infestations have had an impact on people's lives in various ways. Near the city of Port Moresby, an invasion of sewage treatment ponds disrupted the natural sewage treatment processes when water hyacinth covered the aeration ponds. This resulted in raw sewage being emptied into the Waigani Lakes, where a very productive tilapia fishery was a thriving business for unemployed squatter settlers in the poor suburbs of Morata and Moitaka. No one knows whether this affected human health. The weed also invaded parts of the Laloki River

45 km east of Port Moresby, which is used for hydroelectricity generation. This caused regular power stoppages and necessitated cleaning of the turbines and intake areas, all of which could have cost the Papua New Guinea Electricity Commission thousands of kina from lost production and from burning expensive diesel fuel at a backup power plant.

In May 1985, 500 adults of the South American weevil *Neochetina eichhorniae* were provided by CSIRO and released at Madang, from where a handful were re-collected in July 1986. These were taken to Angoram for mass rearing and eventual liberation into the Sepik River by the Department of Primary Industry's Aquatic Weed Control Unit. Low-level rearing and releases of the biocontrol agent continued in the Sepik until December 1989 (P. Pandau, Papua New Guinea Department of Agriculture and Livestock (DAL), unpublished data). However, continuation of the work against water hyacinth ceased in 1988 when support by the Papua New Guinea government ceased and personnel were transferred. Fortunately, with funding from the Australian and Papua New Guinean governments from January 1993 to December 1998, the most focused and structured biological control programme ever undertaken against the weed in the Pacific islands brought it under control (Fig. 6.6b). The programme cost over US$1.5 million and required the intensive rearing and redistribution of *N. eichhorniae* and the related species *N. bruchi* (Fig. 6.6c) in the Sepik and at 136 locations throughout Papua New Guinea (Julien and Orapa, 1999, 2001). Two additional classical biological control agents, the moths *Xubida infusellus* and *Niphographta albiguttalis*, were also released at selected locations, but only the former species has become established – at Waigani Lake near Port Moresby – where the weed cover has remained below 30% since mid-1996 (Julien and Orapa, 2001); Figs 6.6d–f show the weed cover before the release of these agents, and during and after release). Water hyacinth in Papua New Guinea is now under permanent biological control at most of the sites where the two weevils were released. At any time of the year, the infestation levels at these sites fluctuate between 10 and 30% without causing serious persistent problems. At Bulolo where water hyacinth was first introduced, infestations at six gold-dredging ponds were kept to under 5% cover by the two weevils during the 1990s (M. Julien and author's unpublished data), and in 2016 only very small highly damaged plants can be seen.

Institutional responses to plant invasions

Regionally, the SPC has been involved with agriculturally related plant pest management work since the 1950s as part of the Plant Protection Service provided to 22 member developing island states and territories. More recently, SPC has acknowledged the threat of invasive species to the well-being of Pacific peoples and their fragile environments, and this is recognized in its Land Resources Division's Strategic Plan for 2013–2017 (SPC, 2013), the regional Millennium Development Goals and, more recently, the regional Sustainable Development Goals.

Through the Plant Protection in the Pacific Project (SPC, 2008), work related to addressing invasive alien plant problems or threats promoted and implemented by the SPC has been considered under three broad areas. Termed the '3-P' Strategy, these areas of work are: preparedness, prevention and pest management.

- Preparedness: improvement of national capacities in risk assessment, development of emergency response plans for any new incursions, raising of travellers' awareness and ensuring that biosecurity concerns are addressed as part and parcel of international trade. The simple object is to keep alien species risks offshore from the Pacific islands.
- Prevention: field surveys, eradication of new weed incursions, providing diagnostic services and information, and fostering early public awareness of new alien species threats to prevent them from becoming widespread problems.
- Pest management: development of appropriate pest management packages and implementing them in collaboration with PICTs to manage widespread problems. Biological control, one of the main techniques preferred in the Pacific islands for the control of invasive plants affecting agroecosystems and natural areas, falls under this category.

The other regional agency involved in IAS issues in the Pacific in relation to biological diversity under the CBD is the SPREP. This agency currently lacks the institutional capacity to undertake on-the-ground invasive alien plant management work, but it has significant strengths in bringing regional and international agencies together to form partnerships and networks – such as the PILN, PIP and other nationally based conservation-based action groups – for training, creating awareness and sharing learning experiences about invasive species in general. The SPREP and SPC have also collaborated to publish a handbook, *Guidelines for Invasive Species Management in the Pacific* (Tye, 2009), which was compiled in consultation with the agriculture and conservation sectors of all the PICTs. These guidelines aim to provide support and information on the regional and international agencies that can assist (Pacific Invasives Initiative, 2012).

On the national scene, some island states have attempted to contain or eradicate some weeds of limited distribution. The introduced plants that are subject to containment or eradication programmes and under various stages of development are listed in Table 6.4. Most of these activities are happening in the Federated States of Micronesia, Palau, Vanuatu, Papua New Guinea, Fiji, American Samoa and Niue. In the small Republic of Palau in Micronesia, where the impacts of invasive species have been recognized, an active National Invasive Species Committee operates directly under the Office of the President. As a result, several active invasive plant awareness and eradication efforts have been initiated there in natural areas; these also are listed in Table 6.4. In American Samoa, the National Parks and Wildlife Service has been actively running campaigns with communities to control invasive plants such as the leguminous 'tamalingi' (*Falcataria moluccana*) in its parks and nature reserves. The Kosrae Conservation and Safety Organization of the Federated States of Micronesia has undertaken a feasibility study on the management of invasive plants in a proposed protected area funded by the Critical Ecosystem Partnership Fund (Pacific Invasives Initiative, 2012).

Summary and Outlook

The Pacific islands, with their oceanic isolation acting as natural barriers, once enjoyed relatively low levels of exotic plant introductions, but such natural barriers are no longer impediments.

Globalization has made the world a smaller place, and the once-isolated Pacific islands are now under increasing threat of invasion by many troublesome alien plant species. While many small islands are relatively isolated from major shipping and trade routes, new introductions will still occur for various reasons, such as tourism and travel by local people. The impacts of invasive alien plants can range from causing local nuisance and affecting human health to widespread losses or costs in primary productivity which could hurt both the socio-economic status of local communities or small island countries, and their unique biological diversity. Island biodiversity has strong links to the socio-economic status of Pacific island societies, whose existence continues to be dependent on protecting vulnerable island ecosystems and their scarce biological resources from negative impacts, such as those posed by alien plant invasions.

Scores of invasive plants are already of concern to island countries. The plant families Asteraceae, Poaceae and Fabaceae stand out as containing some of the most common weedy plants in the Pacific, but the members of many other plant groups are also important. Major regional weeds, such as merremia, mikania, elephant grass and the African tulip tree to name but a few, are already causing socio-economic problems. As we have discussed in this chapter, some invasive plants, for example chromolaena and mikania, have recently attracted increased attention from weed managers, and others such as the water weeds from the New World have been brought under successful biological control, but the challenges for the region lie in how effectively these and many other highly adapted species are prevented from reaching the Pacific and spreading further. Some immediate threats caused by the occurrence of the wetland invaders *M. pigra* and *L. flava* in Papua New Guinea, the occurrence of the purple scourge (*Miconia calvescens*) in French Polynesia and New Caledonia, and the silent spread of parthenium in New Caledonia, Vanuatu and French Polynesia are of great concern. Some 'sleeper' weeds may be waiting to become invasive and cause difficulties or losses for islanders if nothing is done to detect and control them. Others not present in the region or localized in one or a few islands threaten to spread because effective border control measures are still deficient.

However, all is not gloomy. Already efforts by some national governments and regional intergovernmental agencies such as the SPC are making a difference by increasing awareness, and the development of legislation and standards and protocols that address biosecurity issues such as the threat of invasive plants to the socio-economic situation as well as the environment of Pacific island countries and territories. For example, a workshop was held in 2010 by the Cooperative Islands Initiative, under the World Summit for Sustainable Development and the CBD, to develop a regional action plan to combat invasive species on islands, in order to preserve biodiversity and support adaptation to climate change (Sheppard and Singleton-Cambage, 2010). Meyer (2014) considers that more effort is needed across the region to document the ecological and socio-economic impacts of invasive alien plants, and emphasizes the importance of ecosystem functioning, resilience and climate change on island communities.

References

Allen, B.J. and Bourke, R.M. (2001) The 1997 drought and frost in PNG: overview and policy implications. In: Bourke, R.M., Allen, M.G. and Salisbury, J.G. (eds) *Food Security in Papua New Guinea: Proceedings of the Papua New Guinea Food and Nutrition Conference,* Lae, Papua New Guinea, 26–30 June 2000. ACIAR Proceedings No. 99. Australian Centre for International Agricultural Research, Canberra, pp. 155–163.

Aterrado, E.D. and Bachiller, N.S.J. (2002) Biological control of *Chromolaena odorata*: preliminary studies on the use of the gall forming fly *Cecidochares connexa* in the Philippines. In: Zachariades C., Muniappan R. and Strathie L.W. (eds) *Proceedings of the Fifth International Workshop on Biological Control and Management of* Chromolaena odorata, *Durban, South Africa, 23–25 October 2000.* Agricultural Research Council-Plant Protection Research Institute (ARC-PPRI), Pretoria, pp. 137–139.

Auld, B. and Seniloli-Nagatalevu, M. (2003) African tulip tree in the Fiji Islands. In: Labrada, R (ed.) *Weed Management for Developing Countries.* Food and Agriculture Organization of the United Nations (FAO), Rome.

Barnard, P. and Waage, J.K. (2004) *Tackling Biological Invasions around the World: Regional Responses to the Invasive Alien Species Threat.* Global Invasive Species Programme, Cape Town, South Africa.

Blanfort, V., Balent, G., Julien, M.P. and Guervilly, T. (2008). Invasive plants and pasture management in New Caledonia. In: Blanfort, V. and Orapa, W. (eds) *Proceedings of the Regional Workshop on Invasive Plant Species in Pastoral Areas, Koné, New Caledonia, 24–28 November 2003.* Secretariat of the Pacific Community, Suva, pp. 16–27.

Bofeng, I., Donnelly, G., Orapa, W. and Day, M. (2004) Biological control of *Chromolaena odorata* in Papua New Guinea. In: Day, M.D. and McFadyen, R.E. (eds) *Chromolaena in the Asia-Pacific Region. Proceedings of the Sixth International Workshop of Biological Control and Management of Chromolaena, Cairns, Australia, 6–9 May 2003.* ACIAR Technical Report No. 55. Australian Centre for International Agricultural Research, Canberra, pp. 14–16.

Brinon, L. (2008) Cactus (*Acanthocereus pentagonus*) control trials on the south-west coast of New Caledonia. In: Blanfort, V. and Orapa, W. (eds) *Proceedings of the Regional Workshop on Invasive Plant Species in Pastoral Areas, Koné, New Caledonia, 24–28 November 2003.* Secretariat of the Pacific Community, Suva, pp. 65–66.

Burke, J.M. and DiTommaso, A. (2011) Corallita (*Antigonon leptopus*): intentional introduction of a plant with documented invasive capability. *Invasive Plant Science and Management* 4, 265–273.

CABI (2013) *Spathodea campanulata* (African tulip tree). Invasive Species Compendium datasheet. CAB International, Wallingford, UK. Available at: www.cabi.org/isc/datasheet/51139 (accessed 2 March 2017).

Cock, M.J.W., Ellison, C.A., Evans, H.C. and Ooi, P.A.C. (2000) Can failure be turned into success for biological control of mile-a-minute weed

(*Mikania micrantha*)? In: Spencer, N.R, (ed.) *Proceedings of the X International Symposium on Biological Control of Weeds, Bozeman, Montana, 4–14 July 1999.* USDA Forest Service, Forest Health Technology Enterprise Team, Morgantown, West Virginia, pp. 155–167.

Davis, R.I., Lomavatu-Fong, M.F., Ruabtete, T.K., Kumar, S. and Turaganivalu, U. (2005) *Cucumber mosaic virus* infection of kava (*Piper methysticum*) and implications for cultural control of kava dieback disease. *Australasian Plant Pathology* 34, 377–384.

Day, M. (2014) Parthenium weed invasion in the Pacific: an update from Vanuatu and PNG. *International Parthenium News* 9, 2–3.

Day, M.D. and Bofeng, I. (2007) Biocontrol of *Chromolaena odorata* in Papua New Guinea. In: Lai, P.-Y., Reddy, G.V.P. and Muniappan, R. (eds) *Proceedings of the Seventh International Workshop on Biological Control and Management of* Chromolaena odorata *and* Mikania micrantha, *Pingtung, Taiwan, September 2006.* National Pingtung University of Science and Technology, Pingtung, Taiwan, pp. 53–67.

Day, M.D. and Winston, R.L. (2016) Biological control of weeds in the 22 Pacific island countries and territories: current status and future prospects. *NeoBiota* 30, 167–192.

Day, M.D., Kawi, A., Kurika, K., Dewhurst, C.F., Waisale, S., Saul-Maora, J., Fidelis, J., Bokosou, J., Moxon, J., Orapa, W. and Senaratne, K.A.D. (2012) *Mikania micrantha* Kunth (Asteraceae) (mile-a-minute): its distribution and physical and socioeconomic impacts in Papua New Guinea. *Pacific Science* 66, 213–223.

Day, M.D., Kawi, A.P., Fidelis, J., Tunabuna, A., Orapa, W., Swamy, B., Ratutini, J., Saul-Maora, J. and Dewhurst, C.F. (2013a) Biology, field release and monitoring of the rust *Puccinia spegazzinii* de Toni (Pucciniales: Pucciniaceae), a biocontrol agent of *Mikania micrantha* Kunth (Asteraceae) in Papua New Guinea and Fiji. In: Wu, Y., Johnson, T., Sing, S., Rhagu, R., Wheeler, G., Pratt, P., Warner, K., Center, T., Goolsby J. and Reardon, R. (eds) *Proceedings of the XIII International Symposium on Biological Control of Weeds, Waikoloa, Hawaii, 11–16 September 2011.* USDA Forest Service, Forest Health Technology Enterprise Team, Morgantown, West Virginia, pp. 211–217.

Day, M.D., Brito, A.A., De Costa Guterres, A., Da Costa Alves, A.P., Paul, T. and Wilson, C.G. (2013b) Biocontrol of *Chromolaena odorata* (L.) King and Robinson (Asteraceae) in Timor Leste. In: Zachariades, C., Strathie, L.W., Day, M.D. and Muniappan, R.(eds.) *Proceedings of the Eighth International Workshop on Biological Control and Management of* Chromolaena odorata, *Nairobi, Kenya, November 2010.* Agricultural Research Council-Plant Protection Research Institute (ARC-PPRI), Pretoria, pp. 134–140.

de Chenon, D.R., Sipayung, A. and Sudharto, P. (2002) A decade of biological control against *Chromolaena odorata* at the Indonesian Oil Palm Research Institute in Marihat. In: Zachariades, C., Muniappan, R. and Strathie, L.W. (eds) *Proceedings of the Fifth International Workshop on Biological Control and Management of* Chromolaena odorata, *Durban, South Africa, 23–25 October 2000.* Agricultural Research Council-Plant Protection Research Institute (ARC-PPRI), Pretoria, pp. 46–52.

Dodd, S., Hayes, L. and Paynter, Q. (2011) *Recent Initiatives to Develop Biocontrol for the Pacific: Strategy Workshop and Weed Prioritisation Exercise.* Biodiversity Conservation Lessons Learned Technical Series 5, Critical Ecosystem Partnership Fund (CEPF) and Conservation International Pacific Islands Program (CI-Pacific), Apia. Available at: www.cepf.net/Documents/5_Pacific_Biocontrol.pdf (accessed 23 March 2016).

Esguerra, N.M. (1998) The Siam weed infestation in the Federated States of Micronesia – seven years of attempting to control it. In: Ferrar, P., Muniappan, R. and Jayanth, K.P. (eds) *Proceedings of the Fourth International Workshop on Biological Control and Management of* Chromolaena odorata, *Bangalore, India, 14–18 October 1996.* Agricultural Experiment Station, University of Guam, Mangilao, Guam, pp. 80–81.

Gopal, B. (1987) *Water Hyacinth.* Elsevier, Amsterdam, The Netherlands.

Government of Tonga (2004) Half yearly budget review 2003–2004. Press Release, Thursday 6 May 2004. Ministry of Finance, Government of the Kingdom of Tonga, Nuku'alofa.

Hartemink, A.E. and O'Sullivan, J.N. (2001) Leaf litter decomposition of *Piper aduncum*, *Gliricidia sepium* and *Imperata cylindrica* in the humid lowlands of Papua New Guinea. *Plant and Soil* 230, 115–124.

Henty, E.E. and Pritchard, P.H. (1973) *Weeds of New Guinea and their Control.* Botany Bulletin No. 7, Department of Forests, Lae, Papua New Guinea.

Julien, M.H. and Orapa, W. (1999) Structure and management of a successful biological control project for water hyacinth. In: Hill, M.P., Julien, M.H. and Center, T.D. (eds) *Proceedings of the First IOBC Working Group Meeting for the Biological and Integrated control of Water Hyacinth, Harare, Zimbabwe, 16–19 November 1998.* Agricultural Research Council-Plant Protection

Research Institute (ARC-PPRI), Pretoria, pp. 123–134.

Julien, M.H. and Orapa, W. (2001) Insects used for biological control of the aquatic weed water hyacinth in Papua New Guinea. *Papua New Guinea Journal of Agriculture, Forestry and Fisheries* 44, 49–60.

Keppel, G. and Watling, D. (2011) Ticking time bombs – current and potential future impacts of four invasive plant species on the biodiversity of lowland tropical rainforests in south-east Viti Levu, Fiji. *South Pacific Journal of Natural and Applied Sciences* 29, 43–45.

Kuniata, L.S. and Korowi, K.T. (2004) Bugs offer sustainable control of *Mimosa invisa* and *Sida* spp. in the Markham Valley, Papua New Guinea. In: Cullen, J.M., Briese, D.T., Kriticos, D.J., Lonsdale, W.M., Morin, L. and Scott, J.K. (eds) *Proceedings of the XI International Symposium on Biological Control of Weeds, Canberra, Australia, 27 April–3 May 2003.* CSIRO (Commonwealth Scientific and Industrial Research Organisation) Entomology, Canberra, pp. 567–573.

Laup, S. (1985) The Sepik salvinia problem is beaten. *Harvest* 11, 49–52.

Laup, S. (1986) Biological control of water lettuce: early observations. *Harvest* 12, 41–43.

Li, Y., Cheng, Z., Smith, W.A., Ellis, D.R., Chen, Y., Zheng, X., Pei, Y., Luo, K., Zhao, D., Yao, Q. *et al.* (2004) Invasive ornamental plants: problems, challenges, and molecular tools to neutralize their invasiveness. *Critical Reviews in Plant Sciences* 23, 381–389.

Mack, R.N. and Lonsdale, M.W. (2001) Humans as global plant dispersers: getting more than we bargained for. *Bioscience* 51, 95–102.

Manzan, T.K.W. (2000) Review of the impact of weeds on coffee, their management and research implications in Papua New Guinea. *PNG Coffee* 40, 1–13.

McFadyen, R.E.C. (2002) *Chromolaena* in Asia and the Pacific: spread continues but control prospects improve. In: Zachariades, C.R., Muniappan, R. and Strathie, L.W. (eds) *Proceedings of the Fifth International Workshop on Biological Control and Management of* Chromolaena odorata, *Durban, South Africa, 23–25 October 2000.* Agricultural Research Council-Plant Protection Research Institute (ARC-PPRI), Pretoria, pp. 13–18.

McGregor, A. (1999) *Linking Market Developments to Farming Systems in the Pacific Islands.* SAPA Publication No. 1999/2, FAO (Food and Agriculture Organization of the United Nations) Sub-Regional Office for the Pacific, Apia.

McWilliam, A. (2000) A plague on your house? Some impacts of *Chromolaena odorata* on Timorese livelihoods. *Human Ecology* 28, 451–469.

Meyer, J.-Y. (2000) Preliminary review of the invasive plants in the Pacific islands (member countries). In: Sherley, G. (ed.) *Invasive Species in the Pacific: A Technical Review and Draft Regional Strategy.* Secretariat of the Pacific Regional Environment Programme, Apia, pp. 85–114.

Meyer, J.-Y. (2004) Threat of invasive alien plants to native flora and forest vegetation of eastern Polynesia. *Pacific Science* 58, 357–375.

Meyer, J.-Y. (2014) Critical issues and new challenges for research and management of invasive plants in the Pacific islands. *Pacific Conservation Biology* 20, 146–164.

Miller, I.L., Nemestothy, L. and Pickering, S.E. (1981) Mimosa pigra *in the Northern Territory.* Technical Bulletin No. 51, Northern Territory Department of Primary Productions, Darwin, Australia.

Mitchell, D.S. (1979a) *The Incidence and Management of* Salvinia molesta *in Papua New Guinea.* Office of Environment and Conservation, Waigaini, Papua New Guinea.

Mitchell, D.S. (1979b) Aquatic weeds in Papua New Guinea. *Science in New Guinea* 6, 154–160.

Mitchell, D.S., Petr, T. and Viner, A.B. (1980) The water fern *Salvinia molesta* in the Sepik River, Papua New Guinea. *Environmental Conservation* 7, 115–122.

Mittermeier, R.A. and Mittermeier, C.G. (1997) *Megadiversity: Earth's Biologically Wealthiest Nations.* CEMEX, Conservation International and Agrupación Sierra Madre, Monterrey, Mexico.

Mittermeier, R.A., Myers, N., Robles Gil, P. and Mittermeier, C.G. (1999) *Hotspots: Earth's Biologically Richest and Most Endangered Terrestrial Ecoregions.* CEMEX, Conservation International and Agrupación Sierra Madre, Monterrey, Mexico.

Mittermeier, R.A., Gil, P.R., Hoffman, M., Pilgrim, J., Brooks, T., Mittermeier, C.G., Lamoreux, J. and Fonseca, G.A.B. (2005) *Hotspots Revisited: Earths Biologically Rich and Most Endangered Terrestrial Ecoregions.* CEMEX, Conservation International and Agrupación Sierra Madre, Monterrey, Mexico.

Muniappan, R., Englberger, K., Bamba, J. and Reddy, G.V.P. (2004) Biological control of chromolaena in Micronesia. In: Day, M.D. and McFadyen, R.E. (eds) *Chromolaena in the Asia-Pacific Region. Proceedings of the Sixth International Workshop on Biological Control and Management of Chromolaena, Cairns, Australia, 6–9 May 2003.* ACIAR Technical Reports No.

55, Australian Centre for International Agricultural Research, Australia, pp. 11–12.

Muniappan, R., Englberger, K. and Reddy, G.V.P. (2007) Biological control of *Chromolaena odorata* in the American Pacific Micronesian Islands. In: Lai P.-Y., Reddy, G.V.P. and Muniappan, R. (eds) *Proceedings of the Seventh International Workshop on Biological Control and Management of* Chromolaena odorata *and* Mikania micrantha*, Pingtung, Taiwan, September 2006.* National Pingtung University of Science and Technology, Pingtung, Taiwan, pp. 49–52.

Napompeth, B. (1983) Background, threat, and distribution of *Mimosa pigra* L. in Thailand. In: Robert, G.L. and Habeck, D.H. (eds) *Proceedings of an International Symposium on* Mimosa pigra *Management, Chiang Mai, Thailand, 22–26 February 1982.* Document 48-A-83, International Plant Protection Centre, Corvallis, Oregon, pp. 15–26.

Orapa, W. (2001) Impediments to food security: the case of exotic weeds in Papua New Guinea. In: Bourke, R.M., Allen, M.G. and Salisbury, J.G. (eds) *Food Security for Papua New Guinea. Proceedings of the Papua New Guinea National Food Security Conference, Lae, Papua New Guinea, 26–30 June 2000.* Australian Centre for International Agricultural Research (ACIAR), Canberra, pp. 308–315.

Orapa, W. (2004) Chromolaena. Pest Advisory Leaflet No. 43,Secretariat of the Pacific Community (SPC), Suva.

Orapa, W. and Julien, M.H. (1996) Incidence of *Mimosa pigra* in Papua New Guinea. *Harvest* 18, 20–25.

Orapa, W., Donnelly, G. and Bofeng, I. (2002) The distribution of Siam weed (*Chromolaena odorata*) in Papua New Guinea. In: Zachariades, C., Muniappan, R. and Strathie, L.W. (eds) *Proceedings of the Fifth International Workshop on Biological Control of* Chromolaena odorata*, Durban, South Africa, 23–25 October 2000.* Agricultural Research Council-Plant Protection Research Institute (ARC-PPRI), Pretoria, pp. 19–25.

Orapa, W., Englberger, K. and Lal, S.N. (2004) Chromolaena and other weed problems in the Pacific islands. In: Day, M.D. and McFadyen, R.E. (eds) *Chromolaena in the Asia-Pacific Region. Proceedings of the Sixth International Workshop of Biological Control and Management of Chromolaena, Cairns, Australia, 6–9 May 2003.* ACIAR Technical Reports No. 55, Australian Centre for International Agricultural Research, Canberra, p. 13. [Abstract]

Pacific Invasives Initiative (2012) *Feasibility Study for the Management of Invasive Plants within a Proposed Protection Area, Olum Watershed, Kosrae, Federated States of Micronesia.* Pacific Invasives Initiative, Auckland, New Zealand.

Papineau, C. and Blanfort, V. (2008) Invasive plants, a threat to New Caledonia's dry forest. In: Blanfort, V. and Orapa, W (eds) *Proceedings of the Regional Workshop on Invasive Plant Species in Pastoral Area, Koné, New Caledonia, 24–28 November 2003.* Secretariat of the Pacific Community (SPC), Suva, pp. 70–74.

Paynter, Q. (2010) *Prioritisation of Targets for Biological Control of Weeds on Pacific Islands.* Landcare Research New Zealand Ltd, Auckland, New Zealand.

Paynter, Q. (2014a) Cook Islands project becomes a reality. *What's New in Biological Control of Weeds?* 67, 4. Landcare Research New Zealand Ltd. Available at: www.landcareresearch.co.nz/publications/newsletters/biological-control-of-weeds/issue-67/cooks-islands-project (accessed 18 March 2016).

Paynter, Q. (2014b) New Zealand and the Cook Islands tackle weed biocontrol. *Biocontrol News and Information* 35, 10N.

PIER (2013) Pacific Island Ecosystems at Risk (PIER): Plant threats to Pacific ecosystems. Available at: www.hear.org/pier/ (accessed 5 May 2016).

Reaser, J.K., Meyerson, L., Cronk, Q., De Poorter, M., Eldrege, L.G., Green, E., Kairo, M., Latasi, P., Mack, R.N., Mauremootoo, J. *et al.* (2003) *Paradise Invaded: The Ecological and Socioeconomic Impacts of Invasive Alien Species in Island Ecosystems. Report of an Experts Consultation.* Global Invasive Species Programme. Available at: www.cbd.int/doc/ref/ais-gisp-report-en.pdf (accessed 18 March 2016).

Richards, A.H. (1979) Salvinia in the Sepik River. *Harvest* 5, 239–243.

Room, P.M. and Thomas, P.A. (1986a) Nitrogen, phosphorus and potassium in *Salvinia molesta* in the field: effects of weather, insect damage, fertilizers and age. *Aquatic Botany* 24, 214–232.

Room, P.M. and Thomas, P.A. (1986b) Population growth of the floating weed *Salvinia molesta*: field observations and a global model based on temperature and nitrogen. *Journal of Applied Ecology* 23, 1013–1028.

Samouth, C. (2004) *Mimosa pigra* infestations and the current threat to wetlands and floodplains in Cambodia. In: Julien, M., Flanagan, G., Heard, T., Hennecke, B., Paynter, Q. and Wilson, C.

(eds) *Research and Management of* Mimosa pigra. *Proceedings of the Third International Symposium on the Management of* Mimosa pigra, *Darwin, Australia, 23–25 September 2002.* CSIRO (Commonwealth Scientific and Industrial Research Organisation) Entomology, Canberra, pp. 29–32.

Shabbir, A. (2012) A report on the International Parthenium Weed Workshop, September 27, 2011, Cairns, Australia. *International Parthenium News* 5, 11–13.

Sheppard, B. and Singleton-Cambage, K. (eds) (2010) *Helping Islands Adapt. A Workshop on Regional Action to Combat Invasive Species on Islands to Preserve Diversity and Adapt to Climate Change, 11–16 April 2010, Hyatt Regency Auckland, New Zealand: Proceedings.* Available at: www.cbd.int/invasive/doc/proceedings-workshop-helping-island-en.pdf (accessed 18 March 2016).

Sinden, J., Jones, R., Hester, S., Odom, D., Kalisch, C., James, R. and Cacho, O. (2004) *The Economic Impact of Weeds in Australia. Report to the CRC for Australian Weed Management.* CRC for Australian Weed Management Technical Series No. 8, Cooperative Research Centre for Australian Weed Management, Glen Osmond, Australia.

Smith, A.C. and Carr, G.D. (1991) *Asteraceae. Flora Vitiensis Nova, Vol. 5.* National Tropical Botanical Garden, Kauai, Hawaii, pp. 254–320.

Space, J.C. and Flynn, T. (2002) *Report to the Government of Samoa on Invasive Plant Species of Environmental Concern.* USDA Forest Service, Pacific Southwest Research Station, Institute of Pacific Islands Forestry, Honolulu.

Space, J.C., Waterhouse, B.M., Miles, J.E., Tiobech, J. and Rengulbai, K. (2003) *Report to the Republic of Palau on Invasive Plant Species of Environmental Concern.* USDA Forest Service, Pacific Southwest Research Station, Institute of Pacific Islands Forestry, Honolulu.

Space, J.C., Waterhouse, B.M., Newfield, M. and Bull, C. (2004) *Report to the Government of Niue and the United Nations Development Programme: Invasive Plant Species on Niue following Cyclone Heta.* UNDP NIU/98/G31 – Niue Enabling Activity. Available at: http://www.hear.org/pier/reports/niue_report_2004.htm (accessed 27 February 2017).

SPC (2004) *Pacific Islands Millennium Development Goals Report.* Prepared by the Secretariat of the Pacific Community in collaboration with the United Nations and UN/CROP MDG Working Group, Noumea. Secretariat of the Pacific Community, Noumea.

SPC (2008) *Plant Protection in the Pacific (8.ACP. RPA.08 & 8.PTF.REG.03) Project. Final Report.* Land Resources Division, Secretariat of the Pacific Community, Noumea. Available at: www.spc.int/lrd/publications/doc_download/1374-pppterminal2008 (accessed 18 March 2016).

SPC (2013) *Strategic Plan 2013–2017.* CRGA 43 (13)/Paper 2.2.6 ANNEX 2, Land Resources Division, Secretariat of the Pacific Community, Noumea. Available at: www.spc.int/lrd/publications/doc_download/1374-pppterminal2008 (accessed 22 August 2017).

SPREP (2004) *Action Strategy for Nature Conservation in the Pacific Region, 2003–2007.* Secretariat of the Pacific Regional Environment Programme, Apia.

SPREP (2009) *Action Strategy for Nature Conservation and Protected Areas in the Pacific Region, 2008–2012.* Secretariat of the Pacific Regional Environment Programme, Apia.

SPREP (2014) *Framework for Nature Conservation and Protected Areas in the Pacific Islands Region 2014–2020.* Secretariat of the Pacific Regional Environment Programme, Apia.

Sulistyawan, B.S. and Hartono (2004) Mapping of *Mimosa pigra* on Maro River in Wasur National Park, Papua-Indonesia. In: Julien, M., Flanagan, G., Heard, T., Hennecke, B., Paynter, Q. and Wilson, C. (eds) *Research and Management of* Mimosa pigra. *Third International Symposium on the Management of* Mimosa pigra, *Darwin, Australia, 23–25 September 2002.* CSIRO (Commonwealth Scientific and Industrial Research Organisation) Entomology, Canberra, pp. 52–56.

Swarbrick, J.T. (1997) *Weeds of the Pacific Islands.* Technical Paper No. 209, South Pacific Commission [now Secretariat of the Pacific Community], Noumea.

Thomas, P.A. and Room, P.M. (1986) Taxonomy and control of *Salvinia molesta. Nature* 320, 581–584.

Tye, A. (2009) *Guidelines for Invasive Species Management in the Pacific: A Pacific Strategy for Managing Pests, Weeds and Other Invasive Species.* Secretariat of the Pacific Regional Environment Programme (SPREP), Apia.

Williams, P.A., and Lee, W.G. (2001) Why screen for weediness? *New Zealand Garden Journal* 4, 19–23.

Wilson, G.C. and Widayanto, E.B. (2002) The biological control programme against *Chromolaena odorata* in eastern Indonesia.

In: Zachariades, C., Muniappan, R. and Strathie, L.W. (eds) *Proceedings of the Fifth International Workshop on Biological Control of* Chromolaena odorata*, Durban, South Africa, 23–25 October 2000.* Agricultural Research Council-Plant Protection Research Institute (ARC-PPRI), Pretoria, pp. 53–57.

Winston, R.L., Schwarzländer, M., Hinz, H.L., Day, M.D., Cock, M.J.W. and Julien, M.H. (eds) (2014) *Biological Control of Weeds: A World Catalogue of Agents and Their Target Weeds*, 5th edn. USDA Forest Service, Forest Health Technology Enterprise Team, Morgantown, West Virginia.

7 Understanding the Impact of Invasive *Mikania micrantha* in Shifting Agriculture and Its Management through Traditional Ecological Knowledge

P.S. Ramakrishnan*

School of Environmental Sciences, Jawaharlal Nehru University, New Delhi, India

Introduction

Introduced as a cover crop to check soil erosion in the tea gardens of north-eastern India, *Mikania micrantha* (mikania weed), has become a serious threat to both natural and human-managed ecosystems in humid tropics of the subcontinent (Ramakrishnan, 1991). It is an important component of the early stages of the fallow phase of the shifting agricultural ('jhum') landscape of the north-eastern hill regions (Ramakrishnan, 1992a), but one with mixed fortunes. Thus, as a competing weed on croplands, it poses a serious threat, whereas in the context of the large-scale land degradation that has occurred under the prevailing shortened jhum cycles, it plays an important role in nutrient conservation as well.

The north-eastern region of India, covering about 0.26 million km^2 of hills, valleys and plateaus, is ethnically and culturally very distinct from other areas of the country. The hills where jhum is the chief land use constitute about 70% of the region's land area. In the absence of reliable estimates, the actual area under jhum is a matter for debate. According to the *Forest Sector Review of Northeast India* (Poffenberger, 2006), of the total forested area of 163,799 km^2 in north-eastern India in 1999, an estimated 44,000 km^2 (~27%) was affected by jhum agriculture. Given the large-scale timber extraction for export that has also occurred in the past 100 years, including escalations during the World War periods and in the decades following the country's independence from colonial rule, the total area now deforested is actually much higher than this, and deforestation is continuing today (Forest Survey of India, 2013).

This chapter first discusses some of the dynamics of the patterns and processes determining the presence and role of exotic weeds in general, and mikania in particular, in the jhum shifting agricultural system, and then explores possible control measures through a community participatory landscape management plan.

The Problem of Shifting Agriculture in North-east India

For centuries, forest farmers in many tropical areas have managed a range of traditional shifting (slash and burn) agricultural

E-mail: psr@mail.jnu.ac.in

systems, including those known in India as jhum. Under this system, the vegetation is cut, the debris burnt and the land cropped for a year or more before being abandoned when the soil is exhausted. A natural fallow is allowed to develop, which in time restores the fertility of the soil, while the farmer moves on to another site, and then another – eventually returning to crop each site again. In the past, the small-scale perturbations of jhum agriculture helped to enhance forest biological diversity and the crops and associated organisms benefited from the extra nutrients released when the fallows were burnt under the long jhum cycles. However, with increasing external pressure on forest resources, larger human populations and declining soil fertility, forest cover has been lost (FAO/UNEP, 1982; FAO, 1995, 2010) and agricultural cycles have shortened, causing marked productivity reduction and other negative impacts on a global scale. The north-eastern hill region of India is no exception. Here, when population pressure was not so high, the traditional fallow period between two successive croppings on the same site was as long as 20–30 years. Nevertheless, even though these longer cycles are still occasionally encountered in more remote, less populated areas, the common situation now is cycles of only 4–5 years (Ramakrishnan, 1992a). Although a global survey of the world's forests reports that India has had a small increase in forested areas in the last 20 years (FAO, 2010), the Forest Survey of India (2013) continues to record decreases in the forested area of north-eastern India, citing biotic pressure and shortened jhum cycles as the major reasons, and the shorter jhum cycles as the main cause of forest loss in the north-eastern state of Nagaland.

Forest Succession, Shortened Jhum Cycles and Invasive Weeds

Secondary forest succession

The process of converting forest to cultivable land, as under jhum, involves several major disturbances: the original vegetation is destroyed, followed by repeated perturbations from burning, the introduction of crop species, weeding and hoeing, and disturbances to the soil when the crops are harvested. In addition, the system acquires less biomass during the cropping phase, with its annual clearance/burning and harvesting of plant biomass, than it does under fallow succession. Together, these practices cause a progressive reduction in fertility and species diversity over the lifetime of the cropping phase. Thus, when the land is first returned to fallow, the soil is depleted of nutrients and the plant community has a relatively simple level of organization, with few species. Gradually, changes take place as the fallow community develops, with some species becoming more abundant, others less so, and some species replacing others (Toky and Ramakrishnan, 1983a). This process is called secondary succession, because it follows the destruction of the previous, established plant community during the burning and subsequent cultivation phases.

There is a steady increase in production in the first 5–6 years of the jhum fallow phase because the first species, the pioneer herbaceous species (Saxena and Ramakrishnan, 1984a), are particularly efficient at extracting nutrients from the soil, use the available light effectively (Saxena and Ramakrishnan, 1983a,b, 1984b), and also reproduce efficiently (Saxena and Ramakrishnan, 1982). As the early pioneer species enhance the soil nutrient content and structure, the developing vegetation begins to create shade and decrease light penetration at ground level, and this provides opportunities for other species. In the north-eastern hill regions of India, the pioneer species give way first to fast-growing bamboos (Rao and Ramakrishnan, 1989) and then to early successional tree species (Toky and Ramakrishnan, 1983a), which are characterized by rapid growth over a short period. The change in community structure over time is very marked at lower altitudes – from the initial weedy herbs, to bamboo forest and, finally, to mixed broadleaved forest.

For about the first 10 years of fallow regrowth, succession is dominated by native and exotic weedy species (Toky and Ramakrishnan, 1983a). At lower elevations, these include native C_4 (see Chapter 2, this volume) weeds such as *Imperata cylindrica*, *Thysanolaena maxima*, *Saccharum spontaneum* and *Panicum* spp., and the exotic C_3 weeds *Chromolaena odorata* (formerly known as *Eupatorium odoratum*) and *M. micrantha*. Between 10 and 30 years of fallow regrowth, many species of bamboo dominate at lower elevations, often forming pure formations and gradually replacing the weeds. Among the more important bamboo species are *Dendrocalamus hamiltonii*, *Bambusa tulda* and *B. khasiana*. These, and many other bamboo species, are of multipurpose value as they can be used for making household utensils and items of aesthetic value, in house construction and even for their tender shoots, which are of food value. Beyond 30 years of fallow regrowth, broadleaved tree species increase in numbers at all elevations. The mixed early successional forests that replace the bamboo forests at lower elevations contain trees such as *Anthocephalus kadamba*, *Duabanga sonneratioides* and *Lagerstroemia parviflora*, which are largely of timber value, and *Alnus nepalensis*, though this species occurs at a range of altitudes (300–1700 m above sea level). At lower altitudes, these species are replaced over time by a variety of others, such as *Castanopsis indica*, *Shorea robusta*, *Engelhardia spicata*, *Actinodaphne angustifolia*, *Myristica latifolia*, *Cinnamomum* sp., *Mesua ferrea* and *Miliusa roxburghiana*, which form the late successional forests. The species composition of these newly constituted forests may not be quite the same as that of the original, pre-clearance ('climax') forest, and they are hence called secondary forests.

Many of the accessible forested areas in the north-eastern hill region are predominantly secondary forest (Ramakrishnan, 1992a; Ramakrishnan and Kushwaha, 2001). In more remote areas, where climax forests are still to be found and the population pressure is not high, secondary forest regeneration is fairly rapid in the sparsely distributed small gaps generated by jhum, and this leads to infilling of the climax forest landscape. As we have seen above, secondary forest development occurs only if the jhum cycle exceeds 30 years, while cycles of 10–30 years lead largely to the formation of bamboo forest. However, with much of the area now under a short agricultural cycle averaging 4–5 years, the land area developing into secondary forests at any given point of time is very limited indeed (Ramakrishnan, 1992a). Furthermore, the tropical rainforest ecosystem is very fragile owing to a highly leached soil on which forest has developed over a long period of time, and it often functions through tight nutrient cycling by a surface root mat. Consequently, in extreme cases, large-scale deforestation has led to site desertification and, ultimately, a completely denuded landscape.

Secondary succession, weed potential and crop productivity

The pattern of secondary succession to a forest community – once cropping ceases and the land is left fallow – depends upon the degree of destruction of the original vegetation and the nature of the bank of reproductive propagules in the soil. The weeds of cultivated sites predominate at the early stage, but there is considerable variation in community composition depending on the length of the jhum cycle, the intensity of weeding, and the composition and size of the bank of reproductive propagules.

If a short jhum cycle is imposed continuously at a site, then succession does not progress beyond the weed stage, so there is a build-up of the propagules of these species in the soil (Saxena and Ramakrishnan, 1984a). Moreover, the plant species that dominate one fallow tend to dominate the fallow that follows the next cropping phase, and thus weed problems tend to escalate. In the humid tropical low elevations of Meghalaya in north-east India, the dominant early successional plant species under short jhum cycles can be either the exotic C_3 species *C. odorata* and *M. micrantha*, or the C_4 natives *I. cylindrica* and *S. spontaneum*. As noted above, if the cycle is longer, the vegetation changes rapidly from weed colonizers to bamboo species, and other shade tolerant/intolerant tree species come into the succession (Ramakrishnan, 1992a).

The known biological traits that facilitate the frequent dominance of *M. micrantha* in short-cycle jhum systems are reviewed in Chapter 2, this volume. However, the presence, and indeed the dominance, of exotic weeds like *M. micrantha* are often accentuated by burning (Swamy and Ramakrishnan, 1987a, 1988a), especially under short cycles. This arises from the ability of *M. micrantha* to take up nutrients efficiently when they are abundant as a result of burning, and also from its reproductive strategy (Box 7.1).

In the first few years of a fallow, exotic species, such as *M. micrantha* and *C. odorata*, contribute a major proportion of the total biomass because of their dominance, and they play a significant role in reducing the losses of soil and nutrients in runoff (Toky and Ramakrishnan, 1981a). The exotics exhibit higher nutrient uptake and so they accumulate larger quantities of nutrients in their tissues than native pioneer species (Saxena and Ramakrishnan, 1983a). Under shortened jhum cycles, nutrient losses may be accelerated in the absence of exotic species. For example, potassium was accumulated rapidly by vegetation for the first 4 years of a fallow in which *M. micrantha* was present (Swamy and Ramakrishnan, 1987a,b). Indeed, what is achieved by *M. micrantha* under jhum fallows less than 5 years old mirrors the role of the bamboo species *D. hamiltonii* in conserving potassium

Box 7.1. Mikania and fire.

Some exotic weeds, such as *Mikania micrantha*, often closely follow fire events, and their presence and indeed their dominance in the community are often related to these events (Swamy and Ramakrishnan, 1987a, 1988a). The ruderal nature of this weed is evident from a number of factors:

- Seedling recruitment occurs exclusively in burnt sites in 2- and 4-year-old fallows. The total absence of seedling recruitment in unburnt plots of 2- and 4-year-old fallows may be due, in part, to rapid vegetative regeneration, which would reduce the available space and light penetration at ground level.
- The higher density of *M. micrantha* in burnt plots than in unburnt plots of the same fallow can partly be related to increased insolation and enriched soil through ash deposits (Ramakrishnan and Toky, 1981).
- In burnt plots, *M. micrantha* may be released from the allelopathic effects of another exotic species, *Chromolaena odorata* (Yadav and Tripathi, 1981), but competition may not be a factor in the increase of *M. micrantha* density in such plots, as the densities of most associated species are either unaffected by fire or increased after the burn, as in an 8-year-old fallow (Ramakrishnan, 1992a).
- The percentage of surviving rosette roots of *M. micrantha* and their turnover rates increase with fallow age in burnt plots, while they decline with fallow age in unburnt plots. Recruitment through ramets is also influenced by burning. Such a fire-related seedling recruitment strategy is in spite of the profuse seed production of *M. micrantha* (Swamy and Ramakrishnan, 1987a, 1988a).
- In an 8-year-old fallow, seedling recruitment occurs even in unburnt plots, largely because of the vertical growth made by the already existing but fewer *M. micrantha* individuals that do not interfere with light penetration to the ground level. However, rosette formation occurs only in burnt plots (Swamy and Ramakrishnan, 1987b).
- The response to fire by *M. micrantha* is most striking with respect to nutrient uptake and use (Swamy and Ramakrishnan, 1987a,b). In burnt sites, the efficiency of nutrient uptake is generally higher than in unburnt plots, and this difference increases with fallow age. Soil fertility status always increases after the burn, and to a greater extent in older fallows because of the higher fuel loads there. Consequently, nutrient use efficiency in burnt plots is generally lower than in unburnt plots.

In summary, fire is an important process at various stages in the life cycle of *M. micrantha*, and the species itself is closely adapted to a ruderal environment subject to frequent perturbation (Swamy and Ramakrishnan, 1987a,b, 1988a,b).

in 10- to 30-year-old fallows (Toky and Ramakrishnan, 1982, 1983b). None the less, the beneficial role of *M. micrantha* in conserving nutrients within the jhum system is largely neutralized by its adverse impact upon the system as a whole, which arises from its extreme weediness and the problem that this causes farmers in short jhum cycles (Swamy and Ramakrishnan, 1987c).

Crop productivity is directly linked to the length of the jhum cycle, and is drastically reduced under short cycles of 4–5 years where the weed intensity is very high (Toky and Ramakrishnan, 1981b). A team led by Ramakrishnan (1992a) was able to establish a direct correlation between weed intensity and crop yield, though individual weed species were not considered separately. Under shortened jhum cycles, the land is cleared before the soil fertility has been restored (a particularly acute problem in leached tropical rainforest soils), which also has a detrimental impact on crop productivity and favours weed species.

The net consequences of the arrested succession of the short jhum cycle are thus reduced stability and resilience in the agricultural and natural ecosystems, adverse impacts on both natural and human-managed ecosystems, increased (both intensive and extensive) biological invasions and a resulting biodiversity decline, with a long-term loss in productivity of the system because of rapid soil fertility decline, which can even lead to site desertification (Ramakrishnan and Toky, 1981; Ramakrishnan, 1992a). It is in this context that finding ways to manage invasive weeds such as *M. micrantha* becomes important. In particular, there are clear lessons we could derive here from the point of view of restoring a degraded landscape dominated by native and exotic weedy species, and in the process better managing, if not controlling, invasives like *M. micrantha*. Stated simply, a formal knowledge-based understanding of species (fast-growing and light-demanding early-successional versus slow-growing and shade-tolerant late-successional tree species) could be a basis for controlling exotics through an assisted succession (Ramakrishnan, 1986).

What all of this implies is that while traditional jhum cycles were in harmony with the environment, jhum has fallen into disrepute because the rapid shortening of the cycle to an average of around 5 years does not allow the natural process of succession to adequately restore the ecosystem (Toky and Ramakrishnan, 1981b; Ramakrishnan, 1992a). In this context, it is worth noting that there is a general tendency among policy planners and developmental agencies to blame the poor farmer for the shortening of the jhum cycle that arises from increasing local population pressure. In reality, there is a general realization from many studies across the globe that there is only an indirect linkage between population and land degradation as a proximal driver, and that policies, governmental incentives and market forces are the real culprits in land degradation, including biological invasion. In the case of the north-eastern hills of India, for example, the shortened jhum cycles arise in part from large-scale timber extraction on to which they are superimposed. However, whatever the cause of the shorter cycles, the net result is an arrested succession of weeds, often dominated at lower elevations by invasive weeds like *C. odorata* and *M. micrantha*.

Obstacles to Controlling *M. micrantha*

Once established, *M. micrantha* is difficult to control under the intensive land-use practices now prevailing in the region of north-east India (Ramakrishnan, 1992a). Several methods to curb the spread and reduce abundance of invasive plants, including *M. micrantha*, have been researched and trialled on a local basis in India. Unfortunately, none of these has resulted in any substantial impact. Herbicides have been tried from time to time, but their use is frequently costly, and they are only effective in controlling the invasives temporarily because the cleared land is quickly reinfested via vegetative shoots and seeds; they are also inappropriate for wide-scale use because of environmental concerns. None the less, herbicides may be appropriate on a spot-application basis and in combination with other techniques. Another approach is to try to clear invasives from new plantation areas and gardens by manual means and burning, but the plants quickly reinvade because of the large seed banks present in the soil.

Biological weed control involving native and exotic insect agents has been attempted against some of the invaders since the early 20th century. Early attempts to control *M. micrantha* were made using insect agents, but the failure of the first agent, *Liothrips mikaniae*, to establish (Cock, 1982; Cock *et al.*, 2000) discouraged further investment in this approach. More recent efforts have looked at the use of a broad range of pathogens; in particular, a rust pathogen, *Puccinia spegazzinii*, was identified as a possible effective candidate for control of the weed, as it had been shown to have high potential in experimental field trials (Ellison *et al.*, 2008). The rust pathogen has since been released with effective impacts on the plant in the Pacific region, and this work is reviewed in Chapter 10, this volume.

Nevertheless, it is clear that invasive alien plant problems are complex to deal with and some of the other approaches to control have been researched in India are described in the rest of this chapter. In most instances, the plants are invasive across a range of habitats – various forest types, plantations and agricultural land. Exotics have a number of characteristics that allow them to succeed in habitats that are subject to perturbation, and to ensure coexistence with native species (see Chapter 2, this volume). However, the exotic species that have been able to get a firm foothold in an alien environment often tend to be more aggressive than the native species, and so they often tend to have an edge over them (Ramakrishnan, 1991). It is these factors, together with a general failure to involve communities in tackling the problems, which explain why previous efforts that were focused on a single-component solution for control have either achieved only partial success at a local scale or failed altogether. What is required is a socioecological systems approach towards understanding the issues linked with biological invasion, and to bring in a mix of the knowledge systems appropriately designed, to generate 'hybrid' technologies for landscape management. This is the context in which the integration of measures derived through formal knowledge-based technologies and those derived through the traditional ecological knowledge (TEK) that exists within local communities will prove to be of great value in achieving a sustainable solution.

Towards Developing Sustainable Solutions for Managing *M. micrantha*

When dealing with traditional societies that are largely confined to the mountains of the developing tropics, integration in the real sense is best achieved through a knowledge system to which these communities are able to relate. This is where community-based TEK, derived from an experiential process of societal interaction with nature and natural resources, becomes important for addressing ecosystem sustainability concerns through community participation. Given this provenance, it follows that TEK has a certain degree of location specificity. In our case (north-east India), this applies to an area with over 100 different ethnic societies, each with its own language, culture and customs, and with a strong human element attached to these that emphasizes social emancipation (Elzinga, 1996). The challenge from a policy perspective has always been to arrive at generalizations on TEK that are applicable across socioecological systems, so that regional developmental concerns can be addressed, while contending with the socioecological diversities that exist within the region. Under these circumstances, knowledge systems form the connecting link between ecological and social processes, as shown in Fig. 7.1.

Traditional knowledge: intangible benefits with tangible outcomes

As they are traditionally animistic in their religious beliefs, jhum farmers view nature

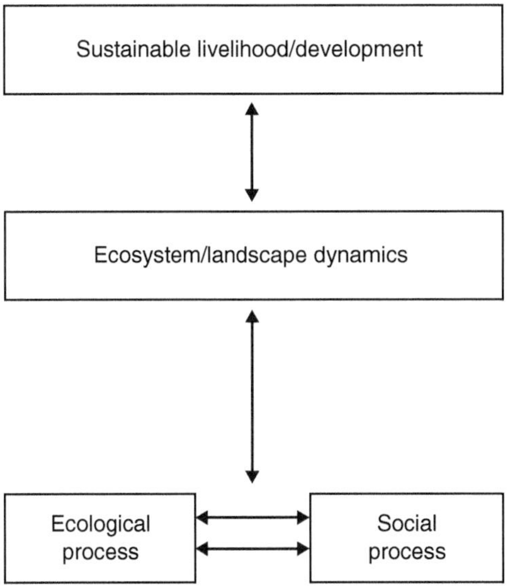

Fig. 7.1. Knowledge systems are the connecting link between ecological and social processes, and have implications for the sustainable natural resource management and livelihood of local communities.

and the natural resources around them with respect and reverence. The intangible benefits that communities perceive extend to all scalar dimensions, ranging from subspecies and species through to ecosystem and landscape levels (Ramakrishnan *et al.*, 1998). These protective impulses cannot often be articulated by traditional societies, but are ingrained in their psyche, and are often linked with TEK, which has been refined and enriched over a long period of time. There are three basic kinds of TEK (Ramakrishnan *et al.*, 1998):

1. Economic TEK: this concerns traditional crop varieties and lesser known plants and animals of food and medicinal value, as well as other wild resources.
2. Ecological/social TEK: this involves the manipulation of biodiversity for coping with environmental uncertainties, hydrological control and soil fertility management.
3. Ethical TEK: this comprises unquantifiable values centred on cultural, spiritual and religious systems, and operates on three scales – sacred species, sacred groves (habitats) and sacred landscapes.

As TEK links ecological and social processes, this discussion focuses on its relevance in soil fertility and water management issues for restoring natural capital with community participation.

TEK in fallow management

TEK can only be deciphered on the basis of an intense participatory mode of research. It operates at a process level, linking the ecological with the social at various scalar dimensions. The challenge before all of us who are concerned with conservation lies in deciphering this knowledge, and validating and integrating it into the modern scientific paradigm. Local communities should be enabled to relate to a value system that they understand and appreciate, and thereby participate in the process of redeveloping and conserving natural resources (Ramakrishnan, 2003).

Arising from these traditional intangible/tangible interconnections between value systems, many species are socially/culturally valued, and we have shown that these species often have ecological keystone value within the ecosystem (Ramakrishnan, 1992a, 2001). In the jhum system itself, early successional species such as Nepalese alder (*Alnus nepalensis*) and many bamboo species come into this category. *Alnus* is one of 15 genera of non-leguminous trees that fix nitrogen, and Nepalese alder can conserve as much as 125 kg N/ha annually in a system (Ramakrishnan 1992a; Rathore *et al.*, 2010). Similarly, many socially valued bamboo species can conserve key elements such as nitrogen, phosphorus or potassium, depending upon the species. Oaks (*Quercus* spp.) are socially valued species right across the Himalayan region, and elsewhere too.

More extensive studies have shown that socially valued species in the Himalayan region are ecological keystone species in other parts of the world as well, and often the same cluster of species (e.g. *Ficus* spp.) are identified as such by communities in completely separate parts of the world (Ramakrishnan *et al.*, 1998). Thus, one or more keystone species within the system may not only determine ecosystem processes but also have an impact upon the structural and functional attributes of the system at the landscape level (see Fig. 7.2). Selecting such species for forest fallow management linked to jhum would ensure an accelerated forest successional process with community participation (Ramakrishnan, 1992b), because local people are then able to relate to a value system that they understand and appreciate. They are, therefore, able to participate in the process of sustainable natural resource management, with its implications for the sustainable livelihoods of the local communities concerned.

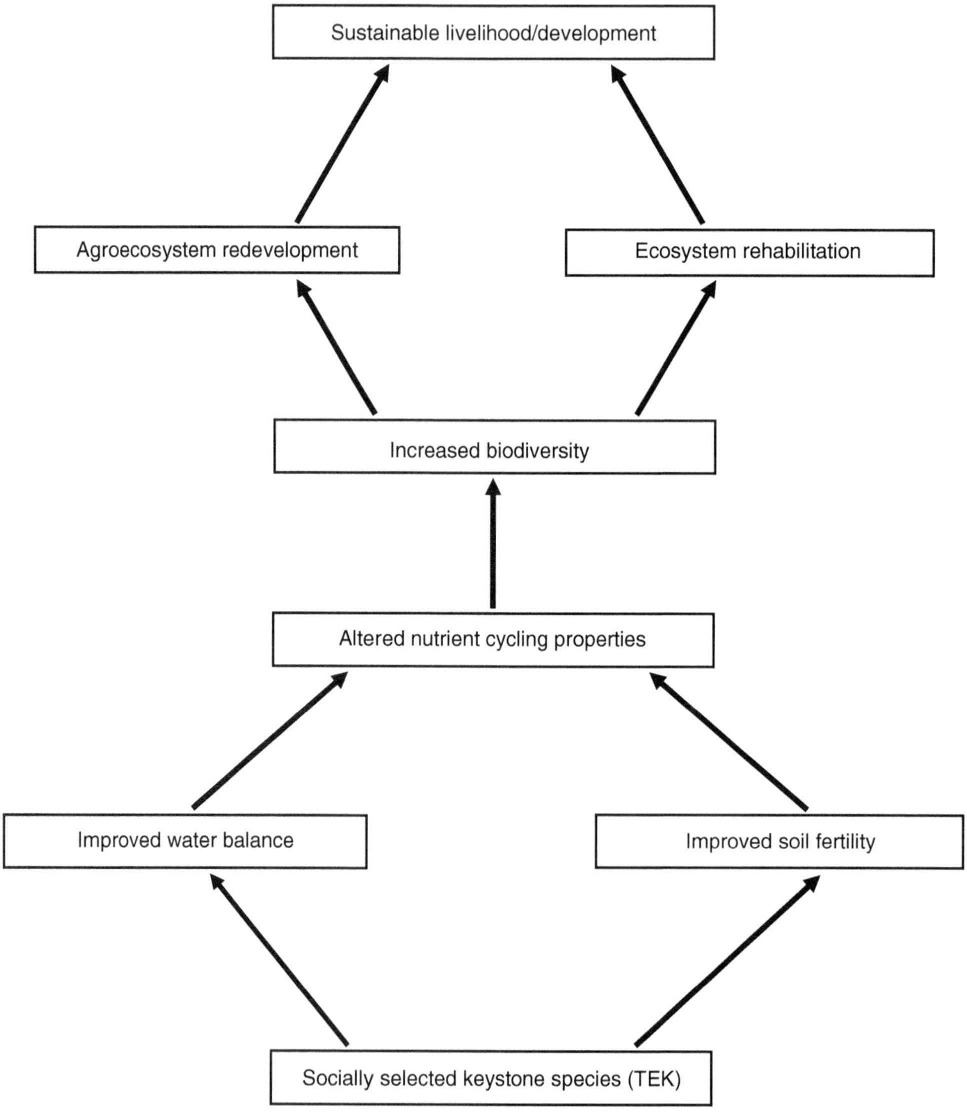

Fig. 7.2. TEK centred around a socially selected keystone species, for example *Quercus* spp., acting as a trigger for rehabilitation of the mountain landscape in the central Himalayas.

Role of formal knowledge in fallow management

We have worked with the existing weakened jhum systems in north-east India with cycle lengths of less than 10 years, and with forest ecosystems at various levels of degradation, and by so doing have gained an understanding of forest successional principles and the functional attributes of species/ecosystems during the successional processes (Ramakrishnan, 1989, 1992a). From these, we have been able to arrive at formal knowledge-based conclusions, at the species, ecosystem and landscape levels, on issues such as population adaptation and the dynamics of change, soil fertility and nutrient cycling dynamics. All of these are relevant to the fallow management practices that are needed to strengthen the weakened jhum systems operating under cycle lengths of less than 10 years (Ramakrishnan, 2001). Natural or assisted regeneration of the forest fallows beyond 10 years would reduce nutrient losses and allow a return to the steady-state cycling characteristics of mature forest (Toky and Ramakrishnan, 1983a,b).

Linked with all of these studies, we have looked at the growth pattern and architectural development of the shoot and root systems of early versus late successional tree species. This has enabled us to identify tree species that have rapid growth and also the right architectural features to be used for fallow management in a redeveloped jhum system, which will at the same time control exotics within the system (Boojh and Ramakrishnan, 1982a,b; Shukla and Ramakrishnan, 1984a,b, 1986). The aim was to look at species that could potentially be used in a condensed succession in a mixed plantation programme for the complementarities of late versus early successional shoot architecture for effective light capture (fast-growing, light-demanding early successionals with tall canopies versus slow-growing, shade-tolerant late successionals at a lower stratum) and root architecture for nutrient capture from the soil profile (shallow-rooted early successional species picking up nutrients from the surface soil versus deep-rooted late-successional species extracting them from lower down the soil profile) (Ramakrishnan, 1986).

At an ecosystem level, we have shown that much of the TEK is centred around sustainable water availability and its use within the system. Indeed, we have found that water is a powerful connecting link between many traditional communities living in the rural environment of the developing tropics (Ramakrishnan, 2001; Ramakrishnan *et al.*, 2005).

Linking knowledge systems towards sustainable fallow management

What do these linkages between formal and traditional knowledge at the species and ecosystem levels imply? They suggest that there is potential for using the two sources of knowledge to form powerful tools for sustainable natural resource management that mesh with the sustainable livelihood concerns of local communities, particularly in the developing tropics (Ramakrishnan, 2001, 2007; Ramakrishnan *et al.*, 2005). In other words, we need to develop 'hybrid' technologies based on linking traditional and formal knowledge systems, and arising from this a socioecological approach towards community participatory fallow management linked to sustainable livelihoods. When these triggers are appropriately linked, there can be a whole range of cascading implications for natural resource management that have an impact upon the livelihoods of local communities (Fig. 7.3). The particular example given here is the role of oak (and bamboo) regeneration in fallow management.

This approach formed the basis for a jhum redevelopment plan through community participatory fallow management, which was put in place by this author through the NEPED (Nagaland Environmental Protection and Economic Development) project – a local governmental initiative in the state of Nagaland in north-eastern India (Box 7.2).

Over a 1000 villages in the state of Nagaland were organized into village

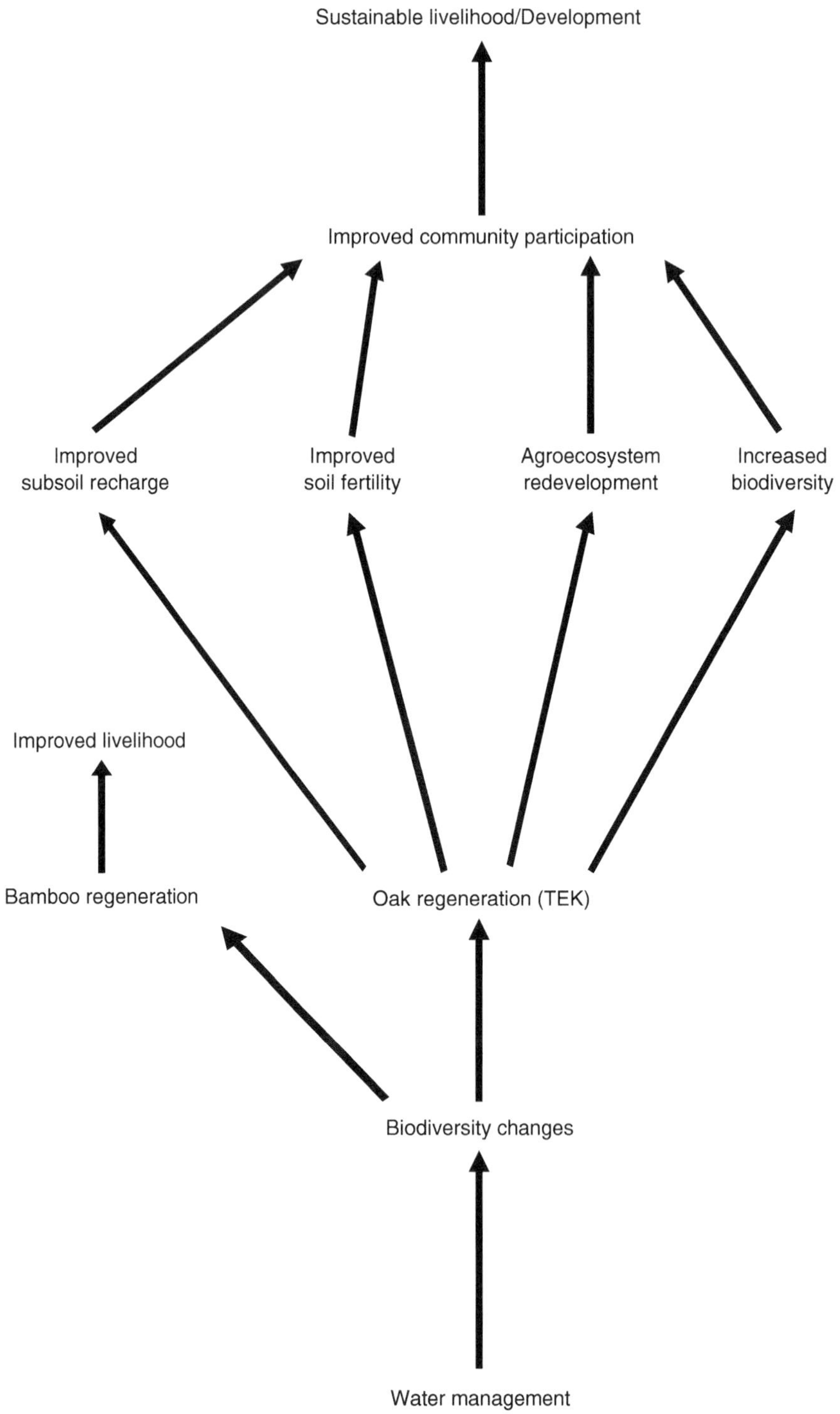

Fig. 7.3. TEK linked with water availability at the landscape level act as drivers for land use/societal development and lead to cascading impacts on ecosystem/landscape functions.

Box 7.2. Building upon jhum in Nagaland.

- The Nagaland jhum redevelopment plan (a project proposal prepared by this author and funded by the India–Canada Environment Facility) was a project of some magnitude, involving all of the villages in the state – about 1200 in all.
- Two key objectives of the plan were: (i) participatory land-use development that improves the quality of life of the people of Nagaland; and (ii) biodiversity conservation, including the control of biological invasions.
- Land-use redevelopment was initiated through participatory extension and dissemination, with gender issues adequately addressed. Village development boards (VDBs), each constituted on the basis of the local value system (see text), were the vehicle for land-use development.
- As part of the community participatory project implementation, about 200 experimental jhum plots in farmers' fields were used to test of farmer-selected tree species for agroforestry technology redevelopment, with a coverage of about 5500 ha of replicated test plots.
- Many farmers, on their own initiative, adopted tree-based strengthened jhum systems based on agroforestry principles for testing in 870 villages, covering a total area of about 33,000 ha (38 ha/village × 870 villages); in these plots, the emphasis was on testing local adaptations and innovations in activities such as soil and water management.
- The incorporation of TEK related to Nepalese alder (*Alnus nepalensis*) during both the cropping and the fallow phases of jhum is widespread throughout the north-eastern region, but has been honed by the Angami tribe of Khonoma Village near Kohima over as much as 500 years. A single tree can fix up to 120 kg nitrogen/ha a year. The significance of this can be seen if we consider that under a 5-year jhum cycle, the system loses as much as 600 kg N/ha each year, of which not more than half gets replaced under a 5-year fallow period of natural succession. With Nepalese alder in the jhum system, all of the 600 kg N lost could be replaced under a 5-year fallow period (Mishra and Ramakrishnan, 1984; Ramakrishnan, 1992a). This was the starting point and the basis for identifying a number of other tree species for a redeveloped jhum system. The system was disseminated to almost all project villages. Farmers living at other altitudes were able to correlate the concept with the trees used there, which included *Schima wallichii* and *Gmelina arborea*.
- Following consultation with local communities, ten tree species used for poles for house construction and fuelwood, which could be harvested 5–10 years after planting, and 20 tree species of value for timber were identified and introduced into jhum plots to strengthen the jhum system. Following the completion of the project, some of the economic benefits to communities are being realized. More than 7.8 million trees were planted in the first 6 years of the project, of which over 4.5 million were estimated to be surviving as the project ended.
- Traditional rainwater harvesting systems using dugout ponds, and erosion control measures using contour trenches and bunds have been incorporated into the redeveloped jhum practices, where appropriate.
- Mixed tree plantations (e.g. *Gmelina* sp. with *Phoebe* sp.; *Gmelina* sp. with *Melia composite* [an ambiguous synonym for *M. azedarach*]) in the jhum plots were shown to be superior to monocultures, and these were recommended.

This account is based on the author's analysis for NEPED (the Nagaland Environmental Protection and Economic Development initiative) and IIRR (the International Institute of Rural Reconstruction) (NEPED/IIRR, 1999), with additional information from a review of alder-based jhum in Nagaland (Cairns *et al.*, 2007) and the project final report (NEPED, 2007).

development boards (VDBs), with the specific purpose of rural development in mind. The process of establishing the VDBs took into consideration traditional institutional arrangements. There are over 35 ethnic societies, each with its own way of constructing village-level traditional institutions, and the individual VDBs were created based on the value system of each one. However, although the VDBs were formed differently depending upon the ethnicity of the community, all of them had the same function, namely, to facilitate decision making for a redeveloped jhum system. Using this institutional mechanism, the highly distorted shifting agricultural system (in essence an agroforestry system now operating at or below subsistence level) was redeveloped by strengthening the tree component that had been weakened through land degradation in the region. Such an incremental pathway for agricultural system development (Swift *et al.*, 1996), as part of the larger landscape, drew on the rich TEK base of these hill societies. The project was implemented by Nagaland government officials through the VDBs created by the Government of Nagaland. It aimed to augment the traditional system of agriculture, rather than attempting to radically change it. In this process the 'formal knowledge'-based technologies were brought in only to the extent required.

Because it was realized that working out the ecological keystone value of the tree species for fallow management is a time-consuming process requiring detailed research analysis, tree selection was done on a participatory basis with local communities, hoping that the species identified by this process would accelerate forest succession. In order to ensure community involvement, it was often found helpful to form village-level institutions following traditional ways of doing this, which varied from one ethnic group to another. Apart from stabilizing the shortened jhum cycle by using tree species for fallow management, the more rapid development of a closed canopy will also control invasive species like *M. micrantha* and other species from the Asteraceae (Eupatoriaceae), and prevent their resurgence during the fallow regrowth phase.

This work aimed to achieve multiple end products:

- A redeveloped jhum system that improves the quality of life for the people through assured food security (Ramakrishnan, 1992a).
- Sustainable fallow management using socially valued early-successional fast-growing tree species with a narrow crown structure (Ramakrishnan, 1986), so ensuring light penetration to ground level for a few years of crop production during the early growth phase of the tree. Then, as the light availability declines, the fallow is left for several more years before tree harvest; slash and burn and cultivation during these years is carried out elsewhere so that each plot goes through another cycle of fallow management before being brought into production again.
- Improved capabilities within local communities for coping with environmental uncertainties arising from 'global change'-related diversification of the production system.
- Assisted successional fallow management in the landscape that provides multipurpose economic benefits to local communities, and at the same time contributes towards carbon sequestration through the introduced tree species. Such an approach to tree fallow management helps to cope with the environmental uncertainties arising from global change, and thus contributes to combating adverse impacts arising from global change in general, and biological invasion in particular.

General Considerations

What we aimed to do over the course of the NEPED project in shifting agriculture-affected Nagaland in north-east India was to use assisted succession to strengthen natural processes of forest succession that had been weakened under shortened jhum cycles. This will help to control weedy species in general and exotics in particular. The fond hope is that the local communities will, over a period of time, not feel the need for 'slash and burn' agriculture, but will move towards a rotational fallow management system, and indeed, an agroforestry system where trees and crops are temporally separated from one another. In this chapter, I have emphasized that understanding 'knowledge systems' is a powerful tool for sustainable natural resource management. Therefore, we need to integrate the traditional forester's way of looking at forestry as a silvicultural activity with a whole variety of other dimensions (as shown in Fig. 7.4). Ecological knowledge from areas such as tree biology and architecture, the ecophysiology of developing forest communities, reproductive biology, ecosystem and landscape level complexities, and nutrient-cycling processes could all be integrated into the current management process and future management options. To ensure community participation, we need to bring in the social and cultural dimensions through an in-depth understanding of the tangible and intangible values of the forest ecosystem through the TEK base available in the given situation, appropriately linking this with the formal knowledge-based understanding of the forest ecosystem function.

Such community-based integrated landscape management approaches can best be complemented with technologies such as biological control through the introduction of exotic agents that can provide a long-term sustainable component in an overall management plan. In the ultimate analysis, a multipronged socioecological systems approach to control weeds through an integrated landscape management plan (Ramakrishnan, 2001) alone can lead to lasting results in controlling exotics like *M. micrantha* and the many other invasive species that will follow it.

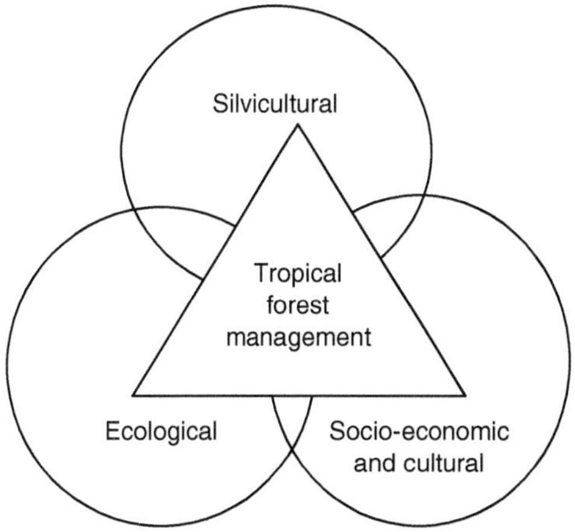

Fig. 7.4. Interdisciplinary interactions called for in tropical forest management and conservation (with particular reference to land affected by shifting agriculture). (From Ramakrishnan, 1992b)

References

Boojh, R. and Ramakrishnan, P.S. (1982a) Growth strategy of trees related to successional status. I. Architecture and extension growth. *Forest Ecology and Management* 4, 355–374.

Boojh, R. and Ramakrishnan, P.S. (1982b) Growth strategy of trees related to successional status. II. Leaf dynamics. *Forest Ecology and Management* 4, 375–386.

Cairns, M., Keitzar, S. and Yaden, T.A. (2007) Shifting forests in northeast India. Management of *Alnus nepalensis* as an improved fallow in Nagaland. In: Cairns, M. (ed.) *Voices from the Forest: Integrating Indigenous Knowledge into Sustainable Upland Farming.* Resources for the Future, Washington, DC, pp. 341–378.

Cock, M.J.W. (1982) Potential biological control agents for *Mikania micrantha* H.B.K. from the Neotropical region. *Tropical Pest Management* 28, 242–254.

Cock, M.J.W., Ellison, C.A., Evans, H.C. and Ooi, P.A.C. (2000) Can failure be turned into success for biological control of mile-a-minute weed (*Mikania micrantha*)? In: Spencer, N.R. (ed.) *Proceedings of the X International Symposium on Biological Control of Weeds, Bozeman, Montana, 4–14 July 1999.* USDA Forest Service, Forest Health Technology Enterprise Team, Morgantown, West Virginia, pp. 155–167.

Ellison C.A., Evans, H.C., Djeddour, D.H. and Thomas, S.E. (2008) Biology and host range of the rust fungus *Puccinia spegazzinii*: a new classical biological control agent for the invasive alien weed *Mikania micrantha* in Asia. *Biological Control* 45, 133–145.

Elzinga, A. (1996) Some reflections on post-normal science. In: Rolen, M. (ed.) *Culture, Perceptions and Environmental Problems: Interscientific Communication on Environmental Issues.* Swedish Council for Planning and Coordination of Research, Stockholm, pp. 32–46.

FAO (1995) *Forest Resources Assessment 1990 – Global Synthesis.* FAO Research Paper No. 124, Food and Agriculture Organization of the United Nations, Rome.

FAO (2010) *Global Forest Resources Assessment 2010. Main Report.* FAO Forestry Paper No. 163, Food and Agriculture Organization of the United Nations, Rome.

FAO/UNEP (1982) *Tropical Forest Resources.* Forestry Paper No. 50, Food and Agriculture Organization of the United Nations, Rome.

Forest Survey of India (2013) *India State of Forest Report 2013.* Forest Survey of India, Dehradun, India.

Mishra, B.K. and Ramakrishnan, P.S. (1984) Nitrogen budget under rotational bush fallow agriculture (jhum) at higher elevations of Meghalaya in north-eastern India. *Plant and Soil* 81, 37–46.

NEPED (2007) *Strengthening Natural Resource Management and Farmers' Livelihoods in Nagaland. Final Technical Report, 2001 to 2007.* Nagaland Empowerment of People through Economic Development, Kohima, India.

NEPED/IIRR (1999) *Building upon Traditional Agriculture in Nagaland.* Nagaland Environmental Protection and Economic Development, Kohima, India and International Institute of Rural Reconstruction, Silang, Philippines.

Poffenberger, M. (2006) *Forest Sector Review of Northeast India.* Background Paper No. 12, Community Forest International, Santa Barbara, California.

Ramakrishnan, P.S. (1986) Morphometric analysis of growth and architecture of tropical trees. In: *L'arbre: Compte-rendu du Colloque International l'Arbre, Montpellier, France, 9–14 Septembre 1985.* Naturalia Monspeliensia, Institut de Botanique, Montpellier, France, pp. 209–222.

Ramakrishnan, P.S. (1989) Nutrient cycling in forest fallows in north-eastern India. In: Proctor, J. (ed.) *Mineral Nutrients in Tropical Forest and Savanna Ecosystems.* Special Publication No. 9, British Ecological Society. Blackwell, Oxford, UK, pp. 337–352.

Ramakrishnan, P.S. (ed.) (1991) *Ecology of Biological Invasion in the Tropics. Proceedings of an International Workshop, Nainital, India, September 1989.* National Institute of Ecology/International Science Publications, New Delhi.

Ramakrishnan, P.S. (1992a) *Shifting Agriculture and Sustainable Development: An Interdisciplinary Study from North-Eastern India.* Man and Biosphere Book Series No. 10, UNESCO (United Nations Educational, Scientific and Cultural Organization), Paris and Parthenon Publishing, Carnforth, UK. (Republished in 1993 by Oxford University Press, New Delhi.)

Ramakrishnan, P.S. (1992b) Tropical forests: exploitation, conservation and management. *Impact of Science on Society* 42, 149–162.

Ramakrishnan, P.S. (2001) *Ecology and Sustainable Development.* National Book Trust, New Delhi.

Ramakrishnan, P.S. (2003) Conserving the sacred: the protective impulse and the origins of modern protected areas. In: Harmon, D. and Putney, A.D. (eds) *The Full Value of Parks: From Economics to the Intangible.* Rowman and Littlefield, Lanham, Maryland, pp. 26–41.

Ramakrishnan, P.S. (2007) Sustainable mountain development: the Himalayan tragedy. *Current Science* 92, 308–316.

Ramakrishnan, P.S. and Kushwaha, S.P.S. (2001) Secondary forests of the Himalaya with emphasis on the north-eastern hill region of India. *Journal of Tropical Forest Science* 13, 727–747.

Ramakrishnan, P.S. and Toky, O.P. (1981) Soil nutrient status of hill agro-ecosystems and recovery pattern after slash and burn agriculture (jhum) in north-eastern India. *Plant and Soil* 60, 41–64.

Ramakrishnan, P.S., Saxena, K.G. and Chandrashekara, U.M. (eds) (1998) *Conserving the Sacred for Biodiversity Management.* UNESCO (United Nations Educational, Scientific and Cultural Organization) and Oxford & IBH Publishing, New Delhi.

Ramakrishnan, P.S., Boojh, R., Saxena, K.G., Chandrashekara, U.M., Depommier, D., Patnaik, S., Toky, O.P., Gangwar, A.K. and Gangwar, R. (2005) *One Sun, Two Worlds: An Ecological Journey.* UNESCO (United Nations Educational, Scientific and Cultural Organization) and Oxford & IBH Publishing, New Delhi.

Rao, K.S. and Ramakrishnan, P.S. (1989) Role of bamboos in nutrient conservation during secondary succession following slash and burn agriculture (jhum) in north-eastern India. *Journal of Applied Ecology* 26, 625–633.

Rathore, S.S., Karunakaran, K. and Prakash, B. (2010) Alder based farming system – a traditional farming practice in Nagaland for amelioration of jhum land. *Indian Journal of Traditional Knowledge* 9, 677–680.

Saxena, K.G. and Ramakrishnan, P.S. (1982) Reproductive efficiency of secondary successional herbaceous populations subsequent to slash and burn of subtropical humid forests in north-eastern India. *Proceedings of the Indian Academy of Sciences (Plant Sciences)* 91, 61–68.

Saxena, K.G. and Ramakrishnan, P.S. (1983a) Growth and allocation strategies of some perennial weeds of slash and burn agriculture (jhum) in north-eastern India. *Canadian Journal of Botany* 61, 1300–1306.

Saxena, K.G. and Ramakrishnan, P.S. (1983b) Growth resource allocation pattern and nutritional status of some dominant annual weeds of slash and burn agriculture (jhum) in north-eastern India. *Acta Oecologica: Oecologia Plantarum* 4, 323–333.

Saxena, K.G. and Ramakrishnan, P.S. (1984a) Herbaceous vegetation development and weed potential in slash and burn agriculture (jhum) in north-eastern India. *Weed Research* 24, 135–142.

Saxena, K.G. and Ramakrishnan, P.S. (1984b) C_3/C_4 species distribution among successional herbs following slash and burn in north-eastern

India. *Acta Oecologica: Oecologia Plantarum* 5, 335–346.

Shukla, R.P. and Ramakrishnan, P.S. (1984a) Biomass allocation strategies and productivity of tropical trees related to successional status. *Forest Ecology and Management* 9, 315–324.

Shukla, R.P. and Ramakrishnan, P.S. (1984b) Leaf dynamics of tropical trees related to successional status. *New Phytologist* 97, 697–706.

Shukla, R.P. and Ramakrishnan, P.S. (1986) Architecture and growth strategies of tropical trees in relation to successional status. *Journal of Ecology* 74, 33–46.

Swamy, P.S. and Ramakrishnan, P.S. (1987a) Weed potential of *Mikania micrantha* H.B.K. and its control in fallows after shifting agriculture (jhum) in north-eastern India. *Agriculture, Ecosystems and Environment* 18, 195–204.

Swamy, P.S. and Ramakrishnan, P.S. (1987b) Effect of fire on population dynamics of *Mikania micrantha* H.B.K. during early succession after slash and burn agriculture (jhum) in north-eastern India. *Weed Research* 27, 397–404.

Swamy, P.S. and Ramakrishnan, P.S. (1987c) Contribution of *Mikania micrantha* during secondary succession following slash-and-burn agriculture (jhum) in north-east India I. Biomass, litterfall and productivity. *Forest Ecology and Management* 22, 229–237.

Swamy, P.S. and Ramakrishnan, P.S. (1988a) Effect of fire on growth and allocation strategies of *Mikania micrantha* H.B.K. under early successional environments. *Journal of Applied Ecology* 25, 653–658.

Swamy, P.S. and Ramakrishnan, P.S. (1988b) Growth and allocation patterns of *Mikania micrantha* H.B.K. in successional environments after slash and burn agriculture. *Canadian Journal of Botany* 66, 1465–1469.

Swift, M.J., Vandermeer, J., Ramakrishnan, P.S., Ong, C.K., Anderson, J.M. and Hawkins, B. (1996) Biodiversity and agroecosystem function. In: Mooney, H.A., Cushman, J.H., Medina, E., Sala, O.E. and Schulz, E.-D. (eds) *Functional Roles of Biodiversity: A Global Perspective. SCOPE 55.* Wiley, New York, pp. 261–298.

Toky, O.P. and Ramakrishnan, P.S. (1981a) Run-off and infiltration losses related to shifting agriculture (jhum) in north-eastern India. *Environmental Conservation* 8, 313–321.

Toky, O.P. and Ramakrishnan, P.S. (1981b) Cropping and yields in agricultural systems of the north-eastern hill region of India. *Agro-Ecosystems* 7, 11–25.

Toky, O.P. and Ramakrishnan, P.S. (1982) Role of bamboo (*Dendrocalamus hamiltonii* Nees and Arn.) in conservation of potassium during slash and burn agriculture (jhum) in north-eastern India. *Journal of Tree Science* 1, 17–26.

Toky, O.P. and Ramakrishnan, P.S. (1983a) Secondary succession following slash and burn agriculture in north-eastern India. I. Biomass, litterfall and productivity. *Journal of Ecology* 71, 737–745.

Toky, O.P. and Ramakrishnan, P.S. (1983b) Secondary succession following slash and burn agriculture in north-eastern India. II. Nutrient cycling. *Journal of Ecology* 71, 747–57.

Yadav, A.W. and Tripathi, R.S. (1981) Population dynamics of the ruderal weed *Eupatorium odoratum* L. and its regulation in nature. *Oikos* 23, 355–361.

8 Prevention and Related Measures for Invasive Alien Plants in India: Policy Framework and Other Initiatives

Ravi Khetarpal,[1]* Kavita Gupta,[2] Usha Dev[2] and Kavya Dashora[3]

[1]CABI South Asia, New Delhi, India; [2]Division of Plant Quarantine, National Bureau of Plant Genetic Resources, New Delhi, India; [3]Indian Institute of Technology, Hauz Khas, New Delhi-110016, India

Introduction

Invasive alien species (IAS) are pathogens, plants and animals that are introduced into new areas where they are not part of the native flora and fauna, and because they are free of the natural enemies and/or competition found in their areas of origin, they spread or reproduce prolifically. Globally, they are a serious impediment to agricultural production and the conservation and sustainable use of biodiversity and other natural resources, often with significant irreversible impacts. Their costs include not only the direct costs of the agricultural and/or biodiversity loss they cause and the management needed to reduce their impacts, but also the indirect costs due to their impacts on ecological services, such as pollination, climate regulation and water purification. On a global scale, some of the most pernicious IAS are invasive alien plants, and many of these species are still spreading between and within countries (CABI, 2016); here, this situation is exemplified by India.

India's people and economy are heavily dependent on agriculture and India's vast natural resources. The country ranks among the world's megadiverse countries. India is one the world's centres of crop diversity, and holds approximately 12% of the world flora, which includes 5725 endemic species of higher plants (Sharma and Brahmi, 2011). With increasing trade and movement of people, as well as agriculture (crops, livestock, forest products), India's natural biodiversity is under continuous threat from accidental or intentional introduction of invasive alien plants and other IAS. Although IAS have been identified as a major factor in natural resource management (Diwakar, 2003), reliable estimates of India's invasive flora are lacking (Khuroo *et al.*, 2012) and the impact of invasive plants on community structure and ecosystem processes is poorly understood (Mandal, 2011). Previous estimates of invasive alien plant numbers have not been based on reliable empirical data, which has serious implications for the management of plant invasions (Khuroo *et al.*, 2012). Two recent studies estimated that: (i) 173 species in 117 genera are invasive alien plants in India, which represents 1% of the vascular flora (Mandal, 2011); and (ii) 225 species (14%) of alien plant species in India are invasive, with another 134 (8%) having the potential to become invasive (Khuroo *et al.*, 2012). Even though these two studies do not

* Corresponding author. E-mail: r.khetarpal@cabi.org

agree completely, they do report figures of the same order of magnitude.

Apart from numbers, the impact of individual invasive alien plant species can be dramatic. Species frequently monopolize the environment and cause health hazards, reduce crop yields, increase labour costs and prevent the re-establishment of native species such as fodder grasses, shrubs and forest trees after land disturbance. In north-eastern India and the Western Ghats, invasive plants, including *Chromolaena odorata*, *Lantana camara* and *Mikania micrantha* infest extensive tracts of agricultural and forest land, displacing native flora and animals and even human activities (on this last point see, for example, Ramakrishnan, 2001). Interference with water flow and availability following invasion by waterweeds such as water hyacinth (*Eichhornia crassipes*) has led to decline in wetlands and associated wildlife. Parthenium weed (*Parthenium hysterophorus*) dominates the vegetation in town and city wastelands and is a potential allergen threatening human health.

In India, as in other countries, the development of appropriate national institutional and policy frameworks to address the increasing tide of invasive plants and other IAS has been complicated because of the traditional divide in responsibilities for the main two sectors at risk: agriculture and the environment (Griffin, 2003). In addition, there has been no commensurate increase in human resources and infrastructure facilities to handle the increased movement of IAS with trade (Sathyanarayana and Sathyagopal, 2013). The status of management of IAS in general in India has been reviewed by Mandal (2011) and Reshi and Khuroo (2012), while the review of IAS in India by Rana *et al.* (2004) includes an account of regulatory mechanisms. Efforts are also continuing to improve the export certification process and standards (Shah, 2008).

In this chapter, the focus is on the history and current status of the regulatory mechanisms and other initiatives in India that relate to preventive and related measures for invasive alien plants; the emphasis is on prevention because this is the most cost-effective means of reducing the threats of these IAS (Wittenberg and Cock, 2001). Naturally, much of this is embedded in or stems from the country's general regulatory framework development and actions. Lastly, the chapter outlines the need for a national strategy on IAS. However, as the national regulatory framework has, by necessity, needed to harmonize with some binding international agreements and conventions that contain provisions on regulation, a brief overview of these is given first.

International Instruments Relating to the Prevention of IAS

There are many binding international instruments that include provisions for preventing the movement (intentional or accidental) of IAS, and these naturally include invasive alien plants; a complete review is provided by Shine *et al.* (2000). Three of the more important instruments to which India is a signatory are outlined here.

For trade, the Agreement on the Application of Sanitary and Phytosanitary Measures (SPS Agreement) is an international treaty that came into force in early 1995 under the World Trade Organization (WTO) and relates to prevention of the establishment or spread of alien pests (including invasive plants), diseases or disease-causing organisms of plants and animals into new geographical areas (WTO, 1995). The SPS Agreement aims to overcome impediments to market access that may result from pests and diseases by encouraging the 'establishment, recognition and application of common sanitary and phytosanitary measures by different Members'. The primary incentive for the use of such common international norms is that they provide the necessary protection based on scientific evidence, and at the same time improve trade flow.

The SPS Agreement recognizes the International Plant Protection Convention (IPPC; https://www.ippc.int/en/) as the organization mandated to develop International Standards for Phytosanitary Measures

(ISPMs), and the IPPC is the most relevant standard-setting body in the context of invasive alien plants. The standards that are relevant in the current context are listed in Table 8.1.

One of the major requirements of the SPS Agreement is for signatories to undertake pest risk analysis (PRA) for all the agricultural commodities traded (Khetarpal and Gupta, 2002). The entire procedure is a blend of inductive, deductive and expert opinion that facilitates the identification and characterization of risk. The guidelines for PRA as per ISPM 2 and ISPM 11, which elaborate the general requirements for a PRA, did not originally cover environmental impacts. However, ISPM 11 was revised in 2003 and 2004 to include the analysis of environmental risks and of living modified organisms (or LMOs; the term used in texts on the Convention on Biological Diversity, which equates, for most purposes, including ours, to the more generally used 'genetically modified organisms', or GMOs), thus widening the scope of the IPPC to include specific guidelines for assessing environmental risks due to the introduction of pests. In 2013, ISPM 11 was revised further to include PRA for plants as quarantine pests. The range of pests covered by the IPPC therefore extends beyond pests directly affecting cultivated plants and includes invasive plants and other species having indirect effects on cultivated plants and wild flora. The indirect effects include competition in natural or semi-natural environments and deleterious effects on other plant species/health in habitats/ecosystems. The inclusion of the analysis of environmental risks in a PRA ensures the protection of the environment, ecosystem and wild flora (Gupta and Khetarpal, 2006).

From an environmental perspective, the Convention on Biological Diversity (CBD), which came into force in 1993, also has provisions on IAS. These are contained in Article 8(h) which states that 'Each contracting Party shall, as far as possible and as appropriate, prevent the introduction of, control or eradicate those alien species which threaten ecosystems, habitats or species' (Secretariat of the CBD, 2005). At the sixth meeting of the Conference of the Parties to the CBD (COP 6) the 'Guiding Principles' for the implementation of Article 8(h) were adopted (Decision V1/23) (CBD, 2002), but the Parties also recognized the contribution of existing international instruments covering standards (such as those of the IPPC) that are relevant to these Guiding Principles. In 2014, the IPPC became a member of the Liaison Group of the Biodiversity-related Conventions, which has facilitated a closer working cooperation between the CBD and the IPPC.

At the ninth meeting of the Conference of Parties to the CBD (COP 9), the Parties called for the consideration of IAS when developing policy frameworks for sustainable production and the use of biofuels (Decision IX/2), forest biodiversity (IX/5) and island biodiversity (IX/21), and set out detailed recommendations for improving IAS information systems through the Global Taxonomy Initiative (IX/22) (CBD, 2008).

Both the CBD and the IPPC broadly call for risk analysis/management that is based on transparency and non-discrimination, and

Table 8.1. International Standards for Phytosanitary Measures (ISPMs) cited in this chapter. Information from www.ippc.int/standards

ISPM No.	Title	Date of notification	Reference
2	Framework for Pest Risk Analysis	March 2007 (revised from 1995 version)	IPPC, 2007
11	Pest Risk Analysis for Quarantine Pests	April 2013 (revised from 2001, 2003 and 2004 versions)	IPCC, 2013a
15	Regulation of Wood Packaging Material in International Trade	June 2013 (revised from 2002, 2006 and 2009 versions; Annex 1 and 2 revised in 2013)	IPCC, 2013b
19	Guidelines on Lists of Regulated Pests	April 2003	IPCC, 2003

places emphasis on measures being minimally restrictive while managing the risks.

Many national agricultural departments may have little experience in dealing with alien crop pests, including plant pests (i.e. weeds), but in the international agricultural sector, the Food and Agriculture Organization of the United Nations (FAO), CABI and regional plant protection organizations (RPPOs) provide information on invasive species problems. For example, CABI provides technical support at the international and national levels, in the form of the identification of new alien species, provision of distribution maps and databases such as the Crop Protection Compendium (www.cabi.org/isc) and the Invasive Species Compendium (www.cabi.org/isc).

History and Current Status of Regulatory Mechanisms for Prevention of IAS in India

The effective prevention of IAS, including invasive alien plants, requires appropriate national laws as well as coordinated international action. The international agreements address components of the IAS problem, but comprehensive national legislation is needed for implementation in each country. As noted above, most regulatory mechanisms for invasive plants are contained in more general regulatory mechanisms.

Awareness of the importance of quarantine measures in India started in the early 20th century when in 1906 the Indian government ordered the compulsory fumigation of imported cotton bales to prevent the introduction of a Mexican species, the cotton boll weevil (*Anthonomus grandis*). As indicated earlier, the policy of using government authority to prevent the entry of dangerous exotic pests is based on the principle that it is preferable to undergo some inconvenience in an effort to exclude pests rather than to accept the expense of controlling them. The quarantine law was enacted for the first time in India in 1914 as the Destructive Insects and Pests (DIP) Act. A gazette notification entitled 'Rules for regulating the import of plants etc. into India' was published in 1936. Over the years, the DIP Act has been revised and amended several times to meet the growing requirements of liberalized trade.

The DIP Act, 1914, provides for the following authorizations by the central and state governments:

- It authorizes the central government to:
  - o prohibit or regulate the import into India or any part thereof or any specific place therein or any article or class of articles; and
  - o prohibit or regulate the export from a state or the transport from one state to another state in India of any plants and plant materials, diseases or insects, likely to cause infection or infestation. It also authorizes the control of transport and carriage and gives power to prescribe the nature of documents to accompany such plants, plant materials and articles.
- It authorizes the state government to:
  - o make rules for the detention, inspection, disinfection or destruction of any pest or class of pests or of any article or class of articles, in respect of which the central government have issued notifications; and
  - o regulate the powers and duties of the officers whom it may appoint on its behalf.

The Act provides a penalty for persons who knowingly contravene the rules and regulations issued under the Act and at the same time protects from prosecution or other legal proceedings anyone acting in good faith.

In 1984, a notification was issued under the DIP Act, namely the Plants, Fruits and Seeds (Regulation of Import into India) Order, popularly known as the PFS Order, which was revised in 1989 after the announcement of the New Policy on Seed Development (NPSD) by the Government of India in 1988, which proposed major modifications for smooth quarantine functioning. The new policy covered the import of seeds/planting materials of wheat, paddy, coarse cereals, oil seeds, pulses, vegetables, flowers, ornamentals and fruit crops. After the enactment of the NPSD in 1988, and the WTO Agreement in 1995, the import of agricultural commodities was allowed more freely. This led to the introduction of several new pests and diseases into the country. While liberalizing import though, care had been taken to ensure that there was absolutely no compromise on plant quarantine requirements. Thus, the PFS Order, 1989, was repealed by the Plant Quarantine (Regulation of Import into India) Order, 2003 (PQ Order) (Government of India, 2003) as a legislative attempt to comply with the SPS Agreement of the WTO.

In particular, the PQ Order, 2003, was brought into force as there was an urgent need to:

- fill the gaps in the existing PFS Order, including those pertaining to the regulation of the import of germplasm/genetically modified organisms (GMOs)/transgenic plant material, live insects/fungi, including biocontrol agents, etc.;
- facilitate safe conduct of global trade in agriculture and thereby fulfil India's legal obligations under the various international agreements;
- protect the interests of the farmers of the country by preventing the entry, establishment and spread of destructive pests, vectors and other noxious alien species, and also the national flora and environment; and
- safeguard national biodiversity from threats from alien species invasions.

Under the PQ Order, 2003, apart from PRA being a precondition for all imports, the scope of plant quarantine activities was widened to incorporate additional definitions. The other salient features of the PQ Order are:

- Agricultural imports classified as follows are included:
  - o prohibited plant species (Schedule IV attached to the Order);
  - o restricted species where import is permitted only by authorized institutions (Schedule V);
  - o restricted species whose import is permitted only with further declarations of freedoms from quarantine/regulated pests and subject to specified treatment certifications (Schedule VI); or
  - o plant material whose import for consumption/industrial processing is permitted with a normal Phytosanitary Certificate (Schedule VII).
- The import of commodities with weed/alien species contamination as per Schedule VIII is prohibited.
- There is a restriction on the import of packaging material of plant origin unless it is treated.
- The provisions include regulating the import of:
  - o soil, peat and sphagnum moss;
  - o germplasm/GMOs/transgenic material for research;
  - o live insects/microbial cultures and biological control agents; and
  - o timber and wooden logs.
- Permit requirement is enforced on imports of seeds, including flower seeds, propagating material and mushroom spawn cultures.
- Additional declarations are specified in the Order for the import of agricultural commodities, specifically listing quarantine pests and 31 weed species.
- Notified points of entry are increased to 130 from the previous 59. (The official entry points are the quarantine stations of the Directorate of Plant Protection, Quarantine and Storage, DPPQS.)
- Certification fees and inspection charges are rationalized.

By June 2017, 42 amendments of the PQ Order, 2003 had been notified to the WTO. These comprised: revising quarantine pest lists; the incorporation of ISPM 15 compliance (see Table 8.1); the recognition of irradiation treatment; pest free areas and cold treatments for fruit flies to allow import of fresh fruits; and revising the lists of crops under Schedules VI and VII to include 694 and 297 crops and commodities, respectively.

In 2007, the National Commission on Farmers recommended developing a National

Agricultural Biosecurity System characterized by high professional, public and political credibility through the integration of plant, animal and fish management systems on biosecurity based on risk analysis and management. It also recommended the establishment of synergies in the requirements of international agreements and national regulations across these sectors to avoid the duplication of resources. In response, a Core Group was constituted in 2008 by the then Department of Agriculture and Cooperation to formulate recommendations for the establishment of an Integrated National Biosecurity System. This Core Group, in its report submitted in 2009, recommended the establishment of the National Agricultural Biosecurity System, which would require new legislation that is more relevant in the context of the present scenario. An Agricultural Biosecurity Bill was thus drafted and submitted in 2013. This provided for the establishment of an Authority for the prevention, eradication and control of pests (including invasive alien plants) and diseases of plants and animals and unwanted organisms. Its task was to ensure agricultural biosecurity, meet the international obligations of India for facilitating the imports and exports of plants, plant products, animals, animal products and aquatic organisms, and the regulation of agriculturally important microorganisms, and deal with all matters connected therewith or incidental thereto. This bill is yet to be approved by the government.

Apart from the plant quarantine regulations, the Indian Biological Diversity Act, 2002, was drafted in line with CBD obligations (Government of India, 2002). There is a need to harmonize the regulations on trade and on access to planting material, particularly those on the assessment of the environmental risk of IAS and LMOs, and those concerning eradication and action plans. An analysis of the implications of both the WTO and the CBD obligations, and efforts at compliance, would contribute significantly to minimizing the introduction and establishment of IAS in the country (Khetarpal and Gupta, 2006).

Beside the above legislation, the Environment (Protection) Act (EPA) was enacted in India in 1986 to protect and improve the environment and prevent hazards to human beings, other living creatures, plants and property (Government of India, 1986). This was a response to the United Nations Conference on the Human Environment held in Stockholm in 1972, which urged all countries to take appropriate steps for the protection and improvement of the environment. Also in 1986, in order to exercise its powers under the EPA, 1986, the central government promulgated the Environment (Protection) Rules with the purpose of protecting and improving the quality of the environment and preventing and abating environmental pollution. In their various schedules, the Rules make relevant provisions for the 'management and handling of hazardous wastes, manufacture, rules for storage and import of hazardous chemicals, and rules for the manufacture, use, import/export and storage of hazardous micro-organisms and genetically-engineered organisms or cells'. They empower the central government to prohibit or restrict the handling of hazardous substances, including their export and import into or from different areas, either in qualitative or in quantitative terms, because of the potential damage to the environment, humans, other living creatures, plants and property. Both IAS and LMOs are covered under the EPA, 1986; however, it does not state in clear terms a framework for the restriction and prohibition of these potential threats to the environment.

Institutions that Currently Cover IAS Issues in India

In India, the Ministry of Environment, Forests and Climate Change (MoEFCC) is the nodal ministry for matters relating to biodiversity, and deals and negotiates with the CBD. The Ministry of Commerce and Industry, in cooperation with the Ministry of Agriculture and Farmers Welfare, is the nodal ministry for the implementation of the phytosanitary aspects of the SPS

Agreement of the WTO. It deals with the quarantine norms and standards to be set up at the national level as per international requirements for minimizing the risks associated with the transboundary movement of pests (plants, insects, nematodes and pathogens), along with agricultural commodities. So far, there is no clear-cut emphasis on IAS, although the subject is dealt with from time to time in several departments of these ministries. However, the MoEFCC, which had earlier accorded low priority to the implementation of the CBD's Article 8(h) in its national report submitted to the CBD (Desai, 1999), instigated initiatives to identify the IAS-related issues at the COP-MOP (Conference of the Parties serving as the Meeting of the Parties to the Biosafety Protocol of the CBD) meetings' series, given the deleterious impacts of IAS on the environment. The eighth meeting of the Conference of the Parties to the CBD (COP 8) in Brazil further highlighted the immediate need for gap analysis and in-depth review of the status of IAS in various member countries.

Prevention and Other Management Initiatives for IAS in India

Even though there is no exclusive national legislation or policy addressing the problem of IAS across sectors, historically there is a suite of enabling policies at the state or other administrative levels that contain elements of prevention and/or other management provisions (Kishwan *et al*, 2007). Examples of these policies and acts are:

- the Madras Agricultural Pests and Diseases Act, 1919;
- the Travancore Plant Pests and Plant Diseases Regulation, 1919;
- the Coorg Agricultural Pests and Diseases Act, 1933;
- the Patiala Destructive Insects and Pests Act, 1943;
- the Bombay Agricultural Pests and Diseases Act, 1947;
- the Rewa State Agricultural Pests and Diseases Act, 1947;
- the East Punjab Agricultural Pests, Diseases and Noxious Weeds Act, 1949;
- the East Punjab Agricultural Pests, Diseases and Noxious Weeds Act as extended to Himachal Pradesh, 1949; and
- the Assam Agricultural Pests and Diseases Act, 1950.

It is of interest that the Coorg Agricultural Pests and Diseases Act, 1933 mentioned above was originally proposed (as the Coorg Noxious Weed Act, 1914) because of the invasion of coffee plantations by *L. camara*, followed by heavy forest fires, but it was not implemented until 1933 owing to World War I and its heavy cost to the British (see Kannan *et al.*, 2013, p. 1163).

A number of meetings have been held at the international level on IAS and related topics in the last 15 years, and those up to 2009 have been reviewed by Sharma *et al.* (2009). Moreover – and in light of the realization of the importance of IAS, the recent increase in volume of trade, and the exigencies posed by the WTO and the CBD – various meetings, workshops and seminars have been held in India to deliberate on the emerging issues and to bring together a number of organizations that have to cope with IAS issues, in order to share experience and to develop a strategy for their management. The proceedings and recommendations of these various meetings could act as an eye-opener for policy makers by bringing to their attention the fact that a large sector of the scientific community is aware of IAS; this may motivate them to take action at a policy level, which would go a long way towards tackling the current IAS problems.

The MS Swaminathan Research Foundation (MSSRF), Chennai, in collaboration with CABI, UK, organized the 'Conference on Management of Alien Invasive Species' on 2–5 December 2000. The conference addressed the growing threat of IAS and emphasized the need for research to accurately measure the socio-economic impacts of IAS, including those at community and landscape levels, to provide background information for their prevention and management. Also, immediate action was proposed to eradicate or control biological invasions before their spread caused greater losses and incurred future costs for control. As a key recommendation of the conference, it was proposed that priorities should be set for research on social, economic and other impacts of IAS; the importance of implementing quarantine standards at the national and the state level was emphasized for preventing new invasions and spread within India. In addition, the conference designated various actions to be taken at local, state, national, regional and international levels.

As a sequel to the MSSRF conference outlined above, a brainstorming session was held at the National Bureau of Plant Genetic Resources (NBPGR), New Delhi on 4 June 2001 to discuss 'Research Prioritization for Management of Alien Invasive Species'. These deliberations were focused mainly on the problems pertaining to different ecological habitats, including freshwater systems, mountains and plains (Khetarpal *et al.*, 2001). The exercise also identified various research institutes in the country that could take up work on different aspects of management of IAS and emphasized that a cross-sectoral approach was needed for developing a management model for them.

Several other meetings and symposia were organized on the subject between 2001 and 2003, including one on the 'Ecology of Biological Invasions' at the School of Environmental Studies, University of Delhi on 4–6 December 2003, which aimed to come up with a strategy for dealing with IAS. A workshop organized on 19–21 August 2004 by Banaras Hindu University, Varanasi, explored whether it was possible to predict invasions based on the susceptibility of an ecosystem, and the possibility of enhancing resistance to or recovery from IAS invasions, and then examined mechanisms for controlling the introduction and spread of IAS. The recommendations of all of these meetings should further help the government to tackle the problem in a holistic manner.

In terms of offering practical solutions to the problem of invasive plants, the 'Workshop on Alien Weeds in Moist Tropical Zones: Banes and Benefits', held at Kerala Forest Research Institute (KFRI) in collaboration with CABI on 2–4 November 1999 (Sankaran *et al.*, 2001), recommended the development of an integrated weed management programme that included classical biological control. In particular, an application was proposed for the introduction of a host-specific exotic rust fungus, *Puccinia spegazzinii*, for the control of *M. micrantha*.

As a follow-up to the above recommendation, the implementation phase of a project on classical biological control of *M. micrantha* with *P. spegazzinii*, funded by the UK Department for International Development (DFID), began in 2003, with Assam Agricultural University (AAU), KFRI, the Indian Council for Agricultural Research (ICAR) and CABI as partners. Under this project, studies on rust multiplication and the screening of a variety of non-target plant species against the rust to assess the safety of its use were conducted at NBPGR, New Delhi, using the CL-4 level containment facility developed for the biosafety evaluation of GMOs (see Fig. 8.1). Based on the results of the combined screening by NBPGR (Sreerama Kumar *et al.*, 2016) and CABI (Ellison *et al.*, 2008) of 108 plant species belonging to 35 plant families, the Plant Protection Advisor to the Government of India granted permission for limited field release of the rust inoculum at two identified sites each in Kerala and Assam. With the release of *P. spegazzinii*, India became the eighth country in the world and the first in mainland Asia to deliberately and scientifically introduce a fungal pathogen for the biological control of an invasive plant (Sankaran *et al.*, 2008; Chapter 10, this volume).

In line with its obligations under the CBD, India established a National Biodiversity Action Plan in 2008. Among the Action Plan's objectives is regulation of the introduction of IAS and their management by: (i) developing a unified national system for the regulation of all introductions, including their quarantine checking, assessment and release; and (ii) improving the management of IAS species and restoring the ecosystems that have been adversely affected (Government of India, 2008). The operational

Fig. 8.1. The containment facility developed for the biosafety evaluation of genetically modified organisms at the NBPGR in New Delhi. Photo courtesy C.A. Ellison.

guidelines of the *Intensification of Forest Management* scheme (Government of India, 2009a) issued by the MoEFCC emphasize the need for the control and eradication of forest invasive species and provision of assistance to state-owned or supported research institutions to carry out research into the management or eradication of IAS. Notably, the legislation of the scheduled tribes and other traditional forest dwellers (Recognition of Forest Rights Act, 2006) also provides an enabling policy space for the participation of local communities in protection, as well as the management of protected areas, forests and biodiversity in general at the national level.

Currently in India, a multi-agency and multiprogramme approach, involving several ministries and agencies, is being followed for regulating introductions of and managing IAS. Major activities include regulation of the introduction of exotic living materials, and their quarantine clearance and release for research and direct use. In general, the Ministry of Agriculture and Farmers Welfare deals with cultivated plants, fish and farm livestock, including poultry. It has sponsored projects on the eradication and management of invasive weedy plants, pathogens, pests and harmful fish. The MoEFCC deals with all forest and wildlife related IAS. In addition, it supports and coordinates programmes on the eradication, control measures and utilization of such species in different forest areas, conducts national surveys on their spread, prepares reports on the damage caused and undertakes restorative measures. There is, however, a need to develop a unified national system for regulation of the introduction and management of all IAS across the jurisdictions of all concerned ministries and relevant sectors.

Action points outlined in India's Fourth National Report to the CBD (Government of India, 2009b) to regulate the introduction of IAS and for the management of those already present include measures to:

- develop a unified national system for the regulation of all introductions and carrying out of rigorous quarantine checks;
- strengthen domestic quarantine measures to contain the spread of invasive species to neighbouring areas;
- promote intersectoral linkages to check unintended introductions and contain and manage the spread of IAS;
- develop a national database on IAS reported in India;
- develop an appropriate early warning and awareness system in response to new sightings of IAS;
- provide priority funding to basic research on managing invasive species;
- support capacity building for managing IAS at different levels with priority on local area activities;
- promote restorative measures of degraded ecosystems, preferably using locally adapted native species for this purpose; and
- promote regional cooperation in the adoption of uniform quarantine measures and containment of invasive exotics.

Developing a National Strategy for the Management of IAS in India

In order to develop a stepwise operational procedure for the management of invasive alien plants and other IAS, it is vital to understand their status, which can be classified as introduced and established, recently introduced, or not yet present but with the potential to be introduced (Khetarpal and Gupta, 2006). Overall, such a comprehensive management plan needs to be underpinned with a national strategy that includes scientific and technical considerations, and institutional and policy frameworks. The ultimate goal is the protection of agriculture and livestock, and the conservation or restoration of ecosystems. The goals and objectives for the national strategy should be realistic and result oriented.

The first step towards such a strategy would be to identify a cross-sectoral group to advocate the development of an IAS programme. Next would be the identification and involvement of all stakeholders to address the IAS problem. Key persons need to be strategically involved, and conspicuous invasive species problems in the country can be used to generate public awareness. The formulation of the national strategy forms the third step after the initial assessment is done and stakeholders have been identified and involved. Ideally, a single nodal agency should be identified or created, but if many agencies are involved, the responsibilities and work involved need to be clearly defined and allocated between the agencies, and each should have complete administrative and technical powers for its own specific remits.

The legal and institutional frameworks for the prevention and management of IAS also need to be considered. Effective management requires appropriate national laws as well as coordinated international action based on jointly agreed standards. Many international agreements address components of the invasive species problem, but national legislation is needed for implementation in each country. Further, there is an urgent need to promulgate existing subnational legislation against important IAS that have been introduced/detected in the country in recent years and that are likely to spread fast. Schedules V and VI of the PQ Order, 2003, pertain to the specific requirements for quarantine pests, but the listing of regulated non-quarantine pests as per ISPM 19 (see Table 8.1) has to be instigated for IAS that have serious economic impacts and can gain entry through planting materials. Thus, the Schedules of the PQ Order, 2003, need to be amended by incorporating the list of regulated non-quarantine pests. A national standard on PRA for regulated non-quarantine pests, as developed for quarantine pests, should also be developed as a priority (Gupta and Khetarpal, 2004).

For the IAS already established in India, such as the plants *L. camara* and *M. micrantha*, there is a need for an intensive official control programme through integrated pest/weed management (IPM), with special emphasis on biological control. However, caution must be exercised in introducing a biological control agent, and its host specificity must be thoroughly confirmed prior to release. In addition, the management of established IAS requires habitat restoration to be accorded importance (Murphy, 2001). In the case of recently introduced IAS (e.g. Singapore daisy, *Sphagneticola trilobata*; CABI, 2013), there is a need for early detection, which necessitates extensive surveys that may be site specific, species specific, or both, as the case may be. Based on its status as revealed by the survey, further actions to mitigate the impact of the IAS may need to be decided upon.

Lastly, a national council or agency needs to be identified to act as the nodal point on matters related to IAS. Such a council or agency would need members and experts from among the diverse stakeholders who would be able to: (i) draft regulations at both central and state levels and harmonize them with the international norms set by the WTO and the CBD; (ii) coordinate the activities of the stakeholders concerned with IAS, including area-specific surveys, and propose mitigatory measures; (iii) prepare a national management plan for IAS and review it periodically; (iv) collaborate with international agencies working on IAS; (v) facilitate the development of a Web-enabled national database on IAS; (vi) help in capacity building in the field of IAS; and (vii) generate funds through national and international sponsors and help in the reorientation of research projects in universities/ICAR institutes, etc.

Some steps have been taken towards the development of a national framework. The Department of Agriculture and Cooperation (DAC) of the Ministry of Agriculture and Farmers Welfare has proposed a National Policy for the Control of IAS under DAC's national Integrated Pest Management Programme to prevent and control the threat posed by IAS within the country. This policy uses the 15 Guiding Principles under the CBD as a basis. It would include the involvement of state governments, non-governmental organizations (NGOs), the private sector, research institutions and farmer self-help groups in the surveillance and detection of invasive plants, pests and diseases, and in taking eco-friendly corrective measures within the IPM scheme.

Conclusion

With increases in trade, transport, travel and tourism, there is greater movement of people and commodities, and consequently a greater risk of the spread of invasive alien plants and other IAS. The developed and developing countries need together to address the global problems presented by these species. Identifying, and where possible, quantifying the importance of the pathways that lead to harmful invasions, and addressing the gaps in plant quarantine measures would help in building national capacity to tackle the problems.

Effective management requires not only a national legal framework but also concerted bilateral, regional or global action based on common objectives and jointly agreed international standards. Legislation and a regulatory mechanism are necessary to implement policy, set principles, rules and procedures, and to provide a foundation for global, regional and national efforts. Presently, there are more than 50 global and regional legal instruments (agreements, codes of conduct and technical guidance documents) dealing in one way or another with alien species (Shine *et al.*, 2000). International instruments are usually general in character, but the national legislation necessary to address invasive plants and other IAS should cover all of the issues concerned. The national framework should harmonize objectives and scope, standardize terminology, implement measures for prevention, support measures for early detection, provide management options that include biodiversity/habitat restoration, and promote compliance and accountability.

In India, at the national level, various aspects of invasive plant and other IAS problems are directly or indirectly considered as cross-sectoral themes. Furthermore, many research institutions and universities are well equipped in terms of both facilities and expertise to take up the challenge posed by these species. To strengthen the drive for IAS management, it is a strategic policy decision to bring them under one umbrella of a nodal agency/council supported by a well-defined legislative mechanism. As in many developed countries, the responsibility for IAS control can be shared between various government agencies, provided there is effective coordination and linkage between these agencies and with clear-cut responsibilities are assigned. The lead responsibility may be given to one agency, such as an environment, agriculture or public health department, or to a specially established body. Apart from this, the legal framework should also promote active participation by the general public, so providing a culture of civil and administrative responsibility and accountability. In this context, the GISP (Global Invasive Species Programme) toolkit for the prevention and management of IAS (Wittenberg and Cock, 2001) gives comprehensive guidelines that can be adapted and adopted to suit the national set-ups in developing countries.

References

CABI (2013) *Sphagneticola trilobata* (wedelia). Invasive Species Compendium datasheet. CAB International, Wallingford, UK. Available at: www.cabi.org/isc/datasheet/56714 (accessed 2 March 2017).

CABI (2016) Invasive Species Compendium. CAB International, Wallingford, UK. Available at: www.cabi.org/isc/ (accessed 30 November 2016).

CBD (2002) COP 6 Decision VI/23. Alien species that threaten ecosystems, habitats or species. Available at: www.cbd.int/decision/cop/?id=7197 (accessed 2 March 2017).

CBD (2008) The Ninth Meeting of the Conference of the Parties (COP), Bonn, Germany, 19–30 May 2008. Available at: www.cbd.int/cop9 (accessed 18 March 2016).

Desai, B.H. (1999) Invasive species in India: a preliminary survey of legal and institutional mechanism. Paper presented at: *IUCN/ELC Workshop on Legal and Institutional Dimensions of Alien Invasive Species Introduction and Control, Bonn, Germany, 10–11 December 1999.*

Diwakar, M.C. (2003) South Asia (continental) perspective: invasive alien species in India. In: Pallewatta, N., Reaser, J.K. and Gutierrez A. (eds) *Prevention and Management of Invasive Alien Species. Proceedings of a Workshop on Forging Cooperation throughout South and Southeast Asia, 14–16 August 2002, Bangkok, Thailand.* Global Invasive Species Programme, National Botanical Institute, Cape Town, pp. 47–51.

Ellison C.A., Evans, H.C., Djeddour, D.H. and Thomas, S.E. (2008) Biology and host range of the rust fungus *Puccinia spegazzinii*: a new classical biological control agent for the invasive, alien weed *Mikania micrantha* in Asia. *Biological Control* 45, 133–145.

Government of India (1986) The Environment (Protection) Act, 1986. Available at: http://envfor.nic.in/legis/env/env1.html (accessed 30 November 2016).

Government of India (2002) The Biological Diversity Act, 2002. Available at: http://nbaindia.org/uploaded/Biodiversityindia/Legal/31.%20Biological%20Diversity%20%20Act,%202002.pdf (accessed 1 September 2017).

Government of India (2003) Plant Quarantine (Regulation of Import into India) Order, 2003. *The Gazette of India*, Extraordinary, Part II, Section 3, Sub-section (ii), No. 1037. Updated and consolidated version available at: http://plantquarantineindia.nic.in/pqispub/pdffiles/pqorder2015.pdf (accessed 30 November 2016.

Government of India (2008) *National Biodiversity Action Plan.* Ministry of Environment and Forests, New Delhi Available at: http://nbaindia.org/uploaded/Biodiversityindia/NBAP.pdf (accessed 30 November 2016).

Government of India (2009a) *Intensification of Forest Management: Operational Guidelines, August 2009.* Forest Protection Division, Ministry of Environment and Forests, New Delhi. Available at: www.moef.nic.in/sites/default/files/Book.pdf (accessed 30 November 2016).

Government of India (2009b) *India's Fourth National Report to the Convention on Biological Diversity.* Ministry of Environments and Forests, New Delhi. Available at: http://nbaindia.org/uploaded/Biodiversityindia/4th_report.pdf (accessed 30 November 2016).

Griffin, B. (2003) The role of the International Plant Protection Convention in the prevention and management of invasive alien species. In: Pallewatta, N., Reaser, J.K. and Gutierrez, A.T. (eds) *Prevention and Management of Invasive Alien Species. Proceedings of a Workshop on Forging Cooperation throughout South and Southeast Asia, 14–16 August 2002, Bangkok,*

Thailand. Global Invasive Species Programme, Cape Town, pp. 65–67.

Gupta, K. and Khetarpal, R.K. (2004) Concept of regulated pests, their risk analysis and the Indian scenario. *Annual Review of Plant Pathology* 4, 409–441.

Gupta, K. and Khetarpal, R.K. (2006) International Standards for Phytosanitary Measures. In: Gadewar, A.V. and Singh, B.P. (eds) *Plant Protection in the New Millennium.* Satish Serial Publishing House, New Delhi, pp. 109–139.

IPPC (2003) *ISPM 19: Guidelines on Lists of Regulated Pests.* International Standards for Phytosanitary Measures, Secretariat of the International Plant Protection Convention, Food and Agriculture Organization of the United Nations, Rome. Available at: https://www.ippc.int/en/core-activities/standards-setting/ispms/ (accessed 2 March 2017).

IPPC (2007) *ISPM 2: Framework for Pest Risk Analysis.* International Standards for Phytosanitary Measures, Secretariat of the International Plant Protection Convention, Food and Agriculture Organization of the United Nations, Rome. Available at: https://www.ippc.int/en/core-activities/standards-setting/ispms/ (accessed 2 March 2017).

IPPC (2013a) *ISPM 11: Pest Risk Analysis for Quarantine Pests.* International Standards for Phytosanitary Measures, Secretariat of the International Plant Protection Convention, Food and Agriculture Organization of the United Nations, Rome. Available at: https://www.ippc.int/en/core-activities/standards-setting/ispms/ (accessed 2 March 2017).

IPPC (2013b) *ISPM 15: Regulation of Wood Packaging Material in International Trade.* International Standards for Phytosanitary Measures, Secretariat of the International Plant Protection Convention, Food and Agriculture Organization of the United Nations, Rome. Available at: https://www.ippc.int/en/core-activities/standards-setting/ispms/ (accessed 2 March 2017).

Kannan, R., Shackleton, C.M. and Uma Shaankar, R. (2013) 1159–1165. Current Science 104, Available at: http://www.currentscience.ac.in/Volumes/104/09/1159.pdf (accessed 2 March 2017).

Khetarpal, R.K. and Gupta, K. (2002) Implications of Sanitary and Phytosanitary Agreement of WTO on plant protection in India. *Annual Review of Plant Pathology* 1, 1–26.

Khetarpal, R.K., Kapoor, M.L., Gupta, K. and Pandey, B.M. (2001) Alien invasive species – an overview. Paper presented to: *Brain Storming Session on Research Prioritization for Management of Alien Invasive Species, NBPGR (National Bureau of Plant Genetic Resources), New Delhi, India, 4–6 June 2001.*

Khetarpal, S. and Gupta, K. (2006) Management of invasive alien species: national strategy. In: Rai, L.C. and Gaur, J.P. (eds) *Invasive Alien Species and Biodiversity in India.* Department of Botany, Centre of Advanced Study, Banaras Hindu University, Varanasi, India, pp. 43–56.

Khuroo, A.A., Reshi, Z.A., Malik, A.H., Weber, E., Rashid, I. and Dar, G.H. (2012) Alien flora of India: taxonomic composition, invasion status and biogeographic affiliations. *Biological Invasions* 14, 99–113.

Kishwan, J., Pandey, D., Goyal, A.K. and Gupta, A.K. (2007) *India's Forests.* Ministry of Environment and Forests, Government of India, New Delhi.

Mandal, F.B. (2011) The management of alien species in India. *International Journal of Biodiversity and Conservation* 3, 467–473.

Murphy, S.T. (2001) Alien weeds in moist forest zones of India: population characteristics, ecology and implications for impact and management. In: Sankaran, K.V., Murphy, S.T. and Evans, H.C. (eds) *Alien Weeds in Moist Tropical Zones: Banes and Benefits. Proceedings of a Workshop, Kerala Forest Research Institute, Peechi, India, 2–4 November 1999.* Kerala Forest Research Institute, Peechi, India and CABI Bioscience UK Centre (Ascot), Ascot, UK, pp. 20–27.

Ramakrishnan, P.S. (2001) Biological invasion as a component of global change: the Indian context. In: Sankaran, K.V., Murphy, S.T. and Evans, H.C. (eds) *Alien Weeds in Moist Tropical Zones: Banes and Benefits. Proceedings of a Workshop, Kerala Forest Research Institute, Peechi, India, 2–4 November 1999.* Kerala Forest Research Institute, Peechi, India and CABI Bioscience, UK Centre (Ascot), Ascot, UK, pp. 28–34.

Rana, R.S., Dhillon, B.S. and Khetarpal, R.K. (2004) Invasive alien species: the Indian scene. *Indian Journal of Plant Genetic Resources* 16, 190–213.

Reshi, Z.A. and Khuroo, A.A. (2012) Alien plant invasions in India: current status and management challenges. *Proceedings of the National Academy of Sciences India, Section B, Biological Sciences* 82(suppl. 2), 305–331.

Sankaran, K.V., Murphy, S.T. and Evans, H.C. (eds) (2001) *Alien Weeds in Moist Tropical Zones: Banes and Benefits. Proceedings of a Workshop, Kerala Forest Research Institute, Peechi, India, 2–4 November 1999.* Kerala Forest Research Institute, Peechi, India and CABI Bioscience, UK Centre (Ascot), Ascot, UK.

Sankaran, K.V., Puzari, K.C., Ellison, C.A., Sreerama Kumar, P. and Dev, U. (2008) Field release of the rust fungus *Puccinia spegazzinii* to control *Mikania micrantha* in India: protocols and awareness raising. In: Julien, M.H., Sforza, R., Bon, M.C., Evans, H.C., Hatcher, P.E., Hinz, H.L. and Rector, B.G. (eds) *Proceedings of the XII International Symposium on Biological Control of Weeds, La Grande Motte, France, 22–27 April 2007.* CAB International, Wallingford, UK, pp. 384–389.

Sathyanarayana, N. and Sathyagopal, K. (2013) Invasive alien species: problems and the way forward. *Pest Management in Horticultural Ecosystem* 19, 85–91.

Secretariat of the CBD (2005) *Handbook of the Convention on Biological Diversity Including its Cartagena Protocol on Biosafety*, 3rd edn. Secretariat to the Convention on Biological Diversity, Montreal, Canada.

Shah, A. (2008) The Plant Quarantine Order in India. In: *Identification of Risks and Management of Invasive Alien Species using the IPCC Framework. Proceedings of the Workshop on Invasive Alien Species and the International Plant Protection Convention, Braunschweig, Germany, 22–26 September 2003.* Food and Agriculture Organization of the United Nations (FAO), Rome, pp. 81–86.

Sharma, S.K. and Brahmi, P. (2011) Gene bank curators. Towards implementation of the International Treaty on Plant Genetic Resources for Food and Agriculture by the Indian National Gene Bank. In: Frison, C., López, F. and Esquinas-Alcázar, J.T. (eds) *Plant Genetic Resources and Food Security: Stakeholder Perspectives on the International Treaty on Plant Genetic Resources for Food and Agriculture.* Food and Agriculture Organization of the United Nations (FAO), Rome, Biodiversity International, Rome and Earthscan, Abingdon, UK/New York.

Sharma, S.K., Khetarpal, R.K., Gupta, K. and Lal, A. (eds) (2009) *Invasive Alien Species – a Threat to Biodiversity.* Indian Council of Agricultural Research, Ministry of Environment and Forests and National Biodiversity Authority, New Delhi.

Shine, C., Williams, N. and Gundling, L. (2000) *A Guide to Designing Legal and Institutional Frameworks on Alien Invasive Species.* Environmental Policy and Law Paper No. 40, IUCN (The World Conservation Union), Gland, Switzerland/Cambridge UK, and IUCN Environmental Law Centre, Bonn, Germany.

Sreerama Kumar, P., Dev, U., Ellison, C.E., Puzari, K.C., Sankaran, K.V. and Joshi, N. (2016) Exotic rust fungus to manage the invasive mile-a-minute weed in India: pre-release evaluation and status of establishment in the field. *Indian Journal of Weed Science* 48, 206–214.

Wittenberg, R. and Cock, M.J.W. (eds) (2001) *Invasive Alien Species: A Toolkit of Best Prevention and Management Practices.* CAB International, Wallingford, UK.

WTO (1995) Agreement on the Application of Sanitary and Phytosanitary Measures. World Trade Organization, Geneva, Switzerland. Available at: https://www.wto.org/english/tratop_e/sps_e/spsagr_e.htm (accessed 22 August 2017).

9 Control Options for Invasive Alien Plants in Agroforestry in the Asia–Pacific Region

K.V. Sankaran*
Kerala Forest Research Institute (KFRI), Peechi, India

Introduction

Agroforestry involves the cultivation and use of trees in farming systems and it is a practical and low-cost means of implementing many forms of integrated land management, especially for small-scale producers (Leakey, 2010). It takes many forms, from traditional shifting cultivation to high-density home gardens and complex hedgerow or tree intercropping. Properly managed, it has three characteristics: (i) it maintains or increases crop production and the productivity of the land, which may include more efficient input and labour use; (ii) it is a sustainable land-use system that maintains biodiversity and soil fertility, and reduces erosion, especially via its perennial component(s); and (iii) it is acceptable to farming communities, as it conforms to local practices (Nair, 1992).

The sustainability of well-managed agroforestry is important for the management of pests, diseases and weeds, because it may confer resilience against incursions by and outbreaks of these problems. While biodiversity alone is not a guarantee of resilience, the good soil fertility of a well-planned agroforestry system may make crops less prone to pest, disease and weed damage than they are in monocultures because crop growth is robust and natural enemy populations thrive. Agroforestry systems therefore present an opportunity to implement preventive measures based on integrated pest management (IPM) rather than relying on applying control when problems arise (Day and Murphy, 1998).

Invasive alien plants pose serious threats to agroforestry systems throughout the world; they cause cost escalation and/or income reduction, as well as changes in vegetation dynamics, nutrient cycling and hydrology, and loss of biodiversity. They are characterized by rapid growth, efficient dispersal capabilities, large reproductive outputs and tolerance to a broad range of environmental conditions (Campbell, 2005).

In the Asia–Pacific region, some of the world's worst invasive alien plants have been present for over a century. Top of the list of terrestrial weeds that affect agroforestry are species such as *Lantana camara* (lantana), *Chromolaena odorata* (chromolaena), *Parthenium hysterophorus* (parthenium), *Mikania micrantha* (mikania) and *Mimosa diplotricha* var. *diplotricha* (spiny mimosa).

This chapter reviews briefly the control options for invasive alien plant management in agroforestry in the Asia–Pacific region, and discusses how preventive IPM measures provide a viable means of minimizing their impact.

* E-mail: sankarankv@gmail.com

Physical Control: Manual and Mechanical Removal

Manual and mechanical removal are the most common strategies followed worldwide for managing invasive plants. The enormous variety of methods includes hand pulling, slashing, knife weeding, uprooting, ploughing, stick raking, chain pulling, trampling, girdling and pushing. The tools and machines used, which vary from the basic to the expensive and sophisticated, include hoes, machetes, bush knives, mowers, brush cutters, tillers, ploughs, bulldozers and tractor-drawn equipment.

This kind of control can be made highly species specific, but it is labour intensive and thus slow and usually uneconomic; hence, it is suitable only for small- or medium-sized infestations. The positive points are that the procedures are generally very simple to carry out, often require only basic equipment and minimum training, and demand no supervision (apart from checking against the accidental removal of non-target species). The measures are environmentally benign in most circumstances, although soil disturbance due to hand pulling, up-rooting and similar methods may lead to leaching and erosion and should not be carried out on steep slopes and in gullies.

The capacity of invasive plants for rapid regeneration and growth, together with their characteristic production of abundant and easily dispersible seeds, present major obstacles to management using mechanical and manual methods. Interventions may actually make infestations worse and, over time, allow invasive alien weeds to dominate in place of more manageable native weeds. Follow-up operations such as herbicidal treatments, burning or grubbing to contain fresh growth may have to be carried out after each mechanical or manual operation in order to achieve satisfactory results – though these can be used to optimize control if the different procedures are properly and deliberately integrated.

Given rapid regrowth by invasive alien plants, the main drawback of mechanical or manual interventions is the limited duration of control that they achieve, with measures typically needed to be applied three to four times a year to keep plants under check. Invasive plants frequently regenerate from tiny stem or root fragments or reinfest a treated area from large seed banks. Vigorous regrowth is observed in species such as parthenium, chromolaena and mikania after mechanical removal. Forest user groups in the buffer zone of Koshi Tappu Wildlife Reserve in Nepal found that manual control of mikania was a 'never-ending task' because weeded land is reinfested by seeds and plant fragments (Siwakoti, 2007). None the less, at the seedling stage, hand weeding is often an easy and effective method for many invasive plants infesting a small area, even though not all are amenable to this method. It is not advisable for parthenium because the plant can cause severe dermatitis and allergic reactions in humans, and mechanical removal is more appropriate, but this is restricted to open areas (Sushilkumar and Saraswat, 2001). Equally, spiny mimosa cannot be removed by hand because of the thorny nature of the stem. Hand weeding is also laborious as labour costs in large areas are prohibitively high, and it needs repeating often.

Slashing the above-ground weed mass, which is quicker, and is an easy and therefore widespread practice, gives apparently satisfactory results, but again it provides only short-term control, particularly for plants that can regenerate from rootstock or stolons. Field studies in Haryana, India, demonstrated how cutting enhances regeneration in parthenium, with sprouting from cut stumps producing flowers only 30 days after the weed had been cut back (Sushilkumar and Saraswat, 2001). Conversely, repeated slashing of a woody weed may drain the resources stored in the root system and weaken the plant or, in some cases, even kill it. This is particularly useful as part of integrated control measures, such as treating the cut stump or new growth with a selective herbicide. Field research with forest user groups in the buffer zone of the Royal Chitwan National Park in Nepal (Rai *et al.*, 2012a) found that two consecutive cuttings of mikania (before flowering) with a 3-week interval eliminated up to 91%

of vines and increased mortality when compared with a single slashing.

Salgado (1972) noted that slashing and mowing were not useful against chromolaena in coconut gardens in Sri Lanka because the plants reacted as they might to pruning. He suggested removing it with a tine cultivator followed by hand digging and cover cropping as more effective. In coconut gardens, *Tephrosia purpurea* planted as a cover crop was the cheapest and most effective method for the control of chromolaena. More extensive removal of plants, including their below-ground parts, can be more effective but it is more time- and labour-consuming. A study in the Western Ghats in the Indian state of Karnataka showed that uprooting chromolaena provided more effective control than cutting plants at ground level (Swaminath and Shivanna, 2001).

Infestation by lantana is best managed by clearing using a tractor, stick raking or stalk removal, but these methods are good only on flat areas or on gentle, accessible slopes (Sood *et al.*, 2001; Day *et al.*, 2003). In inaccessible areas, manual uprooting has to be carried out, which minimizes disturbance to the nearby vegetation and is effective in killing the plants. As noted above, the method is labour intensive and costly, yet it is often the only option available to farmers in developing countries. Also, the control costs of mechanical/manual methods should be considered wherever land is of low value. Mechanical/manual removal requires follow-up treatment to kill regrowth, such as spot spraying with herbicides or continued mechanical removal (Day *et al.*, 2003).

Timing is a crucial element in maximizing the effectiveness of manual, mechanical and chemical control measures. In particular, interventions against invasive plants that reproduce by seed must be made before the plants flower or set seed; if they are made later they run the risk of actually aiding seed dispersal and promoting proliferation. In the studies on chromolaena and parthenium described above, both cutting and the more effective uprooting of chromolaena were most successful in September–November, while successful parthenium control required the plants to be completely uprooted before flowering. In southern China, hand pulling of mikania was most effective before the end of October (i.e. before seed maturity) or during the rainy (March) season as growth begins (Zhang *et al.*, 2004). If mikania vines were cut near the ground once a month for 3 consecutive months in summer and autumn, 90% of the plants were eliminated, but the same procedure was far less successful in winter and spring (Kuo *et al.*, 2002). Rai *et al.* (2012a) suggest that the annual slashing of mikania initially adopted by forest user groups in the Chitwan buffer zone in Nepal mentioned above caused the weed to proliferate because the operation was carried out along with other forestry operations in the winter, during the mikania flowering season.

Manual measures are difficult to implement successfully where labour is costly or scarce, because an important factor in their effectiveness is how thoroughly they are carried out. A study of families with landholdings of various sizes (from <1.5 ha to >5 ha) in Kerala, India, showed that most of them managed mikania manually using either simple weeding or complete weeding (in which the plants were totally uprooted). Either or both methods might be adopted in different circumstances on a single farm, depending on the extent of the invasion, the farm income and the availability of family labour; simple weeding was the more common practice (70% of infested farms) because of the high cost of employing labour (Muraleedharan and Anitha, 2001). It is incorrect to assume that manual control is 'cheap'; sickle weeding (slashing) and uprooting were more expensive than chemical options for mikania in agroforestry systems and plantations in Kerala (Sankaran *et al.*, 2001). Moreover, manual control operations eked out over time may be a false (even if unavoidable) economy; it is better to hit the weed hard initially. The field study in the buffer zone of the Chitwan National Park – where manual control was trialled in part because of the affordable labour available in the forest user groups – showed that the first slashing of mikania is the most expensive, and that the second and third are both cheaper and have greater impact (Rai *et al.*, 2012a). In Papua New Guinea and Fiji, where mikania is widespread, farmers adopt physical methods (pulling or periodic weeding once a week/month) to control mikania in mixed cropping systems, where the yield loss due to the weed is over 30% (Day *et al.*, 2011, 2012).

The disposal of weed waste is a problem in many situations. Heaping it along the boundary of the weeded area allows the still-living plants to continue to grow. Littering of waste from a weeded plot (during transportation for safe disposal off-site or people removing it for fodder) can result in new growth and (perhaps) the spread of the weed to previously uninfested areas. A good example is mikania: a small portion of the stem (a node) will grow into a new plant in a matter of a couple of days (Sankaran *et al.*, 2001). Therefore, all the waste needs to be: (i) removed from the plot; and (ii) burnt as soon as possible.

In summary, manual and mechanical control measures are generally not efficient methods for managing invasive alien plants, and most often they may exacerbate problems. Some of the methods will also cause soil disturbance and leaching. However, despite their shortcomings and high cost, they are still practised widely in small infestations because they are simple to explain and understand and easy to carry out. In addition, they can be excellent tools as part of an integrated management programmes for invasive plants, as we shall see later.

Cultural Control

Sanitation and prevention

Maintaining hygiene in a weed-infested area is of prime importance – it is arguably the single most important factor in preventing the spread of many invasive plants. Routine cleaning and inspection of vehicles, machinery and agricultural implements before they enter or leave a weed-infested area are vital as the long-distance spread of many invasive plants has been along a country's major roads (McFadyen, 2003). Concentrating all washing off at a single site on a farm means that it is easier to monitor for invasive plants and that any emerging weeds can be dealt with. If possible, roadside populations of invasive plants should also be managed to prevent spread.

Grain, fodder, fertilizer, seeds and hay need to be checked for contamination by alien plant parts, including seeds, and the movement and transportation of livestock should be subjected to similar checks. A study in Nepal among community forest user groups who gathered natural resources from the core area of Chitwan National Park found that the communities were very dependent on fodder, but even so that people did not remove the mikania with which the harvested material was heavily infested before carrying it off to their homes (Murphy *et al.*, 2013).

Burning

We still do not know enough about the impact of fire on weeds and their seed banks in agroforestry systems. Slashing and burning the weed, or direct burning, are used extensively as control options for lantana, chromolaena, mimosa and mikania. Fire is often used before or after mechanical or herbicidal measures to enhance their effectiveness. In Karnataka, a year-on-year increase in the severity of chromolaena infestations across much of the state was attributed to the prevalence of hand cutting or burning as control measures, and both of these lead to profuse branching and vigorous growth after the onset of rains (Doddamani *et al.*, 2001a). These authors suggested that utilization is the way forward for dealing with the large biomass. Successful control means targeting the chromolaena root system and impairing its resprouting capacity. Although burning was found to be more effective than hand cutting in having a greater impact on resprouting, burning alone was inadequate and the follow-up use of systemic herbicide (glyphosate or 2,4-D) was needed to provide a satisfactory level of control (Doddamani *et al.*, 2001b).

Burning also provides an instant supply of inorganic fertilizer for the soil in the form of ash instead of a slow release from the gradual natural breakdown of plant matter, but this can be more rapidly lost by leaching. Invasive plants such as mikania take up nutrients more efficiently and accumulate larger quantities in their tissues than native plants, and provide a larger nutrient pool when they are burnt. However, the seeds and underground vegetative parts of some weeds, such as mikania, can survive certain fire regimes (e.g. the low temperature fires of shortened jhum cycles) better than native plants, and in these circumstances, burning can aid invasion by an alien plant. Mikania, which outcompetes native species for several years after sites are burnt, could be seen as the greatest beneficiary of its own burning (see Chapter 7, this volume).

Fire is inexpensive and well suited to dense infestations, but its use involves risks to people, property and non-target species, and these need to be carefully managed. Staging a burn can be particularly difficult in an agroforestry system where invasive weeds are growing among or over both tree and field crops. Some invasive plants themselves present a fire hazard; chromolaena, for example, because of its volatile oil content, burns at a much higher temperature than other plants and its dominance has created a serious fire hazard for villages, infrastructure and forests in Indonesia (Moni and Subramanian, 1960; McFadyen, 1989; Wilson and Widayanto, 2004).

Mulching and cover cropping

Mulching as an option for controlling invasive plants is often labour intensive and expensive to implement because large amounts of mulching material are required to cover rampant and strong-growing infestations. Using the slashed weed itself as a mulch presents dangers when, like mikania, it regenerates from plant fragments.

Cover crops might have promise for invasive plant control, but while a good deal of research has been conducted in plantation crops, there is little published on their potential in agroforestry systems. It has been suggested that the growth of mikania can be reduced by the incorporation of fibrous organic matter from *Polygonum hydropiper*, *Oxalis corymbosa* and *O. repens* (Gogoi, 2001). Nishanathan *et al.* (2013) reported that in the agro-ecological regions of northern Sri Lanka, it was possible to control the population of parthenium by mulching with *Gliricidia sepium* leaves, coupled with manual weeding. Care must be taken to ensure that one weed is not replaced by another when choosing cover crops for weed control. Leguminous climbers such as *Mucuna bracteata* may not be suitable, as they can smother trees unless they are intensively managed (Sankaran *et al.*, 2007).

The vigorous growth of invasive plants such as mimosa has frequently led to their use as a green manure or cover crop. In northern Thailand, a thornless variety of this nitrogen-fixing species (*M. diplotricha* var. *inermis*), which was abundant locally, was adopted by upland farmers to help counter the impact of land shortage and shortened shifting cultivation cycles on soil fertility (Prinz and Ongprasert, 2007). As yields of crops such as rice, peanuts and cassava fell and weed problems increased, improved infrastructure opened the possibility of cash crops and a more diversified agroforestry system. Orange trees were planted, but the light canopy and degraded soil facilitated weedy growth of *Imperata cylindrica* (imperata grass). Farmers found that vigorous growth of the thornless mimosa suppressed the grassy imperata weed, but that it also spread to adjacent crop fields. This could have been disastrous, but the farmers integrated measures for its management into their cropping system to facilitate its use as a cover crop, green manure or mulch in a variety of field crops. Prinz and Ongprasert (2007) note, however, that this thornless mimosa variety is likely to be replaced ultimately by edible legume crops, while the need for its control will continue. Further, they note that the spiny form of the weed (*M. d.* var. *diplotricha*), which occurs on a wider scale in northern Thailand, has a substantial impact on crops.

Mohankumar (1996) reported that in Karnataka, when thornless mimosa was grown for a year and then incorporated into soil suppressed weed growth, it prevented soil erosion and improved the growth of coffee grown as an intercrop in coconut groves. Nevertheless, studies conducted in Kerala have shown that both mimosa varieties (the thorny *M. d. diplotricha* and the thornless *M. d. inermis*) have the potential to become weedy in agricultural and agroforestry systems, with the former proved to be a more aggressive colonizer than the latter (Suresh, 2014). Thus, any plant being considered as a cover crop, green manure or mulch needs to be assessed carefully for its potential weediness, because some of the characteristics of a good cover crop and an invasive plant coincide.

Control by utilization

The utilization of an invasive species means that it attains an economic value, and conflicts of interest can be raised in how it is managed. This is a critical, yet delicate, issue. In the absence of external support, resource-poor farmers will 'make the best of the worst situation' and make use of an invasive weed that has replaced traditional crops and natural resources (Rai *et al.*, 2012b), although, making use of such a plant may provide only short-term benefits. As we have seen above, harvesting and moving the plant may enhance its spread, and the nature of the harvesting may not contribute to effective management. For example, slashing above-ground vegetation is a fast and cheap way to harvest plant material, but the plants may grow back more strongly. Good management practices, such as harvesting before flowering or seed set (so that the seed bank is reduced), or uprooting the plant rather than slashing its above-ground plant parts, may not be necessary for harvesting the economically useful plant parts, in which case they are not likely to be implemented.

In India, attempts to produce compost using mikania were unsuccessful in Kerala owing to the high water content of the plant (K.C. Chacko, KFRI alumnus, 2005, personal communication) but the process seems to have been more successful for parthenium in Nepal, where it was commoditized, albeit with external support (Rai *et al.*, 2012b). Saini *et al.* (2014) have proposed a variety of uses for parthenium in India, such as composting, phytoremediation, oxalic acid production, as a source of edible protein and for biogas production, to name a few.

Given the enormous amount of biomass that invasive alien plants can produce, mulching crops with the weeded material is common. In Karnataka, mulching with cut chromolaena weed in areca nut, coconut and cashew nut gardens considerably increased soil moisture to a depth of 15 cm (Swaminath and Shivanna, 2001). Chromolaena was also found to have NPK (nitrogen, phosphorus, and potassium) levels comparable with those of other popular green manures and could increase the nutrient status and humus content of soil (Doddamani *et al.*, 2001b). In Indonesia, farmers have been reported to view chromolaena as a valuable fallow species, and it has been shown to be a good cover crop and green manure, as well as being effective in controlling other weeds such as imperata grass (Baxter, 1995). The negative impacts of chromolaena outweigh its benefits though, and this is the crucial factor to consider in evaluating the utilization of any invasive plant (CABI, 2011).

Some invasive alien plants are used as animal fodder, especially where they have outcompeted normal browse and have thus led to a shortage. Mikania has been used as fodder in India and Indonesia (Soerjani, 1977; Sankaran and Sreenivasan, 2001). In Nepal, lantana is used for fuelwood and fodder, and chromolaena for fuelwood, composting, fodder and fencing (Rai *et al.*, 2012b). Adverse effects of mikania as a fodder have been reported (see also Chapters 3 and 4, this volume). The people in the Koshi Tappu forest buffer zone in Nepal, for instance, used fresh mikania as fodder for cattle and buffalo, but they mixed it with other fodders because it reduced milk yield and caused abdominal disorders (Siwakoti, 2007). Chromolaena has also been reported as unpalatable and toxic to cattle and goats (McWilliam, 2000; McFadyen, 2003). Parthenium, however, has been evaluated for silage production in India using a process that reduces its toxins and destroys seed viability, but this has not yet apparently reached the adoption stage (Narasimhan *et al.*, 1993).

Invasive species biomass is a common fuel for burning on-farm but, with growing interest in 'biofuels', its potential for biogas production has come under the spotlight. Both chromolaena and parthenium have been demonstrated to have potential for biogas production in India, but production from chromolaena has been optimized by using dry leaves rather than whole plants, with maximum production limited to 2 months of the year (Swaminath and Shivanna, 2001). Parthenium was trialled as an additive to cattle manure for methane production and, subsequently, as sole fuel for family-sized biogas fermenters (Gunaseelan, 1987, 1994). There seems to be no evidence that either of these schemes has progressed beyond the experimental stage since then, but Singh and Poudel (2013) have reported promising results using dried biomass of mikania (after cutting) to produce briquette fuel in Nepal.

Material from invasive plants such as lantana is used on-farm for construction purposes, for example for fences and small huts, as well as in basket making, in north-eastern India (K.C. Puzari, Assam Agricultural University, 2005, personal communication). There is a market for other products, especially paper and handicrafts (furniture, baskets, utensils, toys) made using lantana stem in place of traditional timber, cane and rattan (Priyanka and Joshi, 2013). Parthenium has properties suitable for papermaking, and technologies have been standardized in India to make paper boards using lantana stems (Gujral and Vasudevan, 1983).

Such activities have been suggested as a means of conserving indigenous species also used in home and large-scale industries, such as the utilization of lantana by communities living near the Chitwan National Park to conserve rattan and the medicinal species *Wrightia tinctoria* in Karnataka (Kannan *et al.*, 2008). With external support, invasive plant-based small-scale industries may transform communities: Lachhiwala village near Dehradun in northern India, supported by the Dehradun-based non-governmental organization (NGO) HESCO (the Himalayan Environmental Studies and Conservation Organization), has developed an array of lantana-based products from furniture to medicinal remedies that contribute considerably to the community's economy, and has gained Lachhiwala the name 'Lantana Village'. ATREE (the Ashoka Trust for Research in Ecology and the Environment), an NGO based in Bangaluru, India, has put efforts into promoting the use of lantana for making furniture and handicrafts (Kannan *et al.*, 2014). Using the phrase 'adaptive management' as a proxy for the 'management-by-utilization' of lantana weed in India, Bhagwat *et al.* (2012) lauded local communities that had adapted to the presence of lantana weed, describing how cottage industries had sprung up in areas of abundance. In reality, such enterprises are localized and, more importantly, use only parts of the plant (e.g. the thicker stems), meaning that harvesting does not manage the invasive plant effectively and, furthermore, sometimes aids its vigorous regrowth and spread (Witt *et al.*, 2012).

Another mimosa species, *Mimosa pigra*, which is widespread in countries like Australia, Cambodia, Indonesia, Malaysia and Vietnam, is widely used as a green manure, as fodder and firewood, and for fencing purposes. Although it is inefficient in controlling the population of the plant, its harvesting could help in the integrated control of selected populations (Miller, 2004). Another case in point is *Prosopis juliflora*, which is one of the worst invasive trees species in the semi-arid and arid regions of several countries in the Asia–Pacific region. This species is treated as a potential agroforestry crop in hot and arid regions in India, and it has been suggested that the invasiveness of the tree could be controlled by putting it to various uses, such as fuelwood, fodder and timber (Tewari *et al.*, 2006).

Leaving aside their impacts on invasive plant populations, how far such efforts at the utilization of the plants are economic or sustainable without external support is questionable. Fifteen years ago, Ramakrishnan (2001) pointed out that despite many feasibility and pilot-scale studies on the economic utilization of invasive plants in India, none had taken off at the field level. Notwithstanding cases like Lachhiwala that have external support, this seems to hold true even today. Ramakrishnan (2001) argued for 'effective natural resource management strategies to enhance biodiversity in the landscape' in order to keep invasive alien plant invasions under check. This possibility, whether at the landscape level or for individual farmer agroforestry systems, seems unlikely to integrate with economic utilization under any current model.

Chemical Methods

Since their development in the 1940s, synthetic herbicides have been used extensively to control invasive alien weeds. The common herbicides used are glyphosate, triclopyr, picloram, 2,4-D amine, imazapyr, metasulfuron-methyl, atrazine, diuron, hexazinone, tebuthiuron and fluroxypyr. Their ability to control invasive species has meant that synthetic herbicides, more than any other form of weed management, have allowed the agricultural productive potential of infested sites to be maximized. Herbicides are commonly used against invasive alien plants in the Asia–Pacific region, though not so much in India, China and Nepal because of ecological, environmental and social concerns, such as toxicity to users and consumers, water contamination from runoff, and persistence in the soil and crops.

It is beyond the scope of this chapter to review herbicides in detail, although it is useful to have a grasp of how they are applied and how they work in order to understand the issues surrounding their use. However, a guide to the terminology associated with chemical weed control is given in Box 9.1, and information about the herbicides used in agroforestry is presented in Table 9.1.

While chemical methods may achieve the control of invasive weeds quickly, regrowth can be rapid, and repeated spraying may be required even within one growing season to prevent renewed weed growth and seed production. The need to apply herbicides repeatedly against invasive alien plants means that they are not seen as an economically feasible option by smaller scale farmers. They tend to be used more by larger landholders or people leasing land to grow cash crops (Day *et al.*, 2012; Chapter 3, this volume), especially when labour is scarce or expensive. In some situations herbicides can be economic: application of glyphosate was cheaper than manual weeding for controlling mikania infestation in forest plantations in Kerala (Sankaran *et al.*, 2001; Sankaran and Pandalai, 2004; Chapter 4, this volume).

The complexity of agroforestry systems makes herbicide use in them particularly problematic. When farmers resort to heavy pesticide use, which destroys the ecological balance of the system, natural IPM is threatened (Day and Murphy, 1998). If this happens, the weed problem will escalate, other pest outbreaks may also be facilitated, more pesticide will have to be applied, and farmers may find themselves on a 'pesticide treadmill'. Thus, where pesticides are used, they need to be selected and applied so that they optimize control and have minimum non-target impacts, an approach known as rational pesticide use (Bateman, 2003). The potential benefits of this approach, especially where volume application rates are reduced substantially, include cost reduction (for both pesticides and labour), improved safety and reduced environmental impact (through more efficient use of sprays and the use of specific agents). Critical factors in rational pesticide use are, therefore, product efficacy and selectivity, and precision of application and timing.

Herbicide choice

Safety is a key factor in selecting a herbicide, but assessing the comparative safety to

Box 9.1. Terminology used in chemical control.

Contact herbicides destroy the tissues they come into contact with. They generally have fast-acting but short-lived impacts, and do not kill below-ground roots and tubers; they are **non-systemic**.

Systemic herbicides are absorbed by plants and **translocated** (moved in the plant's xylem or phloem vessels), so their impact is not restricted to the point of application. They may be sprayed on to the plant or the soil (in the latter case absorption is by the roots). In the plant, they may be translocated upward (**acropetally**: usually in the xylem) or downward (**basipetally**: in the phloem), or both; herbicides translocated downward can damage or kill below-ground roots and tubers.

Most contact herbicides have a **broad spectrum** of activity; they damage/kill most plants they come into contact with, regardless of taxonomic group. Systemic herbicides may be broad spectrum, but because they interfere with specific biochemical pathways, many have **selective** activity against particular groups of plants, or against specific plant growth stages.

Because of basic differences in plant biochemistry and life history, herbicides are often selective against one of the two many classes of plants: **broadleaved** or dicotyledonous species; and **grassy** or monocotyledonous species. This is why herbicidal grass control in sugarcane is difficult, and likewise control of *Mikania* in tea.

Non-target effects describe the adverse impacts of herbicides on species other than the target flora. Herbicide damage to non-target (e.g. crop) plants is often described in terms of **phytotoxicity**.

The timing of herbicide application is frequently critical. Note that in herbicide terminology, the distinction between pre-emergence and **post-emergence** application (i.e. before or after a plant has emerged from the ground) generally refers to the growth stage of the crop, not the weed, unless it specifically says otherwise (as is the case in Table 9.1, where reference is made to pre- and post-weed emergence).

Herbicides are formulated as products suitable for application in various situations: e.g. emulsifiable concentrates, wettable powders, granules, dusts, oil solutions, etc. A **carrier** is a substance used to dilute or suspend a herbicide during its application, but substances beside the herbicide may be added. Emulsified oils can be used as **adjuvants** to improve the absorption and activity of foliar herbicides. These include **surfactants**, which aid spreading, sticking or wetting, and may also aid leaf penetration by the herbicide. Wetting agents are surfactants that, when added to a spray solution, cause it to cover plant surfaces more thoroughly (common examples include Teepol, Plantowet, Tritone, etc.). Modern herbicide products may include **safeners** (e.g. fenclorim), which are molecules that improve selectivity between the crop plants (for which they reduce the effect of the herbicide) and the target weeds.

Herbicides have varying fates: some are quickly metabolized in the plant or soil; and some are retained in the plant and exhibit residual activity. For food crops, **residual activity** is monitored through the use of **maximum residue limits** (**MRLs**), which are based on the analysis of the quantity of a given chemical remaining on food product samples. Some herbicides are very soluble and may be hazardous because they contaminate groundwater. Some are adsorbed in the soil and remain active for prolonged periods; these **persistent** herbicides exhibit residual soil activity.

Although the majority of herbicide treatments are targeted at the growing plant, in some situations, **cut stem/stump** treatment is appropriate: herbicides are applied/painted on to the cut surfaces after physical or mechanical clearance. A more expensive option is **stem/trunk injection**, in which selected herbicides are injected directly into the plant.

Herbicides often work well in combination. They may be formulated and sold as mixed products under a specific brand name by an agrochemical company. Separate herbicide products can also, in some circumstances, be successfully **tank mixed** by the user, i.e. mixed in the spray tank, but there should be a specific reason for doing this.

Table 9.1. Herbicides commonly used in agroforestry.[a]

Class/mode of action (HRAC Group[b])	Herbicide (WHO class[c])	Target, application and activity	Typical agroforestry/ plantation crops
Contact action			
Bipyridyliums Inhibit photosynthesis (D)	Diquat (II)	A non-residual bipyridyl contact herbicide and crop desiccant. Broad spectrum: contact action only, annual broadleaved and grassy weeds	Banana, cocoa, coconut, coffee, forestry, oil palm, orchards, rubber, tea
Diphenylethers Prevent chlorophyll formation; enzyme inhibitors (E)	Oxyfluorfen (U)	Annual grassy and broadleaved weeds Post-weed emergence application Residual activity	Fruit trees
Nitriles Inhibit photosynthesis (C3)	Ioxynil (II)	Annual broadleaved weeds Post-weed emergence application	Sugarcane
Oxadiazoles Enzyme inhibitors; disrupt cell membranes (E)	Oxadiazon (U)	Annual grassy and broadleaved weeds Post-weed emergence application Absorbed by foliage/shoots more readily than roots	Bananas, conifer seedbeds, nuts, tree fruits
Systemic action			
Benzoic acids Synthetic plant hormones: growth regulators (O)	Dicamba (II)	Annual, perennial and woody broadleaved weeds Applied to the soil or post-weed emergence to the plant Absorbed by roots and leaves Translocated throughout the plant via the xylem and phloem	Sugarcane
Chlorocarbonic acids Inhibit lipid synthesis (N)	Dalapon (acid and sodium salt) (U)	Selective for annual and perennial grasses, including couch and imperata grass Applied directly to weeds Absorbed by roots and leaves Translocated in xylem and phloem Residual activity: 3–4 months	Banana, coffee, forestry, orchards, sugarcane, tea, woodland
Imidazolinones Enzyme inhibitors; disrupt protein, DNA synthesis and cell growth (B)	Imazapyr (U)	Broad spectrum, annual and perennial grassy, broadleaved and woody weeds Applied pre- or post-weed emergence; also to cut stems and by stem injection Absorbed by foliage and roots: kills underground storage organs Rapid translocation to growing regions, where it accumulates Residual activity	Forestry, oil palm, rubber
Nitriles Enzyme inhibitors; block photosynthesis (C2)	Tebuthiuron (II)	Broad spectrum incl. woody weeds Applied to soil pre- or early post-weed emergence Absorbed by roots Readily translocated Soil acting: absorbed mainly by the roots	Sugarcane

continued

Table 9.1. *continued.*

Class/mode of action (HRAC Group[b])	Herbicide (WHO class[c])	Target, application and activity	Typical agroforestry/ plantation crops
Phenoxycarboxylic acids Synthetic plant hormones: growth regulators (O)	2,4-D (various esters and salts) (II)	Annual, perennial and woody broadleaved weeds Applied post-weed emergence Salts readily absorbed by the roots, esters by foliage Translocated; accumulates mostly in growing regions Residual activity	Cocoa (use with great care only), forestry, orchards, oil palm, rubber (when devoid of legume cover), sugarcane
Phosphanoglycine Enzyme inhibitors; block protein synthesis (G)	Glyphosate (various esters and salts) (III)	Broad spectrum, annual, perennial and woody weeds; probably the most widely used herbicide Applied post-weed emergence Absorbed from leaves Rapid translocation throughout plant Residual activity	Forestry, orchards, agricultural areas
Pyridine carboxylic acids Synthetic plant hormones: growth regulators (O)	Fluroxypyr (U)	Annual and perennial broadleaved weeds Applied post-weed emergence Absorbed mostly by leaves; roots to a lesser extent Translocated rapidly Residual activity	Conifers, oil palm, orchards, rubber
	Picloram (U)	Annual, perennial and especially woody broadleaved weeds Applied pre- and post-weed emergence Can be used for killing by trunk injection in the replanting of oil palm and rubber Absorbed by roots and foliage; best results from foliar application Translocated acropetally and basipetally; accumulates in new growth Residual activity	Forestry, sugarcane
	Triclopyr (II)	Annual, perennial and woody broadleaved weeds Applied and rapidly absorbed from foliage, shoots/bark, roots; also applied to cut stumps and by trunk injection Translocation throughout plant; accumulates in growth regions	Conifers, oil palm, rubber (last two including for cover crop maintenance)
Sulfonylureas Enzyme inhibitors; block protein synthesis, hence plant growth (B)	Metsulfuron methyl (U)	Broadleaved weeds: potent at relatively low dosage rates Applied pre- and post-weed emergence Absorbed through roots and foliage Rapid translocation acropetally and basipetally; rapid action with post-weed emergence use	Forestry

continued

Table 9.1. *continued.*

Class/mode of action (HRAC Group[b])	Herbicide (WHO class[c])	Target, application and activity	Typical agroforestry/ plantation crops
Triazines Inhibit photosynthesis (C1)	Simazine (U)	Annual broadleaved and grassy weeds Applied to many germinating annuals grasses and broadleaved weeds; phytotoxic to a number of annual crops Absorbed mainly by the roots, also foliage Translocated acropetally; accumulates in leaves and growing shoots Residual activity	Coffee, oil palm, pineapple, rubber, sugarcane, tea, tree fruits
Triazinones Inhibit photosynthesis (C1)	Hexazinone (II)	Non-selective for annual and perennial grassy and broadleaved weeds Applied post-weed emergence; primarily contact action; some acropetal translocation Absorbed by leaves and roots	Conifers, but not areas planted with deciduous trees, pineapple, sugarcane
Ureas Enzyme inhibitors; block photosynthesis (C2)	Diuron (III)	Annual and perennial broadleaved and grassy weeds Applied to soil, pre- or post-weed emergence Soil acting: absorbed mainly by roots Translocated acropetally in xylem Residual activity	Banana, pineapple, sugarcane, tree fruits

[a]Information in this table is indicative and incomplete and should not be used for guidance in selecting or applying herbicides. The inclusion of a compound does not indicate endorsement. Users should check country registration of herbicides and MRL (maximum residue limit) status. Main sources: Bakar (2004); Turner (2015); the DROPDATA website (www.dropdata.org/RPU/pesticides_MoA.htm).
[b]HRAC Group, as designated by the Herbicide Resistance Action Committee (http://hracglobal.com/).
[c]WHO Class, as designated by the World Health Organization: II, moderately hazardous; III, slightly hazardous; U, unlikely to present acute hazard in normal use (WHO, 2009).

users and the environment of the herbicides currently in use in agroforestry is hampered by the lack of long-term impact studies for these systems. The herbicide that instantly comes to mind in terms of operator exposure is paraquat, which can be toxic and even fatal (especially by inhalation) if proper application procedures are not followed. However, most deaths from paraquat are suicides – presumably made possible owing to poor formulations, to which no emetic had been added, and paraquat is being phased out. Another herbicide of concern is 2,4-D amine, which is toxic to animal life and has a relatively long and persistent residual action, though it has also been associated with severe non-target damage (Wang *et al.*, 1994). Environmental issues are of frequent concern with herbicide use, including the effects of residues, water contamination, the impact on bees and aquaculture, etc. These are particularly important in smallholder agroforestry where the land has multiple uses, including livestock production, and farmer families live close by and share the farm water sources.

Herbicide choice should be based on what provides the best control of the target weed with the least adverse impacts on non-target species, including crops. Given the complexity of agroforestry systems, the use of selective herbicides (see Box 9.1) is strongly warranted, but these are not universally available in all countries. To borrow an example from plantation forestry, Sankaran *et al.* (2001) and Sankaran and Pandalai (2004) found that one of the most effective treatments for mikania in teak and eucalypt plantations in Kerala was Grazon DS, a mixture of triclopyr and picloram, but it was not commercially available in India at that time (Chapter 4, this volume). They suggested that as the mixture was of lower toxicity and was more easily degraded in soils than other available compounds, it should be imported.

Mixtures of herbicides are an important tool for improving the degree and spectrum of weed control, and allowing doses to be reduced, which is beneficial to the environment and saves on costs. Ideally, a herbicidal control strategy would involve the use of a 'knock-down' herbicide to kill weeds that are growing and a residual herbicide to control those that subsequently germinate from seed or grow vegetatively from surviving fragments.

Application and timing

The choice of appropriate herbicide(s) is important but not sufficient. How and when they are applied is critical to the success of the control operation, and also affects user and environmental safety. Important elements to consider are timing (before or after weed, or crop emergence, at what stage of weed growth) and method of application (to soil, foliage, cut stems/stumps, by injection) (see Box 9.1). There may also be efficacy versus cost considerations; for example cut stump and stem injection techniques are labour intensive and time-consuming and thus more expensive than other methods. A good choice of application equipment and the optimization of nozzle/spray droplet parameters in order to reduce volume application and increase work rates, and applying the correct dosage (i.e. knowing when to stop spraying and move on) all increase the efficiency of application as well (Bateman, 2003).

Herbicide treatments are considered to be most effective when the target weeds are small and actively growing, and all treatments need be carried out before flowering and seed setting to ensure maximum efficacy. Treating when the plants are still small is particularly important because strong growth of invasive alien plants can make the treatment of large plants difficult, especially when, like mikania, they grow over and smother crops (Holm *et al.*, 1991) or, like lantana or chromolaena, they grow into dense, tall thickets that are difficult to penetrate. In the latter case, the appropriate herbicides may be applied on to the bark of the lower part of the stems or painted on to freshly cut stumps (i.e. cut-stump treatment). Foliar spraying is highly effective, particularly on regrowth, probably owing to the more efficient penetration of young leaves by the herbicide (Day *et al.*, 2003). As with other control methods, treating before the plants mature is absolutely crucial for invasive weeds that produce vast quantities of easily dispersed seed. Further, it has been reported that chemical control is often more effective after fire or mechanical control (Day *et al.*, 2003).

Success also depends on environmental conditions, such as rainfall, temperature, humidity, wind, etc. It is always advisable to carry out spraying operations on dry days as rain may wash herbicides off the plant before they can be absorbed. Moreover, different herbicides vary in how they are affected by rain; in trials in Malaysia, paraquat was rainfast 15 min after spraying, but glyphosate and glufosinate-ammonium gave poor results when it rained 6 h after spraying (Lee and Ngim, 1997). Finding the best application 'window' is complicated too. For mikania treated with foliar applications of imazapyr, rainfall preceding treatment led to better herbicide uptake, but applications needed to be made at least 6 h before the rain to obtain more than 50% control (Ipor and Price, 1992). In Queensland, Australia, best results were obtained when fluroxypyr was sprayed on to lantana 6 weeks after good rains, and when the minimum temperature exceeded 15°C (Hannan-Jones, 1998). Additionally, this work showed that taking factors beside rain into account makes decisions more complicated; both day length and rainfall in the weeks preceding fluroxypyr treatment affected the mortality of *L. camara* cv. Helidon White. Light may affect the uptake of herbicide; for example, better uptake of paraquat by mikania was recorded in poor light (Ipor and Price, 1994). Plant maturity and stress are other factors that affect success rates; herbicides are often not effective, for example, on lantana plants under stress (Hannan-Jones, 1998).

'Additives' are an important tool for improving herbicide application and efficacy. Adjuvants facilitate the activity of herbicides or modify the characteristics of herbicide formulations and so enhance their efficacy, as shown for sodium dodecyl sulfate with 2,4-D and dicamba against mikania (Huang *et al.*, 2004). Sankaran and Pandalai (2004) reported that the addition of ammonium sulphate increased the weed control efficacy of glyphosate applied to mikania in Kerala. Likewise, the addition of urea to the diuron formulation was found to be better at controlling mikania than the application of diuron alone, and foliar herbicides can be applied with wetting agents to improve their 'stickability'.

Limitations to chemical control

The shortcomings of herbicides that have emerged in this section can be summarized as follows:

- They are not economically viable against invasive alien weeds in many situations, particularly when the weed is covering extensive areas and growing into the canopy of large trees. Conversely, in situations such as high-value agriculture, where intensive mechanical control is needed and would be expensive, herbicides can be useful.
- Their effects tend to be short lived (although this is also the case with mechanical weeding), and in most cases repeated applications may be necessary for long-term control, while the continuous application of herbicides is environmentally damaging.
- Over a long period, continuous seed production, the long-term viability of some seeds (e.g. mimosa and parthenium) and their wind dispersal (e.g. mikania, chromolaena) promote the reinvasion of areas where herbicidal control has been achieved. It is difficult to prevent reinvasion unless the entire seed source is controlled, which is impractical. However, the use of more specific herbicides that, for example, target broadleaved weeds only, so allowing for the survival of grasses, may help limit reinvasion in some situations.
- There may be impacts on non-target species.
- Older compounds, such as atrazine (banned in the European Union since 2004), had a reputation for contaminating groundwater. Many active substances have now been banned in the OECD (Organisation for Economic Co-operation and Development) countries and new compounds are subject to screening and modelling to prevent this potentially disastrous pollution from occurring.
- Poor application technology and techniques, the use of ineffective herbicides or low concentrations of effective herbicides, incorrect and untimely applications, and lack of follow-up action will all frustrate the efficacy of chemical control methods.

As already indicated, the human and environmental impact of herbicide application may be particularly high in an agroforestry situation where the farm family lives on the farm and the source(s) of drinking water and food is (are) close to the sprayed areas. Also, while we have outlined good practices in rational pesticide use, the reality is that there is a lack of training in applying herbicides; in developing/least developed countries, low-paid and poorly educated workers are usually engaged for the field application of herbicides. In most situations, they are unaware of the toxicity problems associated with the herbicides and the stipulated safety devices are not provided for them, thereby increasing unintentional exposure and health risks.

Public fears about environmental and human health risks associated with herbicides are hard to allay despite the accumulated scientific documentation on their safety and benign environmental effects when they are used according to the safety instructions. There is an urgent need, therefore, to minimize the environmental impact of herbicides and to make them more effective and efficient, and more appropriate to the soil, site and species (Little *et al.*, 2006). It is also necessary to create awareness among the public of the suitability of herbicides for managing invasive plants and the environmental hazards that may result from their use. As with mechanical methods, chemical methods alone will not serve for the efficient management of invasive alien plants; but their use can be complementary in an integrated management approach.

Biological Control

The classical biological control of invasive plants employing natural enemies has elicited wide interest as an environmentally benign, cost effective, self-sustaining and permanent solution for the control of invasive alien plants. There are many success stories from around the world that illustrate this (Evans, 2001; Winston *et al.*, 2014), and a large number of biocontrol agents (mostly insects) have been introduced into different countries in the Asia–Pacific region during the last century for managing invasive alien plants of concern in agroforestry, though with varying degrees of success. The agents that have been used for control of the five main species of invasive plants in the Asia–Pacific region are summarized below, along with discussion of other agents that have been used in the region and the impact of this type of control.

Lantana

Over 30 species have been introduced to manage lantana in Australia, of which 16 have established. Four of these species, *Teleonemia scrupulosa* (a sap-sucking bug), *Uroplata girardi* (a leaf-mining beetle), *Octotoma scabripennis* (another leaf-mining beetle) and *Ophiomyia lantanae* (a seed-feeding fly), have shown promise (Day *et al.*, 2003). In India, even though eight species were introduced, including the above four, there has been limited success for several reasons (Sushilkumar, 2001). Three more species, including the flower-feeding moth *Lantanophaga pusillidactyla*, were introduced into India accidentally. Among these introduced insects, *T. scrupulosa* has spread far and wide in the country and appears to have had the most impact. However, defoliation by this insect alone is unlikely to be sufficient to result in a significant reduction in the infestation of lantana. When the bug population falls owing to environmental stresses, the weed regains its vigour and, in this situation, even subsequent increases in the population of the bug are unable to severely reduce the vigour of lantana bushes. The poor impact of other insect agents is mostly related to variation in climatic factors and partial resistance by different biotypes of the weed. Although the recovery of *O. scabripennis* and *U. girardi* in many of the introduced areas gives some hope of their establishment, there appears to be no real lasting solution for the biological control of lantana (Sushilkumar, 2001). Some of India's indigenous natural enemies (e.g. the gall fly, *Asphondylia lantanae*, and the defoliating moth, *Hypena laceratalis*) do show promise as biocontrol agents of the weed. It would thus appear worthwhile to augment the populations of these insects and disseminate them to newer infested areas, along with fresh introductions of some exotic insects that have shown some potential.

Chromolaena

A gall fly, *Cecidochares connexa*, which was introduced into Indonesia in 1995, has proved to be an efficient agent there (McFadyen, 2002; Wilson and Widayanto, 2004). It is also reported to have established in Palau, Guam and Papua New Guinea (Bofeng *et al.*, 2004; Muniappan *et al.*, 2004) and India (Bhumannavar and Ramani, 2007). This species is considered to be the most effective agent for the control of chromolaena.

A moth, *Pareuchaetes pseudoinsulata*, was introduced into India from Trinidad in the 1970s to manage chromolaena, but the control achieved has been largely insignificant. In most cases, the larvae did not establish, mainly because of predation by ants. In 1984, about 40,000 larvae and 400 adults were released in the campus of the Kerala Agricultural University in Thrissur (Joy *et al.*, 1987). Although this release resulted in the establishment of the insect and partial control of the weed on the campus, further spread of the insect was not reported until its recent rediscovery in Walayar, Kerala, about 70 km from the original release site (Varma *et al.*, 2006). This instance provides some hope for the moth as a biocontrol agent. The same agent has shown variable impacts in other countries in the region, such as Indonesia, the Philippines, Papua New Guinea, Malaysia and Sri Lanka (Zachariades *et al.*, 2009; Winston *et al.*, 2014). Satisfactory control of chromolaena using *Pareuchaetes* has been reported from Guam and the Commonwealth of the Northern Mariana Islands (CNMI) (Seibert, 1989; Muniappan *et al.*, 2004).

Other exotic agents introduced to India for the biocontrol of chromolaena, e.g. *Apion brunneonigrum* and *Mescinia parvula*, failed to establish (Singh, 2001). *Acalitus adoratus*, an eriophyid mite, which was accidently introduced in to South-east Asia, has been established in South and South-east Asia and in the East Carolinas (Federated States of Micronesia) and CNMI, but the damage that it causes appears to be slight (Muniappan *et al.*, 2004; Winston *et al.*, 2014).

No fungal pathogens have so far shown potential as biocontrol agents of chromolaena, though Prashanthi and Kulkarni (2005) and Chetti *et al.* (2001) reported the potential of the fungus *Aureobasidium pullulans*, a floral pathogen, as a mycoherbicide for the biocontrol of the weed in India. The potential of the foliar pathogens *Colletotrichum gloeosporioides* and *Alternaria alternata* as biocontrol agents of chromolaena has also been reported (Chetti *et al.*, 2001). There have been no reports on the use of these pathogens in the field.

Parthenium

The biocontrol agents introduced to control parthenium in the region include *Zygogramma bicolorata* (a leaf-feeding beetle), *Epiblema strenuana* (a stem-boring moth) and *Puccinia abrupta* var. *partheniicola* and *P. melampodii* f. sp. *parthenium hysterophorae* (rust fungi). Of these, the two insects and *P. abrupta* have shown promise as biocontrol agents in Australia. In India, the two species *Z. bicolorata*, *E. strenuana* and *Smicronyx lutulentus* (a flower-feeding weevil) were imported into quarantine, but only *Z. bicolorata* was permitted to be released because *S. lutulentus* could not be reared in the laboratory and *E. strenuana* was found to complete its life cycle on an oil crop (niger, *Guizotia abyssinica*). *Z. bicolorata* has established and is causing damage in some areas in the country, especially in northern India, and the beetle is continuing to be released. Even though several pathogens have been recorded on parthenium in India, including *P. abrupta* var. *partheniicola*, the biocontrol efficacy of these, except for *Sclerotium rolfsii* [preferred name *Athelia rolfsii*], was low (Sushilkumar and Saraswat, 2001; Atkins and Shabbir, 2014).

Mimosa

Biocontrol attempts against mimosa (*M. d.* var. *diplotricha*) have included the introduction of a sap-feeding bug, *Heteropsylla spinulosa*, which causes growing-tip distortion, brittle stems and stunted plants, and has given good control of the weed in Australia, Papua New Guinea and, more recently, in at least an additional eight Pacific Island countries and territories (Willson and Garcia, 1992; Day and Winston, 2016). A fungus, *Corynespora cassiicola*, which causes defoliation and dieback of the plant in Queensland, is considered to be a potential biocontrol agent in Australia. *Fusarium pallidoroseum*, a fungus isolated from diseased mimosa in the Philippines, has been shown to provide excellent control of mimosa seedlings when sprayed as crude culture filtrate or cell-free filtrate. Further results on the suitability of the fungus to control mimosa are awaited (APFISN, 2008). Biocontrol agents for the plant are yet to be introduced into India. Promising biocontrol agents for *M. pigra* included the cerambycid *Milothris irrorata*, the larvae of which girdle the stems of the weed, resulting in the death of the shoots (Napompeth, 1983).

Mikania

The biocontrol of mikania was attempted in Malaysia and the Solomon Islands using an introduced insect natural enemy, *Liothrips mikaniae*, but this failed to establish in the field mainly because of predation (Cock *et al.*, 2000). Research by CABI in collaboration with several research institutions in India has resulted in understanding the potential of a rust fungus, *Puccinia spegazzinii*, in controlling *M. micrantha* (Ellison *et al.*, 2008; Sankaran *et al.*, 2008). The rust fungus has established and is having a significant impact on the weed in a range of ecosystems, including agroforestry, in Taiwan, Papua New Guinea and a number of Pacific Island countries and territories (Ellison and Day, 2011; Day *et al.*, 2013a,b). The work on this biological control agent is fully covered in Chapter 10, this volume.

Use of plants as biocontrol agents

Cuscuta spp. (dodders) are known to parasitize and control populations of lantana and mikania in India (Pundir, 1985; author's personal observation) and China (Yu *et al.*, 2009). Observations reported by Pundir (1985) indicated that *C. santapaui* killed numerous lantana plants in Dehradun. The parasitic plant spreads quickly and has been recommended against lantana in heavily infested areas. Unfortunately, prospects for using *Cuscuta* as a biological control agent are seriously limited because dodders are not host specific and can be crop pests (Chapter 4, this volume).

Plants such as *Cassia tora*, *C. caesia*, *Khaya senegalensis* and *Crotalaria mucronata* were observed to suppress growth of chromolaena in Karnataka (Swaminath and Shivanna, 2001). The potential of using these plants as biocontrol agents is yet to be assessed. Likewise, Sushilkumar and Saraswat (2001) noted the use of *Cassia uniflora*, *C. tora*, *Abutilon indicum* and a host of other plants for the competitive replacement of parthenium in India. As noted earlier, climbing legumes such as *Mucuna* and *Pueraria* are promising aggressive plants for competing with and controlling the growth of mikania (Dutta, 1978). However, some of the species considered for use as cover crops may also have the potential to be invasive.

Impact of biological control

Classical biological control can have a very positive benefit–cost ratio and the benefits normally outweigh the drawbacks. Indeed, even partial control has been shown to provide significant savings to agriculture and the environment in terms of reducing the need for other control methods to be implemented (McFadyen, 2000). When properly conducted, this type of control represents the cheapest and safest option for managing invasive alien plants. The main disadvantages are: (i) the lack of certainty about the level of control that will be achieved; and (ii) the delay until the agents, once established, achieve their full impact.

As described above, the history of biocontrol agents introduced into the Asia–Pacific region shows that some agents have been very useful and some have not, and that the results varied significantly between countries. A number of biocontrol agents have been introduced into India, mainly for the control of chromolaena, lantana and parthenium, but success has been limited and, at best, successful only at a local level (Evans, 2001; Singh, 2001). The reasons for these recurrent failures are linked to the genetic diversity of the invasive plant species, their ability to adapt to a wider range of ecosystems than the biocontrol agents and a lack of expertise in handling the agents. Therefore, it is essential to strengthen the capabilities in classical biological control within countries of the region.

Classical biological control is not the only approach. Barbora (2001) advocated augmentative biocontrol, ideally with a bioherbicide, for mikania control in tea in northern India. He argued that this would allow the implementation to be stopped if necessary, and that it could also be integrated with chemical herbicide usage.

Integrated Management

The foregoing review of different control methods of invasive alien plants in agroforestry has shown that no single method is efficient enough to manage a weed population on its own. Hence, the complexity of agroforestry systems and the robust growth and reinvasion potential of invasive alien weeds provides an environment where an integrated management approach can yield the best results. Where possible, biological control should provide the cornerstone, and the appropriate mechanical, chemical, cultural control methods are then integrated as required.

For example, as we have seen above, burning and follow-up herbicide treatment gave the best control of chromolaena in Karnataka (Doddamani *et al.*, 2001b). The biodiversity and soil health promoted by agroforestry supports the implementation of preventive IPM, and rational pesticide use is critical to safeguarding the natural enemies that are supported by the diverse crops and have an impact on weed growth (Bateman, 2003). In turn, this leads to greater resilience in the agroforestry system. It may be that the greater the variety of techniques employed, the smaller their individual importance and, by the same token, the smaller their potential for any impact on the environment or, in the case of herbicides, for accumulation and the development of resistance (Harr, 1992; Bakar, 2004). To cite an example, in taro farms in Fiji, in plots treated by ploughing combined with the frequent application of herbicides, there was successful control of mikania infestation compared with plots which received minimum soil tillage, hand weeding and occasional herbicide application (Macanawai *et al.*, 2010).

The participation of the local community needs to be ensured in implementing integrated management because they are the main stakeholders who will continue to put the validated methods into practice. Awareness of the usefulness of different methods, especially the use of biocontrol agents, needs to be raised through media such as television, radio and newspapers (Sankaran *et al.*, 2008). Scientists should interact and work with foresters, farmers, planters and other stakeholders to spread the message on the integrated management of invasive alien species. These issues were highlighted by Sushilkumar and Saraswat (2001), who reviewed current measures to control parthenium in India, and concluded that no single measure would be effective. They presented a scheme for its integrated management, including mechanical control, chemical control, utilization and biological control, with a time frame for the various activities on an annual and long-term basis. They stressed the importance of participation by the local community in all of the activities, and also the crucial need for relevant legislation to be enforced to ensure that control measures are implemented.

Likewise, Sushilkumar (2001) recommended integrated management as the only likely way to control lantana in India. The strategy he proposed involves biological management using introduced exotic insects in association with silvicultural (an appropriate felling system), mechanical (uprooting) and chemical (glyphosate alone or in combination with oxadiazon and fluroxypyr) management. Similar suggestions were put forth by Kumar *et al.* (2007), who successfully practised the mechanical removal of lantana (cutting and uprooting), followed by herbicidal treatment of the regrowth and the planting of grass species in a reclaimed area in Himachal Pradesh, India.

The potential for the integrated weed management of invasive alien plants has been proposed and discussed widely by the scientific community, policy makers, foresters and other stakeholders. Regrettably, concerted efforts have not been made for its effective implementation, especially in the Asia–Pacific region. The reasons perhaps are: (i) a resistance to changing the methods already in practice; (ii) an inability to appreciate the relatively simple technology; and (iii) the fact that economic returns are not perceived to be worth the additional resources needed to implement this type of management (Little *et al.*, 2006).

Acknowledgments

The author thanks Dr J.K. Sharma, former Director, Kerala Forest Research Institute, for encouragement and Dr Roy Bateman, International Pesticide Application Research Consortium (IPARC), UK, for advice on herbicides (Table 9.1).

References

Adkins S. and Shabbir, A. (2014) Biology, ecology and management of the invasive parthenium weed (*Parthenium hysterophorus* L.). *Pest Management Science* 70, 1–7.

APFISN (2008) *Mimosa diplotricha.* Invasive Pest Fact Sheet, Asia-Pacific Forest Invasive Species Network Secretariat, Kerala Forest Research Institute, Peechi, Thrissur, Kerala. Available at: http://apfisn.net/sites/default/files/mimosa.pdf (accessed 12 December 2014).

Bakar, B.H. (2004) Invasive weed species in Malaysian agro-ecosystems: species, impacts and management. *Malaysian Journal of Science* 23, 1–42.

Barbora, A.C. (2001) Weed control in tea plantations: current scenario in northeast India. In: Sankaran, K.V., Murphy, S.T. and Evans, H.C. (eds) *Alien Weeds in Moist Tropical Zones: Banes and Benefits. Proceedings of a Workshop, Kerala Forest Research Institute, Peechi, India, 2–4 November 1999.* Kerala Forest Research Institute, Peechi, India and CABI Bioscience, UK Centre (Ascot), Ascot, UK, pp. 107–111.

Bateman, R. (2003) Rational pesticide use: spatially and temporally targeted application of specific products. In: Wilson, M.F. (ed.) *Optimising Pesticide Use.* Wiley, Chichester, UK, pp. 131–159.

Baxter, J. (1995) *Chromolaena odorata*: weed for the killing or shrub for the tilling? *Agroforestry Today* 7, 6–8.

Bhagwat, S.A., Breman, E., Thekaekara, T., Thornton, T.F. and Willis, K.J. (2012) A battle lost? Report on two centuries of invasion and management of *Lantana camara* L. in Australia, India and South Africa. *PLoS ONE* 7(3): e32407. Available at: http://dx.doi.org/10.1371/journal.pone.0032407.

Bhumannavar, B.S. and Ramani, S. (2007) Introduction of *Cecidochares connexa* (Macquart) (Diptera: Tephritidae) into India for the biological control of *Chromolaena odorata.* In: Lai, P.-Y., Reddy, G.V.P. and Muniappan, R. (2007) *Proceedings of the Seventh International Workshop on Biological Control and Management of* Chromolaena odorata *and* Mikania micrantha, *Pingtung, Taiwan, 12–15 September 2006.* National Pingtung University of Science and Technology, Pingtung, Taiwan, pp. 38–48.

Bofeng, I., Donnelly, G., Orapa, W. and Day. M. (2004) Biological control of *Chromolaena odorata* in Papua New Guinea. In: Day, M.D. and McFadyen, R.E. (eds) *Chromolaena in the Asia-Pacific Region. Proceedings of the Sixth International Workshop on Biological Control and Management of Chromolaena, Cairns, Australia, 6–9 May 2003.* ACIAR Technical Report No. 55, Australian Centre for International Agricultural Research, Canberra, Australia, pp. 14–16.

CABI (2011) *Chromolaena odorata* (Siam weed). Invasive Species Compendium datasheet. CAB International, Wallingford, UK. Available at: www.cabi.org/isc/datasheet/23248 (accessed 14 December 2014).

Campbell, S. (2005) A global perspective on forest invasive species: the problem, causes and consequences. In: McKenzie, P., Brown, C., Su, J., and Wu, J. (eds) *The Unwelcome Guests. Proceedings of the Asia-Pacific Forest Invasive Species Conference, Kunming, China, 17–23 August 2003.* FAO Regional Office for Asia and the Pacific, Bangkok, Thailand, pp. 9–10.

Chetti, M.B., Hiremath, S.M., Prashanthi, S.K., Mummigatti, U.V. and Kulkarni, S. (2001) Survey and screening of various pathogen for biological control of *Chromolaena odorata*. In: Sankaran, K.V., Murphy, S.T. and Evans, H.C. (eds) *Alien Weeds in Moist Tropical Zones: Banes and Benefits. Proceedings of a Workshop, Kerala Forest Research Institute, Peechi, India, 2–4 November 1999*. Kerala Forest Research Institute, Peechi, India and CABI Bioscience, UK Centre (Ascot), Ascot, UK, pp. 146–149.

Cock, M.J.W., Ellison, C.A., Evans, H.C. and Ooi, P.A.C. (2000) Can failure be turned into success for biological control of mile-a-minute weed (*Mikania micrantha*)? In: Spencer, N.R. (ed.) *Proceedings of the X International Symposium on Biological Control of Weeds, Bozeman, Montana, 4–14 July 1999*. USDA Forest Service, Forest Health Technology Enterprise Team, Morgantown, West Virginia, pp. 155–167.

Day, M.D. and Winston, R.L. (2016) Biological control of weeds in the 22 Pacific island countries and territories: current status and future prospects. In: Daehler, C.C., van Kleunen, M., Pyšek, P. and Richardson, D.M. (eds) *Proceedings of the 13th International EMAPi Conference, Waikoloa, Hawaii. NeoBiota* 30, 167–92.

Day, M.D., Wiley, C.J., Playford, J. and Zalucki, M.P. (2003) *Lantana: Current Management Status and Future Prospects*. ACIAR Monograph 102, Australian Centre for International Agricultural Research, Canberra, Australia.

Day, M.D., Kawi, A., Tunabuna, A., Fidelis, J., Swamy, B., Ratutuni, J., Saul-Maora, J., Dewhurst, C.F. and Orapa, W. (2011) The distribution and socio-economic impacts of *Mikania micrantha* (Asteraceae) in Papua New Guinea and Fiji and prospects for its biological control. In: *23rd Asian-Pacific Weed Science Society Conference, The Sebel Cairns, Australia, 26–29 September 2011: Weed Management in a Changing World. Conference Proceedings, Volume 1*. Asia-Pacific Weed Science Society, pp. 146–153. Available at: http://apwss.org/apwss-publications.htm (accessed 14 July 2017).

Day, M.D., Kawi, A., Kurika, K., Dewhurst, C.F., Waisale, S., Saul-Maora, J., Fidelis, J., Bokosou, J., Moxon, J., Orapa, W. and Senaratne, K.A.D. (2012) *Mikania micrantha* Kunth (Asteraceae) (mile-a-minute): its distribution and physical and socio economic impacts in Papua New Guinea. *Pacific Science* 66, 213–223.

Day, M.D., Kawi, A.P., Fidelis, J., Tunabuna, A., Orapa, W., Swamy, B., Ratutini, J., Saul-Maora, J. and Dewhurst, C.F. (2013a) Biology, field release, and monitoring of the rust fungus *Puccinia spegazzinii* (Pucciniales: Pucciniaceae), a biological control agent of *Mikania micrantha* (Asteraceae) in Papua New Guinea and Fiji. In: Wu, Y., Johnson, T., Singh, S., Raghu, S., Wheeler, G, Pratt, P., Warner, K., Center, T., Goosby, J. and Reardon, R. (eds) *Proceedings of the XIII International Symposium on Biological Control of Weeds, Waikoloa, Hawaii, 11–16 September 2011*. USDA Forest Service, Forest Health Technology Enterprise Team, Morgantown, West Virginia, pp. 211–217.

Day, M.D., Kawi, A.P. and Ellison, C.A. (2013b) Assessing the potential of the rust fungus *Puccinia spegazzinii* as a classical biological control agent for the invasive weed *Mikania micrantha* in Papua New Guinea. *Biological Control* 67, 253–261.

Day, R.K. and Murphy, S.T. (1998) Pest management in agroforestry: prevention rather than cure. *Agroforestry Forum* 9, 11–14.

Doddamani, M.B., Mummigatti, U.V., Nadagoudar, B.S. and Chetti, M.B. (2001a) *Chromolaena* in Karnataka: problems and prospects. In: Sankaran, K.V., Murphy, S.T. and Evans, H.C. (eds) *Alien Weeds in Moist Tropical Zones: Banes and Benefits. Proceedings of a Workshop, Kerala Forest Research Institute, Peechi, India, 2–4 November 1999*. Kerala Forest Research Institute, Peechi, India and CABI Bioscience, UK Centre (Ascot), Ascot, UK, pp. 42–45.

Doddamani, M.B., Mummigatti, U.V., Nadagoudar, B.S. and Chetti, M.B. (2001b) Influence of chemical and conventional methods of weed control on *Chromolaena odorata*. In: Sankaran, K.V., Murphy, S.T. and Evans, H.C. (eds) *Alien Weeds in Moist Tropical Zones: Banes and Benefits. Proceedings of a Workshop, Kerala Forest Research Institute, Peechi, India, 2–4 November 1999*. Kerala Forest Research Institute, Peechi, India and CABI Bioscience, UK Centre (Ascot), Ascot, UK, pp. 139–142.

Dutta, T.R. (1978) The present status of *Eupatorium* and *Mikania* problems in the NE hills region and projections on their control. In: Borthakur, D.N. and Gosh, S.P. (eds) *Studies on Weeds and their Control*. Meghalaya Science Society, Shillong, India, pp. 17–21.

Ellison, C.A. and Day, M. (2011) Current status of releases of *Puccinia spegazzinii* for *Mikania micrantha* control. *Biocontrol News and Information* 32, 1N.

Ellison, C.A., Evans, H.C., Djeddour, D.H. and Thomas, S.E. (2008) Biology and host range of the rust fungus *Puccinia spegazzinii*: a new classical biological control agent for the

invasive, alien weed *Mikania micrantha* in Asia. *Biological Control* 45, 133–145.
Evans, H.C. (2001) Classical biological control: a tailor made strategy for the management of alien weeds. In: Sankaran, K.V., Murphy, S.T. and Evans, H.C. (eds) *Alien Weeds in Moist Tropical Zones: Banes and Benefits. Proceedings of a Workshop, Kerala Forest Research Institute, Peechi, India, 2–4 November 1999.* Kerala Forest Research Institute, Peechi, India and CABI Bioscience, UK Centre (Ascot), Ascot, UK, pp. 35–41.
Gogoi, A.K. (2001) Status of *Mikania* infestation in north-eastern India: management options and future research thrust. In: Sankaran, K.V., Murphy, S.T. and Evans, H.C. (eds) *Alien Weeds in Moist Tropical Zones: Banes and Benefits. Proceedings of a Workshop, Kerala Forest Research Institute, Peechi, India, 2–4 November 1999.* Kerala Forest Research Institute, Peechi, India and CABI Bioscience, UK Centre (Ascot), Ascot, UK, pp. 77–79.
Gujral, G.S., and Vasudevan, P. (1983) *Lantana camara* L. – a problem weed. *Journal of Scientific and Industrial Research* 42, 281–286.
Gunaseelan, V.N. (1987) Parthenium as an additive with cattle manure in biogas production. *Biological Wastes* 21, 195–202.
Gunaseelan, V.N. (1994) Methane production from *Parthenium hysterophorus* L., a terrestrial weed, in semi-continuous fermenters. *Biomass and Bioenergy* 6, 391–398.
Hannan-Jones, M.A. (1998) The seasonal response of *Lantana camara* to selected herbicides. *Weed Research* 38, 413–423.
Harr, J. (1992) The role of industry in the future of weed science. *Weed Technology* 6, 177–181.
Holm L.G., Plucknett, D.L., Pancho, J.V. and Herberger, J.P. (1991) *The World's Worst Weeds: Distribution and Biology.* Kreiger Publishing, Malabar, Florida.
Huang, H.-Z., Zhao, J.-B., Huang, B.-Q., Zhang, Y.-X. and Yan, L. (2004) Phenoxy-hydroxy-acid herbicides for controlling the weed *Mikania micrantha. Journal of South China Agricultural University* 25, 52–55. [In Chinese, English abstract.]
Ipor, I.B. and Price, C.E. (1992) The effect of simulated rain on the activity and uptake of imazapyr on *Mikania micrantha* H.B.K. *Pertanika* 15, 99–103.
Ipor, I.B. and Price, C.E. (1994) Uptake, translocation and activity of paraquat on *Mikania micrantha* H.B.K. grown in different light conditions. *International Journal of Pest Management* 40, 40–45.
Joy, P.J., Satheesan, N.V. and Lyla, K.R. (1987) Biological control of weeds in Kerala. In: Joseph, K.J. and Abdurahiman, U.C. (eds) *Advances in Biological Control Research in India. Proceedings of the First National Seminar on Entomophagous Insects and other Arthropods and their Potential in Biological Control, Calicut University, Kerala, India, 9–11 October 1985.* University of Calicut, Kerala, India, pp. 247–251.
Kannan, R., Gladwin, J. and Shaanker, R.U. (2008) Conserving rattan and *Wrightia tinctoria* by utilizing the invasive weed *Lantana camara* as a substitute. In: *Abstract Book of the Proceedings of the 22nd Annual Conference of the Society for Conservation Biology, From the Mountains to the Sea, 13–17 July 2008, Chattanooga, Tennessee, USA.* Society for Conservation Biology, Washington, DC, Abstract 18. Available at: http://conbio.org/images/content_conferences/2008_Abstract_Book.pdf (accessed 6 March 2017).
Kannan, R., Shackleton, C.M. and Shaanker, R.U. (2014) Invasive alien species as drivers in socio-ecological systems: local adaptations towards use of *Lantana* in southern India. *Environment, Development and Sustainability* 16, 649–669.
Kumar, N., Chander, N., Sood, B.R. and Bhandari, J.C. (2007) Effect of weed management techniques and planting of improved grasses on forage production, economics and nutrient uptake from lantana infested wasteland. *Range Management and Agroforestry* 28, 176–178.
Kuo, Y.-L., Chen, T.-Y. and Lin, C.-C. (2002) Using a consecutive-cutting method and allelopathy to control the invasive vine, *Mikania micrantha* H.B.K. *Taiwan Journal of Forest Science* 17, 171–181. [In Chinese, English abstract.]
Leakey, R.R.B. (2010) *Should We Be Growing More Trees on Farms to Enhance the Sustainability of Agriculture and Increase Resilience to Climate Change?* ISTF News, Special Report, February 2010, International Society for Tropical Forestry, Bethesda, Maryland. Available at: http://www.istf-bethesda.org/specialreports/leakey/Agroforestry-Leakey.pdf (accessed 6 March 2017).
Lee, L.J. and Ngim, J. (1997) Evaluating non-selective herbicides for rain fast properties in Malaysia. *Planter* 73, 65–68.
Little, K.M., Willoughby, I., Wagner, R.G., Adams, P., Frochot, H., Gava, J., Gous, S., Lautenschlager, R.A., Orlander, H., Sankaran, K.V. and Wei, R.P. (2006) Towards reduced herbicide use in forest vegetation management. *South African Journal of Forestry* 207, 63–79.
Macanawai, A.R., Day, M.D., Tumaneng-Diete, T. and Adkins, S.W. (2010) Frequency and density of *Mikania micrantha* and other weeds in taro and banana systems in eastern Viti Levu, Fiji. In: Zydenbos, M.N. (ed.) *Proceedings of the 17th*

Australasian Weeds Conference. New Frontiers in New Zealand: Together We Can Beat the Weeds, Christchurch, New Zealand, 26–30 September 2010. New Zealand Plant Protection Society, Auckland, New Zealand, pp. 116–119.

McFadyen, R.E.C. (1989) Siam weed: a new threat to Australia's north. *Plant Protection Quarterly* 4, 3–7.

McFadyen, R.E.C. (2000) Successes in biological control of weeds. In: Spencer, N.R. (ed.) *Proceedings of the X International Symposium on Biological Control of Weeds, Bozeman, Montana, 4–14 July 1999.* USDA Forest Service, Forest Health Technology Enterprise Team, Morgantown, West Virginia, pp. 3–14.

McFadyen, R.E.C. (2002) *Chromolaena* in Asia and the Pacific: spread continues but control prospects improve. In: Zachariades, C.R., Muniappan, R and Strathie, L.W. (eds) *Proceedings of the Fifth International Workshop on Biological Control and Management of* Chromolaena odorata, *Durban, South Africa, 23–25 October 2000.* Agricultural Research Council-Plant Protection Research Institute (ARC-PPRI), Pretoria, pp. 13–18.

McFadyen, R.C. (2003) *Chromolaena* in Southeast Asia and the Pacific. In: da Costa, H., Piggin, C., da Cruz, C.J. and Fox, J.J. (eds) *Agriculture: New Directions for a New Nation – East Timor (Timor Leste). Proceedings of a Workshop, Dili, East Timor, 1–3 October 2002.* ACIAR Proceedings No. 113, Australian Centre for International Agricultural Research, Canberra, pp. 130–134.

McWilliam, A. (2000) A plague on your house? Some impacts of *Chromolaena odorata* on Timorese livelihoods. *Human Ecology* 28, 451–469.

Miller, I.L. (2004) Uses for *Mimosa pigra.* In: Julien, M., Flanagan, G., Heard, T., Hennecke, B., Paynter, Q. and Wilson, C (eds) *Research and Management of* Mimosa pigra. *Proceedings of the Third International Symposium on* Mimosa pigra, *Darwin, Australia, 23–25 September 2002.* CSIRO (Commonwealth Scientific and Industrial Research Organisation) Entomology, Canberra, pp. 63–67.

Mohankumar, H.T. (1996) Benefit of growing green manure crop in coffee under coconut. *Indian Coffee* 60, 7–8.

Moni, N.S. and Subramoniam, R. (1960) Essential oil from *Eupatorium odoratum* – a common weed in Kerala. *Indian Forester* 86, 209.

Muniappan, R., Englberger, K., Bamba, J. and Reddy, G.V.P. (2004) Biological control of chromolaena in Micronesia. In: Day, M.D. and McFadyen, R.E. (eds) *Chromolaena in the Asia-Pacific region. Proceedings of the Sixth International Workshop on Biological Control and Management of Chromolaena, Cairns, Australia, 6–9 May 2003.* ACIAR Technical Report No. 55, Australian Centre for International Agricultural Research, Canberra, Australia, pp. 11–12.

Muraleedharan, P.K. and Anitha, V. (2001) The economic impact of *Mikania micrantha* in the agroforestry production system in the Western Ghats of Kerala. In: Sankaran, K.V., Murphy, S.T. and Evans, H.C. (eds) *Alien Weeds in Moist Tropical Zones: Banes and Benefits. Proceedings of a Workshop, Kerala Forest Research Institute, Peechi, India, 2–4 November 1999.* Kerala Forest Research Institute, Peechi, India and CABI Bioscience, UK Centre (Ascot), Ascot, UK, pp. 80–85.

Murphy, S.T., Subedi, N., Jnawali, S.R., Lamichhane, B.R., Upadhyay, G.P., Kock, R. and Amin, R. (2013) Invasive mikania in Chitwan National Park, Nepal: the threat to the greater one-horned rhinoceros, *Rhinoceros unicornis,* and factors driving the invasion. *Oryx* 47, 361–368.

Nair, P.K.R. (1992) *An Introduction to Agroforestry.* Kluwer, Dordrecht, The Netherlands with the International Centre for Research in Agroforestry, Nairobi. Available at: www.worldagroforestry.org (accessed 17 July 2017).

Napompeth, B. (1983) Preliminary screening of insects for biological control of *Mimosa pigra* L. in Thailand. In: *Proceedings of an International Symposium on* Mimosa pigra *Management, Chiang Mai, Thailand, 22–26 February 1982.* Document 48-A-83, International Plant Protection Centre, Corvallis, Oregon, pp. 121–127.

Narasimhan, T.R., Murthy, B.S.K. and Rao, P.V.S. (1993) Nutritional evaluation of silage made from the toxic weed *Parthenium hysterophorus* in animals. *Food and Chemical Toxicology* 31, 509–515.

Nishanathan, K., Sivachandiran, S. and Marambe, B. (2013) Control of *Parthenium hysterophorus* L. and its impact on yield performance of tomato (*Solanum lycopersicum* L.) in the northern province of Sri Lanka. *Tropical Agricultural Research* 25, 56–68.

Prashanthi, S.K. and Kulkarni, S. (2005) *Aureobasidium pullulans*, a potential mycoherbicide for biocontrol of eupatorium (*Chromolaena odorata* (L) King and Robinson) weed. *Current Science* 88, 18–21.

Prinz, K. and Ongprasert, S. (2007) Management of *Mimosa diplotricha* var. *inermis* as a simultaneous fallow in northern Thailand. In: Cairns, M. (ed.) *Voices from the Forest: Integrating Indigenous Knowledge into Sustainable Upland*

Farming. Resources for the Future, Washington, DC, pp. 214–225.
Priyanka, N. and Joshi, P.K. (2013) A review of *Lantana camara* studies in India. *International Journal of Scientific and Research Publications* 3, 1–11.
Pundir, Y.P.S. (1985) Preliminary observations on the control of *Lantana camera* by *Cuscuta santapaui* Banerji and Das in Dehradun valley. *Acta Botanica Indica* 13, 298–300.
Rai, R.K., Sandilya, M. and Subedi, R. (2012a) Controlling *Mikania micrantha* HBK: how effective manual cutting is? *Journal of Ecology and Field Biology* 35, 235–242.
Rai, R.K., Scarborough, H., Subedi, N. and Lamichhane, B. (2012b) Invasive plants – do they devastate or diversify rural livelihoods? Rural farmers' perception of three invasive plants in Nepal. *Journal for Nature Conservation* 20, 170–176.
Ramakrishnan, P.S. (2001) Biological invasion as a component of global change: the Indian context. In: Sankaran, K.V., Murphy, S.T. and Evans, H.C. (eds) *Alien Weeds in Moist Tropical Zones: Banes and Benefits. Proceedings of a Workshop, Kerala Forest Research Institute, Peechi, India, 2–4 November 1999.* Kerala Forest Research Institute, Peechi, India and CABI Bioscience, UK Centre (Ascot), Ascot, UK, pp. 28–34.
Saini, A., Aggarwal, N.K., Sharma, A., Kaur, M. and Yadav, A. (2014) Utility potential of *Parthenium hysterophorus* for its strategic management. *Advances in Agriculture* 2014, Art. ID 381859. Available at: http://dx.doi.org/10.1155/2014/381859.
Salgado, M.L.M. (1972) *Tephrosia purpurea* (pila) for the control of *Eupatorium* and as a green manure on coconut estates. *Ceylon Coconut Planters' Review* 6, 160–174.
Sankaran, K.V. and Pandalai, R.C. (2004) Field trials for controlling mikania infestation in forest plantations and natural forests in Kerala. KFRI Research Report No. 265, Kerala Forest Research Institute, Peechi, India.
Sankaran, K.V. and Sreenivasan, M.A. (2001) Status of *Mikania* infestation in the Western Ghats. In: Sankaran, K.V., Murphy, S.T. and Evans, H.C. (eds) *Alien Weeds in Moist Tropical Zones: Banes and Benefits. Proceedings of a Workshop, Kerala Forest Research Institute, Peechi, India, 2–4 November 1999.* Kerala Forest Research Institute, Peechi, India and CABI Bioscience, UK Centre (Ascot), Ascot, UK, pp. 67–76.
Sankaran, K.V., Muraleedharan, P.K. and Anitha, V. (2001) Integrated management of the alien invasive weed *Mikania micrantha* in the Western Ghats. KFRI Research Report No. 202, Kerala Forest Research Institute, Peechi, India.
Sankaran, K.V., Mendham, D.S., Chacko, K.C., Pandalai, R.C., Grove, T.S. and O'Connell, A.M. (2007) Increasing and sustaining productivity of tropical eucalypt plantations over multiple rotations. *Journal of Sustainable Forestry* 24, 109–121.
Sankaran, K.V., Puzari, K.C., Ellison, C.A., Sreerama Kumar, P. and Dev, U. (2008) Field release of the rust fungus *Puccinia spegazzinii* to control *Mikania micrantha* in India: protocols and awareness raising. In: Julien, M.H., Sforza, R., Bon, M.C., Evans, H.C., Hatcher, P.E., Hinz, H.L. and Rector, B.G. (eds) *Proceedings of the XII International Symposium on Biological Control of Weeds, La Grande Motte, France, 22–27 April 2007.* CAB International, Wallingford, UK, pp. 384–389.
Seibert, T.F. (1989) Biological control of the weed *Chromolaena odorata* (Asteraceae) by *Pareuchaetes pseudoinsulata* in Guam and northern Mariana islands. *Entomophaga* 34, 531–539.
Singh, R.M. and Poudel, M.S. (2013) Briquette fuel – an option for management of *Mikania micrantha. Nepal Journal of Science and Technology* 14, 109–114.
Singh, S.P. (2001) Biological control of invasive weeds in India. In: Sankaran, K.V., Murphy, S.T. and Evans, H.C. (eds) *Alien Weeds in Moist Tropical Zones: Banes and Benefits. Proceedings of a Workshop, Kerala Forest Research Institute, Peechi, India, 2–4 November 1999.* Kerala Forest Research Institute, Peechi, India and CABI Bioscience, UK Centre (Ascot), Ascot, UK, pp. 11–19.
Siwakoti, M. (2007) Mikania weed: a challenge for conservationists. *Our Nature* 5, 70–74.
Soerjani, M. (1977) Weed management and weed science development in Indonesia. In: Day, M.D. and McFadyen, R.E. (eds) *Chromolaena in the Asia-Pacific Region. Proceedings of the Sixth Asian-Pacific Weed Science Society Conference, Jakarta Indonesia, 11–17 July 1977, Vol. 1.* Asian-Pacific Weed Science Society, pp. 31–41. Available at: http://apwss.org/apwss-publications.htm (accessed 14 July 2017).
Sood, K.G., Rawat, G.S., Kumar, D. and Singh, R.P. (2001) Uprooting of lantana – a comparative study. *Indian Forester* 127, 512–518.
Suresh, T.A. (2014) Ecology and invasion dynamics of the giant weed complex in the Western Ghats. PhD thesis, Forest Research Institute, Dehradun, India.
Sushilkumar (2001) Management of lantana in India: trend[s], prospects and need of integrated

approach. In: Sankaran, K.V., Murphy, S.T. and Evans, H.C. (eds) *Alien Weeds in Moist Tropical Zones: Banes and Benefits. Proceedings of a Workshop, Kerala Forest Research Institute, Peechi, India, 2–4 November 1999.* Kerala Forest Research Institute, Peechi, India and CABI Bioscience, UK Centre (Ascot), Ascot, UK, pp. 95–106.

Sushilkumar and Saraswat, V.N. (2001) Integrated management: the only solution to suppress *Parthenium hysterophorus.* In: Sankaran, K.V., Murphy, S.T. and Evans, H.C. (eds) *Alien Weeds in Moist Tropical Zones: Banes and Benefits. Proceedings of a Workshop, Kerala Forest Research Institute, Peechi, India, 2–4 November 1999.* Kerala Forest Research Institute, Peechi, India and CABI Bioscience, UK Centre (Ascot), Ascot, UK, pp. 150–168.

Swaminath, M.H. and Shivanna, M. (2001) The ecological impact of *Chromolaena odorata* in the Western Ghat forests of Karnataka and the management strategies to minimize the impact. In: Sankaran, K.V., Murphy, S.T. and Evans, H.C. (eds) *Alien Weeds in Moist Tropical Zones: Banes and Benefits. Proceedings of a Workshop, Kerala Forest Research Institute, Peechi, India, 2–4 November 1999.* Kerala Forest Research Institute, Peechi, India and CABI Bioscience, UK Centre (Ascot), Ascot, UK, pp. 112–114.

Tewari, J.C., Harsh, L.N., Sharma, A.K. and Khan, M.A. (2006) *Prosopis juliflora* debate: is it a valuable resource or an invasive weed in India? *APANews* 28, 3–4.

Turner, A.J. (ed.) (2015) *The Pesticide Manual: A World Compendium,* 17th edn. British Crop Protection Council, Alton, UK.

Varma, R.V., Shetty, A., Swaran, P.R., Paduvil, R. and Shamsudeen, R.S.M. (2006) Establishment of *Pareuchaetes pseudoinsulata* (Lepidoptera: Arctiidae), an exotic biocontrol agent of the weed, *Chromolaena odorata* (Asteraceae) in the forests of Kerala, India. *Entomon* 31, 49–51.

Wang, Y.S., Jaw, C.G. and Chen, Y.I. (1994) Accumulation of 2,4-D and glyphosate in fish and water hyacinth. *Water, Air and Soil Pollution* 74, 397–403.

WHO (2009) *The WHO Recommended Classification of Pesticides of Hazard and Guidelines to Classification 2009.* World Health Organization, Geneva, Switzerland. Available at: http://www.who.int/ipcs/publications/pesticides_hazard_2009.pdf (accessed 3 March 2017).

Willson, B.W. and Garcia, C.A. (1992) Host specificity and biology of *Heteropsylla spinulosa* (Hom.: Psyllidae) introduced into Australia and Western Samoa for the biological control of *Mimosa invisa. Entomophaga* 37, 293–299.

Wilson, C.G. and Widayanto, E.B. (2004) Establishment and spread of *Cecidochares connexa* in eastern Indonesia. In: Day, M.D. and McFadyen, R.E. (eds) *Chromolaena in the Asia-Pacific Region. Proceedings of the Sixth International Workshop on Biological Control and Management of Chromolaena, Cairns, Australia, 6–9 May 2003.* ACIAR Technical Report No. 55, Australian Centre for International Agricultural Research, Canberra, pp. 39–44.

Winston, R.L., Schwarzländer, M., Hinz, H.L., Day, M.D., Cock, M.J.W. and Julien, M.H. (eds) (2014) *Biological Control of Weeds: A World Catalogue of Agents and Their Target Weeds,* 5th edn. USDA Forest Service, Forest Health Technology Enterprise Team, Morgantown, West Virginia.

Witt, A., Day, M., Urban, A., Sankaran, K.V. and Shaw, D. (2012) Lantana: the battle can be won. *Biocontrol News and Information* 33, 13N–15N.

Yu, H., He, W.-M., Liu, J., Miao, S.-L. and Dong, M. (2009) Native *Cuscuta campestris* restrains exotic *Mikania micrantha* and enhances soil resources beneficial to natives in the invaded communities. *Biological Invasions* 11, 835–844.

Zachariades, C., Day, M., Muniappan, R. and Reddy, G.V.P. (2009) *Chromolaena odorata* (L.) King and Robinson (Asteraceae). In: Muniappan, R., Reddy, G.V.P. and Raman, A. (eds) *Biological Control of Tropical Weeds Using Arthropods.* Cambridge University Press, Cambridge, UK. pp. 130-162.

Zhang, L.Y., Ye, W.H., Cao, H.L. and Feng, H.L. (2004) *Mikania micrantha* H.B.K. in China – an overview. *Weed Research* 44, 42–49.

10 Classical Biological Control of *Mikania micrantha*: the Sustainable Solution

Carol A. Ellison* and Matthew J.W. Cock
CABI, Egham, UK

Introduction – Concepts

Invasive alien plants pose a serious threat to agricultural production, biodiversity of the natural environment and the sustainability of human societies (Wittenberg and Cock, 2001; Pimentel, 2002; Diamond, 2005; Perrings *et al.*, 2010). Physical, cultural and chemical control options (Chapter 9, this volume) can help to reduce the impact of these introduced species in limited areas of agriculture and forestry, but they are both expensive and unsustainable. Moreover, the non-target effects can be environmentally undesirable, and there is now concern throughout the world about the environmental impact and toxic effects of the widespread use of chemical methods of pest control (e.g. Diamond, 2005). This concern has, in part, fuelled the increasing global interest in developing more sustainable and environmentally friendly methods of pest management in agriculture (Bale *et al.*, 2008; Willer *et al.*, 2013). The Global Invasive Species Programme (GISP) helped countries to catalyse action against invasive alien species by developing national and regional control and prevention strategies from 1997 to 2011. Under this programme, the first global best practice guidelines were produced (Wittenberg and Cock, 2001); these advocate classical biological control (CBC) as a sustainable control strategy available for invasive weeds, particularly as part of an integrated management approach. CBC fits well into an integrated, biologically based approach to pest management in agro-ecosystems (Labrada, 1996; Charudattan, 2001). Increasingly, it is the only viable long-term option for the control of invasive alien weeds in rangeland and natural environments (McFadyen, 1998).

Principles of classical biological control (CBC)

CBC is based on the principle that plant species tend to be moved around the world, either deliberately or accidentally, and without most or all of the natural enemies that attack them in the countries they originate from, i.e. their native range. Once freed from these controlling agents, some plant species, therefore, become more competitive than the native plants around them that are subject to natural control, and are able to multiply rapidly (Wilson, 1969; Blossey and Nötzold, 1995; Keane and Crawley, 2002). Fortunately, most introduced plant species do not become weedy once they are established in a new region. However, without their natural enemies to control them, and often aided by human disruption of natural habitats, population explosions of some species occur, with the subsequent development of weed invasions (Mack *et al.*, 2000).

* Corresponding author. E-mail: c.ellison@cabi.org

A study by Mitchell and Power (2003) has provided strong evidence in support of the concept that plants become weedy in a new region because their natural enemies are not there to control them. Their study looked specifically at co-evolved plant pathogens (see below), and found that in the USA, invasive plants originating from Europe have only 23% of the fungal and viral diseases that they have in their native range.

Natural enemies can be insects, mites, fungi, bacteria, viruses and, of course, larger grazing animals. If these natural enemies are to be used to control, or at least to suppress, invasive plants in their new, exotic range, it is essential that those selected for release have the following characteristics:

- They attack a very narrow range of plant species, ideally only the target weed.
- They spread easily between plants and between populations of the weed.
- They are able to develop high-density populations or infestations on the target plants.
- They are damaging to the plant, reducing its ability to reproduce and compete.

Many species of insects, mites and fungi have these characteristics, and these are the groups almost exclusively used for the CBC of invasive alien weeds.

Locating natural enemies that attack only a narrow range of species is critical to successful CBC. It has a strong theoretical base, from which safety testing protocols have also been developed (see later in this section). A natural enemy with a restricted host range has usually co-evolved with its host over their joint evolutionary history, with the host evolving mechanisms to evade the natural enemy, and the natural enemy evolving mechanisms to overcome them – the analogy of an evolutionary arms race is often made. However, these defence and attack mechanisms come at a cost, and resources are limited. In evolving adaptations to overcome the defences of a particular host plant, the natural enemy becomes less able to attack plants that do not share these defences. The relationship between co-evolved natural enemies and their host plants involves many genes. Closely related plants are more likely to share similar genes than unrelated plants, so a co-evolved natural enemy tends to be restricted to, and be able to survive on, only one or a limited number of closely related plant species. If there is no host plant available, then the natural enemy dies.

The implementation of CBC is underpinned by one of the International Standards for Phytosanitary Measures (ISPMs) from the IPPC (International Plant Protection Convention), specifically *ISPM 3: Guidelines for the Export, Shipment, Import and Release of Biological Control Agents and Other Beneficial Organisms* (IPPC, 2005). Selected natural enemies have to pass a comprehensive evaluation and screening programme, following the guidelines described in, for example, Harley and Forno (1992), in order to meet the information requirements set out in ISPM 3 (IPPC, 2005; also see Kairo *et al.*, 2003). Following this, a decision can be made on their introduction and release as biological control agents in the exotic target area. The most critical part of this risk assessment involves testing the capacity of the natural enemy to feed on (insect or mite) or infect (pathogen) other plant species. The number of plants tested can be quite high – often over 100 different species. The test plants are chosen according to a method developed by Wapshere (1974), which is known as the phylogenetic (evolutionary relatedness), centrifugal testing method. Test plant species are selected on the basis of their relatedness to the target weed species, as these are the species most at risk. Starting from the most closely related, and working outwards (genus, tribe, family – hence 'centrifugal') the testing determines how specific the natural enemy is to the weed host. In addition to this scientific approach, local crop species and plants of importance for other reasons (e.g. their medicinal or cultural importance) are screened to allay the understandable concerns of the public. Post-release analyses of many natural enemy introductions for weed biological control have confirmed that only the most closely related plant species are at risk (e.g. Pemberton, 2000). The advantages and disadvantages ('pros' and 'cons') of CBC are summarized in Box 10.1.

Box 10.1. The pros and cons of CBC.

Advantages

Inherently safe: only co-evolved, host-specific natural enemies are considered as potential biological control agents.
Cost effective: once biological control agents are released and established, no further human inputs are required, although redistribution programmes will speed up their impact.
Target specific: the natural enemy selected will only attack one or a limited number of plant species.
Practical: can be implemented in almost any environment, using relatively inexpensive, uncomplicated procedures.
Environmentally benign: has a minimal negative effect on native flora and fauna.
Efficacious: damage to the target weed can be an important factor in the reduction and spread of that weed.
Sustainable: once released and established, the biological control agent is self-perpetuating.
Proven track record: there are many examples of successful projects (Winston *et al.*, 2014), and, since the introduction of scientific testing procedures, virtually no unpredicted non-target effects have been recorded (Marohasy, 1996; Barton, 2012; Suckling and Sforza, 2014).

Disadvantages

Can have long lag-phase: there is often 5–10 years after release before an impact on the weed population is seen.
Long research phase: it can take 5–10 years for a project to reach the stage when the first natural enemy can be released in the field.
Need for opinion of wide range of stakeholders: delays in project implementation can occur during the pre-release consultative phase.
Potential conflicts of interest: some invasive weeds have benefits, e.g. providing pollen for bees, firewood and soil stabilization; thus, objections can be raised to the implementation of a CBC programme.
Does not always work: some natural enemies, on their own, do not have sufficient impact to control a weed; often, a suite or guild of natural enemies, or a combination of control methods, is required.
Non-commercial: funding for CBC has to be almost exclusively through donor organizations, governments or industry groups because a self-perpetuating method of control cannot be exploited for monetary gain.
No eradication: once an agent has been released, there is little chance of removing it from the environment; hence, each agent must be fully risk assessed before release.

How successful is CBC?

CBC has been employed for over a century utilizing insects and mites, and there have been some spectacular success stories (see McFadyen, 2000; Winston *et al.*, 2014). For example, the complete control of the South American water fern (*Salvinia molesta*) in Australia (and subsequently in other parts of the world; see Chapter 6, this volume) was achieved with a weevil (*Cyrtobagous salviniae*) imported from Brazil (Thomas and Room, 1986). In contrast, the exploitation of fungal pathogens is a relatively new, but growing approach (See Box 10.2). Clewley *et al.* (2012) looked at the effectiveness of the CBC of invasive plants by analysing 61 published studies from 2000 to 2011, and found that biological control agents significantly reduced plant size (28±4%), plant mass (37±4%), flower and seed production (35±13% and 42±9%, respectively) and target plant density (56±7%). They also found that the non-target plant diversity significantly increased (88±31%). McFadyen (2012) points out that even a small impact of a CBC agent on a weed can significantly reduce the cost of controlling it using conventional methods. Culliney (2005) reviewed the economics from 32 CBC

Box 10.2. CBC with fungi: success stories.

The first release of a pathogen for the CBC of a weed was made in 1972 in Australia, when a rust fungus (*Puccinia chondrillina*) was introduced from Europe to control skeleton weed, *Chondrilla juncea*, in cereal crops (Cullen *et al.*, 1973; Hasan and Wapshere, 1973). The total estimated saving due to increased crop yields and reduced herbicide usage varied; however, the cost:benefit ratio has been put at 1:112 (Marsden *et al.*, 1980). Mortensen (1986) credited this pathogen with an annual saving of over US$12 million, and Marsden *et al.* (1980) estimated a saving of AU$260 million, projected up to the year 2000. Since 1972, over 28 introductions of fungal pathogens to control alien weeds have been made worldwide, and a significant number of these have either been successful in reducing the impact of an exotic weed or are looking highly promising (Evans, 2002; Morin *et al.*, 2006; Barton, 2012).

For example, the gall-forming rust fungus (*Uromycladium tepperianum*) (see Fig. 10.1 below) was introduced from Australia to control Port Jackson willow (*Acacia saligna*), an invasive alien and damaging weed of the unique fynbos ecosystem of South Africa. After an 8- to 10-year lag phase, the rust is now responsible for a 90–95% reduction in the weed population, and by the turn of the century, the fynbos was in the process of recovery (Morris, 1997).

In Queensland, Australia, the rubbervine weed (*Cryptostegia grandiflora*) was dramatically controlled by a highly specific rust fungus (*Maravalia cryptostegiae*), introduced from the native range of the plant in Madagascar in 1992. Rubbervine is now diminishing in importance over the entire 60,000 km^2 of grazing land and the native plant communities that it had invaded – in riverbank habitats, in particular, the vine literally overgrew the gallery forest, eliminating the native flora and fauna. The rust is having a spectacular effect in reducing the growth and seed production of the weed, eliminating seedling recruitment in many areas and slowing the spread of the weed. The native vegetation is now regenerating under the diminished rubbervine infestations (Tomley and Evans, 2004).

Fig. 10.1. The gall-forming rust *Uromycladium tepperianum*. Photo courtesy A.R. Wood.

projects for which adequate data existed; he found that all of the projects showed an impact on the weeds under CBC, with a mean cost:benefit ratio of 1:200 (range from 1:2.3 to 1:4000).

Key issues in implementing CBC

There are a number of key issues that need to be addressed when considering CBC as a strategy for the control of invasive alien weeds, and these are summarized below.

Funding

As CBC can seldom be a profit-making enterprise, almost all of its funding comes from governments, usually at the national level. The ephemeral nature of governments inevitably means that CBC is rarely funded for adequate periods of time, i.e. for what is a long-term strategy, except where the relevant government agencies are already familiar with this issue.

New environmental agenda

The ratification of the Convention on Biological Diversity (CBD) in 1992 by 150 government leaders was an important event in world history for the protection of global biodiversity (www.cbd.int/intro/default.shtml). However, new access and benefit-sharing (ABS) procedures being developed under the 'Nagoya Protocol on Access to Genetic Resources and the Fair and Equitable Sharing of Benefits Arising from their Utilization to the Convention on Biological Diversity', an international agreement (CBD, 2010), may threaten the future of CBC (Cock *et al.*, 2010). Already, recent applications of the CBD principles, ostensibly for the protection of countries' rights over their genetic resources, have resulted in the curtailment of surveying and export of natural enemies for biocontrol research. If this situation continues, CBC as a safe and effective method of pest management will be put at risk. Cock *et al.* (2010) provide recommendations to help policy makers during the process of developing ABS under the Nagoya Protocol.

Government policy frameworks

Government policy frameworks need to be in place to facilitate the importation of biological control agents. The framework in India, which has a successful history in CBC implementation, is discussed in Chapter 12 (this volume).

Time lag

It needs to be made clear to stakeholders in a project that results will not be instantaneous – it is likely to take 5–10 years before a significant impact of the biological control agent(s) can be recorded. Hence, other control methods being implemented prior to the release of a biological control agent should be continued during the initial years after release, though it may be important to modify these techniques so they do not impact negatively on the biological control agents.

Partial control

Many biological control agents that have been deemed to be 'unsuccessful' in controlling a weed species, may actually be reducing the competitive ability of the target weed. Their impact, although apparently 'insignificant', can in reality decrease the spread of the weed and reduce – while it does not completely eliminate – the need for conventional control measures (McFadyen, 2012).

The remainder of this chapter provides a summary of the progress that has been made in the implementation of a CBC management strategy in Asia for the invasive alien weed *Mikania micrantha* (mikania weed). Options for the control of mikania are discussed in other chapters, but it is clear that conventional methods, i.e. manual and cultural control and herbicide application, provide only limited relief from this prolific weed. Further, as India has a history of CBC implementation using insects (Sreerama Kumar *et al.*, 2008), the policy framework for the import and release of natural enemies was in place before the start of this project in that country (Chapter 12, this volume).

History of the CBC of *M. micrantha*

Insect agents

Research into the CBC of *M. micrantha* was started in 1978, and concentrated on insect natural enemies (Cock, 1982a). Extensive surveys were undertaken in Latin America, and a species of thrips (*Liothrips mikaniae*) from Trinidad was prioritized from several potential agents for the following reasons:

- A closely related species of thrips (*L. urichi*) had successfully controlled another invasive alien weed, *Clidemia hirta*, in Fiji (Simmonds, 1933; Winston *et al.*, 2014).
- *L. mikaniae* is widespread in the Neotropics and, therefore, it is likely to be adaptable to a wide range of habitats in the Old World tropics.
- It attacks and damages the youngest leaves and causes dieback, and so it is likely to have a substantial impact on the aggressive vegetative growth, which is the most weedy aspect of *M. micrantha*.
- It is a locally damaging natural enemy on mikania in the Neotropics, despite being heavily attacked by parasitoids, predators and entomopathogenic fungi, and so freed of these natural enemies, the insect should be more effective in the Old World.

Host-range screening of *L. mikaniae* was carried out using 37 test species: three *Mikania* spp., 13 other Asteraceae, and 21 further species from 15 other plant families. Although there was slight feeding on a few other species, *L. mikaniae* was shown to be host specific to *Mikania*, and fed only on species closely related to *M. micrantha* (and within the genus *Mikania*) (Cock, 1982b). Releases of the thrips were made in the Solomon Islands in 1988, and in west Malaysia in 1990, but neither led to the establishment of the insect (Cock *et al.*, 2000). Possible reasons for this failure have been suggested and include:

- High levels of opportunistic predation by indigenous natural enemies, e.g. ants, spiders, predatory thrips, because *L. mikaniae* is a conspicuous (it has red nymphs and black adults) and sedentary insect.
- Disruption of dispersal, mate location or host-finding behaviour, because of the continuous distribution of invasive *M. micrantha* compared with its distribution in small patches in the area of origin of the insect.
- Unidentified pupal mortality – the thrips do not pupate on the host plants, though the causes of mortality at this stage have not been investigated.
- Host plant incompatibility, whereby the thrips may not have been able to attack all of the biotypes of *M. micrantha* present in the areas of release.
- A suboptimal release strategy because the optimal approach had not been fully investigated prior to release.

From the available information, it would appear that the impact of indigenous natural enemies, particularly predatory thrips, was likely to have been the most significant factor in the failure of *L. mikaniae* in these efforts in CBC (Cock *et al.*, 2000).

Fungal pathogen agents

The failure of *L. mikaniae* in the Solomon Islands and west Malaysia described above discouraged more donor investment, despite many other promising agents having been identified during the survey phase of the project. However, the mikania weed problem did not go away – quite the opposite – the weed continued to proliferate and dominate vast tracts of the moist forest zones of Asia, and was flagged by organizations in many countries as a major, growing problem in need of sustainable management (Sen Sarma and Mishra, 1986; Muniappan and Viraktamath, 1993; Waterhouse, 1994). After specific requests from Indian scientists, a new initiative was born, to look at the potential of fungi as part of an integrated pest management (IPM) strategy for the weed in the Western Ghats of Kerala and in Assam. The project was funded by the UK Department for International Development

(DFID) through the Natural Resources International Limited (NRIL) Crop Protection Programme, and ran from 1996 until the end of 2005. The project was divided into two phases: a broad first phase that considered the mikania weed problem as a whole in India and weed management strategies; and a second phase that focused on CBC with a selected biological control agent identified during Phase 1.

Phase I: development of a biocontrol strategy

Phase I of the project ran between 1996 and 2000, and involved two Indian organizations – Kerala Forest Research Institute (KFRI) and Assam Agricultural University (AAU); Viçosa University, Minas Gerais, Brazil and CABI collaborated in the venture. The work included:

- mapping the distribution and monitoring the spread of mikania weed in Kerala;
- assessing its socio-economic impact on home-garden subsistence agriculture in Kerala;
- investigating control methods; and
- evaluating fungal pathogens as biological control agents – both local (indigenous) fungi and exotic fungi were considered as potential mycoherbicides for possible introduction as CBC agents.

The results of this work are published in Sankaran *et al.* (2001a), apart from the evaluation of the fungal pathogens as CBC agents, which is summarized in Ellison (2001), Evans and Ellison (2005) and Ellison *et al.* (2008). Also, Chapters 2 and 9, this volume provide details of the impact of mikania and of conventional control options for the weed, respectively.

The surveys in India revealed no indigenous natural enemies that could effectively be used to control *M. micrantha*. Conversely, mikania is rarely a weed in its native range in the Neotropics, where natural enemies are seen to exert significant pressure on its occurrence and abundance (Parker, 1972; Cock, 1982a; Cock *et al.*, 2000). Previous literature studies, fungal herbaria records (e.g. the Herb. IMI database; www.herbimi.info/herbimi/home.htm) and field surveys in the Neotropical native range of the plant had revealed a wealth of fungal pathogens on *M. micrantha* (Evans, 1987; Barreto and Evans, 1995). The work that finally led to the selection of the rust *Puccinia spegazzinii* as the prime candidate for the control of *Mikania* in India is summarized below.

SURVEYS IN THE NATIVE RANGE OF *M. MICRANTHA*. Surveys for pathogens of mikania weed were undertaken throughout the Neotropics, and a selection of possible candidate natural enemies was made based on field observations and knowledge of the potential biocontrol value of particular fungal groups. The factors influencing the choice of pathogen were that:

- It should cause substantial damage to *M. micrantha*.
- It should be commonly found and abundant at individual sites.
- There should be a lack of the same symptoms on adjacent vegetation, particularly other *Mikania* spp. (of which there are approximately 200 in the Americas).
- Ideally, it should be a rust fungus – this group of pathogens is almost always considered first as CBC agents, because they are inherently host specific and damaging and they spread well in the field.

From evaluations based on the above criteria, several rust species were selected for further assessment as potential CBC agents against mikania weed in India. Their taxonomic identities were elucidated by Evans and Ellison (2005), who showed that there was a complex of three rusts that caused damage to mikania weed – *Puccinia spegazzinii* (Fig. 10.2a), *Dietelia portoricensis* (Fig. 10.2b) and *Dietelia mesoamericana* (Fig. 10.2c), the last named being a species new to science at the time. All of these species are highly damaging to their host in the field, causing leaf, petiole and stem infections, as shown in Fig. 10.2a–c, which lead to cankering and whole plant death. In contrast, no rust fungi have been found on the weed in its exotic range.

Fig. 10.2. Biological control of *Mikania micrantha*. (a) *M. micrantha* heavily infected with *Puccinia spegazzinii*, showing stem, petiole and leaf infection. Courtesy C.A. Ellison. (b) *Dietelia portoricensis*. Courtesy C.A. Ellison. (c) *Dietelia mesoamericana*. Courtesy C.A. Ellison. (d) sunflower, showing chlorotic spots (arrows) in reaction to being inoculated by *P. spegazzinii*. Courtesy C.A. Ellison. (e) *P. spegazzinii* heavily parasitized in the field in the native range by the mycoparasite *Eudarluca* sp. (arrow). Courtesy H.C. Evans.

PATHOGEN SELECTION. Although all three species of the rust complex described above appeared to be equal in the damage they inflicted on their host, *M. micrantha*, the two *Dietelia* spp. had a narrow distribution, whereas *P. spegazzinii* was widely distributed throughout the Neotropical range of *M. micrantha*, and over a wide range of altitudes (from near sea level to ~1200 m above sea level) (Fig. 10.3). *P. spegazzinii* was therefore considered to be more likely to be (already) adapted to the wide range of environmental conditions it would encounter in Asia. For this reason, the species was subjected to a more detailed assessment, focusing on its tolerance to different environmental conditions and how specific it was to the target weed, i.e. could it attack any other species? Eleven isolates of the rust were collected from six countries (Argentina, Brazil, Costa Rica, Ecuador, Peru and Trinidad and Tobago) and tested against populations of the weeds collected from all over Kerala and Assam. Over the course of the project, populations of mikania from other parts of Asia were also tested for their susceptibility to these isolates.

The situation was found to be more complicated than had been anticipated. The rust isolates differed in their ability to infect and/or cause damage to the genetically different populations of mikania weed (biotypes) that had been identified across the invaded range in Asia. This suggests that there had been multiple introductions of the plant to Asia from different areas of the native range (Ellison *et al.*, 2004). What is more, each rust isolate was found to consistently infect and cause damage to a different selection of weed biotypes; thus, the isolates are referred to as 'pathotypes'. It soon became clear that different pathotypes of

Fig. 10.3. Distribution of rust pathogens infecting *Mikania micrantha* in its native range in Latin America and the Caribbean. The scale given is for latitude.

the rust would need to be used in different countries or regions. Additionally, more than one pathotype would be needed in some areas where the weed was particularly genetically variable (Ellison *et al.*, 2004). For the weed populations in Kerala, and for most of those in Assam, an isolate of the rust from Trinidad was considered to be the most aggressive, and this isolate was selected for full evaluation. The subsequent assessment focused on host-specificity testing and evaluation of the biological parameters of the chosen isolate.

HOST-SPECIFICITY TESTING. Plants for the host-specificity testing were selected based on the procedure developed by Wapshere (1974) (see Introduction). More than 60 non-target plants were challenged with the rust, including 11 other *Mikania* spp., and scored according to the evaluation given in Table 10.1 (Ellison *et al.*, 2008).

The rust was found to be highly specific within the genus *Mikania*; only three species from Africa and one from Asia (*M. cordata*, the only native species in Asia) became infected with the rust, and this was to varying degrees. Three other species, also closely related to *M. micrantha*, developed chlorotic spots in response to the rust (see Fig. 10.2d). In such situations, the interaction between the rust and the test plant is investigated microscopically (Bruzzese and Hasan, 1983); in all three of these cases, it was shown that the attempts by the rust at penetration elicited a resistance response in the plant, which led to a failure of the rust to grow inside the leaf. In addition, some of the chlorotic leaves were observed until they had senesced, and no further symptoms were seen to develop. It was concluded that all three of these closely related test plants were resistant to the rust (Ellison *et al.*, 2008).

EVALUATION OF BIOLOGICAL PARAMETERS. The life cycle and infection process for *P. spegazzinii* was elucidated (Fig. 10.4). Life cycles of rust fungi vary in complexity, with up to five separate stages, each with a distinct spore type. The life cycle of *P. spegazzinii* is relatively simple, with just two stages/spore types, and is referred to as a microcyclic rust. The teliospores are orange–brown and are embedded in the host plant tissue; they can remain viable for many weeks. However, high humidity, such as dew or rain, triggers the teliospores to germinate and produce delicate basidiospores – the dispersal stage of the rust – which are released into air currents. It is these basidiospores that land on fresh plants, and, when there is water present, germinate and infect the young (meristematic) growing tissue at the tips of the mikania vines. Damage in the form of chlorosis is visible 6–9 days after infection, depending on the rust pathotype, and teliospores are visible at 10–14 days.

The biology of the rust in response to key environmental parameters was also investigated, including its temperature tolerance and dew-period requirements. The latter is the length of time that free water (either rain or dew) is required on the plant surface for the infective spores (basidiospores) to germinate and invade the plant tissue. It was found that the rust was able to

Table 10.1. Pathogenicity scoring for the evaluation of *Puccinia spegazzinii*.

Score	Symptoms
0	No macroscopic symptoms
1	Necrotic or chlorotic spots on inoculated leaves – no sporulation
2	Abnormal infection site: chlorotic patches on leaves with very low teliospore production around the edges of chlorosis
3	Abnormal infection site: pustules reduced in size (<4 mm) or with low teliospore production compared with compatible host–pathogen interactions. Lower than compatible infection sites on petioles and stems
4	Fully compatible host–pathogen interaction. Normal pustule formation (4–7 mm diameter) on leaves petioles and stems

Fig. 10.4. Life cycle of *Puccinia spegazzinii*. (a) Mature teliospores in telia (pustules) on the lower leaf surface of *Mikania micrantha*. Under conditions of high humidity, the teliospores germinate and a white bloom can be seen over the telial surface. (b) Microscopic view of a single teliospore germinating and producing basidiospores (arrow). (c) Basidiospores land on and infect the growing point (meristem) of mikania, which is the most susceptible part of the host to infection. (d) Scanning electron microscopic view of a basidiospore germinating on the surface of a leaf. e) Chlorotic spots on the upper leaf surface 6–10 days after inoculation.

infect at temperatures ranging from 15 to 25°C and after less than 10 h of dew (Ellison, 2001; Ellison *et al.*, 2008). Both of these requirements are compatible with the prevailing environmental conditions in the invasive range of *M. micrantha* in Asia.

DOSSIER ON *PUCCINIA SPEGAZZINII*. The data outlined above formed the core of the dossier that was produced by CABI for the Indian project collaborators (Ellison and Murphy, 2001). This was submitted to the Indian Directorate of Plant Protection Quarantine and Storage (DPPQS). A letter was included detailing that permission had been given by the Ministry of Agriculture, Land and Marine Resources, Government of Trinidad and Tobago, for the use of their genetic resources, in line with the intention of the CBD.

IMPACT OF *M. MICRANTHA* AND SETTING UP OF PERMANENT SAMPLE PLOTS. Under this project, the economic impact, spread and potential control options for mikania were assessed in both Kerala and Assam (Sankaran *et al.* 2001a; see also Chapter 3, this volume). This work not only acted as a justification for the CBC management approach, but, importantly, provides the baseline for post-release environmental and economic assessments of the rust in India. It is important to monitor the spread and impact of the rust once it is established in the field. To facilitate this, permanent sample plots, each of 1 ha, were set up in different ecosystems in Kerala (18 plots in forestry) and Assam (four plots in tea), and the plant density and biomass of the mikania were recorded before the rust was released. These data could then be used to measure the impact of the rust over time, post-release.

The experimental design was planned to measure the natural spread of the rust from where the initial rust source plants were placed in the field at the release sites. Mikania plants were to be examined along radial transects (e.g. north, south, east and west from the sample plots) each month over the course of the growing season, and the presence of the rust recorded. Once the rust was established at the sites, impact studies would commence, and continue over a number of years. Within each sample plot, at least 20 subsample quadrats (1 m^2) were to be taken, and the level of rust infection recorded, together with the percentage mikania cover, at regular intervals throughout the season. Changes in plant biodiversity and crop yield or growth were also to be recorded.

PREDICTING THE IMPACT OF *P. SPEGAZZINII*. *P. spegazzinii* is a widespread and often damaging rust on *M. micrantha* throughout its native range. In glasshouse studies, care has to be taken not to over-inoculate plants, because the rust will kill them at high inoculum densities. Studies under controlled conditions on biological parameters such as temperature, dew period and humidity can provide information on what may limit the successful establishment and spread of an agent, and its likely distribution in the invasive range of the weed. However, it is difficult to categorically extrapolate from glasshouse studies and native-range field observations to the field situation in the invasive range.

The impact that CBC can have on the abundance of an invasive alien weed often exceeds expectation based on how the natural enemy/bicontrol agent performs in the native range of the plant. Indeed, although *P. spegazzinii* was common throughout its Neotropical native range, the pustules were smaller, less abundant, and inflicted less apparent damage compared with those that developed during the infection of quarantine-grown plants that had originated from invasive range populations.

The explanation for these observations lies once more in co-evolution, but in this case the evolutionary relationship is with fungi living asymptomatically in the plants (fungal endophytes). More often than not, endophytes do not get transferred with the weed plants to their invasive range, but in the native range they protect plants from, among other things, attack by insects and pathogens. Without their endophytes, the plants are more susceptible to attack by the introduced natural enemies – this is the concept of the 'endophyte-enemy release hypothesis' (Evans, 2008). So we can see that co-evolution leads to a complex web of interactions in native communities, which is left behind when a single member of the community is introduced to a new range, often with multiple impacts. This is taken advantage of in CBC, because fungal (and insect) biological control agents are deliberately introduced free of their own co-evolved natural enemies (e.g. hyperparasites) and so are more able to flourish. For example, Fig. 10.2e shows *P. spegazzinii* heavily hyperparasitized by the damaging fungus *Eudarluca* sp., but strains of species of this genus able to infect *P. spegazzinii* are unlikely to be present in Asia (Evans and Ellison, 2005). Finally, as described above, genetic plant resistance to natural enemies is likely to involve an energy cost to the plant. Hence, without the burden of natural enemies, there will be a selection pressure for plants to lose such genetic resistance mechanisms (Blossey and Nötzold, 1995).

STAKEHOLDERS' WORKSHOP. At the end of the first phase of the programme, an Indian national workshop on alien weeds was held at KFRI at Peechi in Kerala on 2–4 November 1999. This workshop brought together representatives from agriculture, forestry, environment, conservation, industry and the media. Broad support was given for the introduction of *P. spegazzinii* through appropriate government channels, and this formed part of the recommendations of the workshop. Subsequently, the proceedings were published (Sankaran *et al.*, 2001b).

Phase II: Implementation of a CBC strategy

At the end of Phase I, the Project Directorate of Biological Control (PDBC; now incorporated into the National Bureau of Agricultural Insect Resources, NBAIR) of the Indian Council of Agricultural Research (ICAR) entered into the project framework as the nodal point for biological control in India, while the National Bureau of Plant Genetic Resources (NBPGR) provided quarantine facilities for the import of the biological control agent, *P. spegazzini*. Phase II commenced, once permission to import the rust into quarantine in India had been granted, in September 2002, following extensive consultation between DPPQS and PDBC. Before the rust could be considered for release, the authorities deemed it necessary to conduct additional host-specificity tests by Indian scientists (from the NBPGR and PDBC) under Indian conditions. The test plant list was drawn up following consultation with systematic botanists, mycologists and plant pathologists from ICAR, and consisted of 74 test species/varieties of plants; 18 of these were repeats of plants tested by CABI.

The rust was hand carried to the NBPGR on living plants and propagated under quarantine conditions on fresh plants (Sankaran *et al.*, 2008). It was successfully established in quarantine by September 2004, and host-specificity testing was completed by early 2005. A supplementary dossier was submitted with the application for limited field release of the rust to the Plant Protection Adviser to the Government of India, the Ministry of Agriculture (MoA, now the Ministry of Agriculture and Farmers Welfare) (Sreerama Kumar *et al.*, 2016). In June 2005, after extensive consultation between the MoA and PDBC, a permit was granted for the release of *P. spegazzinii* in four identified areas, two each in Kerala and Assam (Sreerama Kumar *et al.*, 2005; Sankaran *et al.*, 2008). The results from the field trials were published by Sankaran *et al.* (2008) and are summarized below.

FIELD RELEASES IN ASSAM. The Trinidad isolate of *P. spegazzinii* was transported to Assam on living plants for field release in 2005. A purpose-built rust propagation facility at AAU (Fig. 10.5) was used to produce the infected potted plants for putting out into the field. The rust was released at two mikania-infested sites in Jorhat in October and November 2005, and again in June 2006. At the first site (the Experimental Garden for Plantation Crops of AAU) the mikania plants were growing at ground level and so pots containing the infected plants were set in the ground, in shady places, within dense mikania stands. The shoot tips of mikania plants growing in the surrounding vegetation were pulled underneath the infected leaves, stems and petioles of the potted plants. The rust-infected plants were kept well watered and the surrounding vegetation was sprayed with water each evening to increase the humidity, and encourage basidiospore release and the infection of surrounding vegetation. At the second site (the Cinnamora Tea Estate), it was necessary to use a modified inoculation strategy, because the mikania leaves were overgrowing the tea bushes. Potted, infected plants were hung in a dense stand of mikania at the level of the tea (plucking) table, suspended from a bamboo pole by a rope.

The rust spread from the infected source plants on to the surrounding vegetation, but did not persist over the dry season. In addition, it became apparent that the infection on the mikania plants in the field in Assam was not a fully susceptible response, as fewer, smaller pustules developed on these plants (Fig. 10.6a) than on a plant genotype from Kerala (Fig. 10.6b). It was established during Phase I of the project that in Assam

Fig. 10.5. The purpose-built rust propagation facility at Assam Agricultural University. Plants were maintained in the net house (right) and inoculated in the air-conditioned mist chamber (attached to left of the net house). The control room for the facility is in a separate building situated to the left of the mist chamber. Photo courtesy C.A. Ellison.

Fig. 10.6. Field infection *Mikania micrantha* by *Puccinia spegazzinii*: (a) *M. micrantha* plants from Assam showing semi-resistant response as indicated by small chlorotic spots on upper leaf surface (see arrows); and (b) *M. micrantha* plants from Kerala showing fully compatible response as shown by large chlorotic spots on upper leaf surface (see arrow). Photos courtesy C.A. Ellison.

(but not in Kerala), there were genotypes of *M. micrantha* present that were semi-resistant to the pathotype of the rust from Trinidad. A pathotype of the rust from Peru was found to be fully infective on these biotypes, but it had been decided by all collaborators that it was best to complete the work with the Trinidad pathotype first. Then, once all of the procedures were in place in India for the import of fungal weed biological control agents, the plan was to submit an application to import a new pathotype for Assam. It was considered that to request two pathotypes at the same time would only confuse issues and be counterproductive, though it was disappointing that the first release site in Assam proved to be within a semi-resistant population. No further releases of the Trinidad isolate were made.

The Peruvian isolate of the rust was imported into NBPGR in February 2006 and checked for its specificity to mikania. It was released in Assam in July 2007. There is no more information available about the performance of this isolate in the field.

FIELD RELEASES IN KERALA. The pathotype of the *P. spegazzinii* from Trinidad was transported to Kerala in 2006 and propagated on to fresh plants in an air-conditioned glasshouse (Fig. 10.7) before field releases in late August–October. There were three release sites: one in a mixed tree cropping system, and two within degraded forest (Fig. 10.8). As in Assam, the rust spread from the initial source plants to field populations of mikania weed, but in Kerala, the field plants proved to be fully susceptible to the rust (Fig. 10.6b). There appeared, however, to be no inter-season survival of the rust. It was only possible to release the rust in 1 year, and quite late in the growing season, by which time the rainfall was starting to decrease. Therefore, there would have been little chance for the rust to establish over a sufficiently wide area, and in a high enough density, to provide the opportunities for over-season survival. The optimum time to release the rust is May/June.

RELEASE OF THE RUST IN CHINA. Towards the end of the India project, funding was secured from the UK Darwin Initiative (Department for Environment, Food and Rural Affairs, Defra) to implement a rust release project in mainland China. The project reached the field-release stage in less than 3 years rather than the 10 years that it had taken in India – a significant example of 'fast-tracking' in CBC. The rust was released in Guangdong Province in 2006 and spread from the initial release site, going through several generations, but whether or not it has survived long term in the field is not known.

WHY DID THE RUST FAIL TO ESTABLISH? Table 10.2 provides a summary of all releases of *P. spegazzinii* made up to 2012, and gives the probable reasons for the successful establishment of the rust on mikania, and

Fig. 10.7. The rust propagation facility at Kerala Forest Research Institute (left); plants ready for infection with rust, prior to placing in the field (right). Photos courtesy C.A. Ellison.

Fig. 10.8. Field release site of *Puccinia spegazzinii* in Kerala, showing a high level of mikania infestation (left) and a rust-infected pot of mikania (right) acting as a source for field infection and showing a rust infected leaf (see arrow). Photos courtesy K.V. Sankaran.

the probable reasons for its failure to establish following the earliest releases in India and mainland China.

There have been many papers published on why biological control agents may fail to establish; for example, insufficient numbers of agents released, climatic conditions not favouring the agent and the resistance of the weed to the strain of the agent released (e.g. Day and Neser, 2000). In the case of *P. spegazzinii*, the restrictions placed on the Indian and mainland Chinese government release permits, which in both cases allowed releases to be made at only a few sites, did not favour success. In addition, only a relatively small number of infected plants were put out into the field. In many areas where mikania is invasive, there is a distinct dry season. In these situations, mikania weed tends to die back in open areas, although plants continue to grow and maintain leaves where there is perennial standing water and along permanent streams. Hence, over most of the mikania-infested areas, the rust cannot maintain a propagating population during the dry season, and so it remains dormant. Evidence from the native range of *M. micrantha*, and from glasshouse studies, suggests that the rust will survive in living stems as cankers in open habitats, and on all aerial parts of plants surviving where there is permanent water. These 'rust refuges' could act as the inoculum source to initiate the rust epidemic as the rains begin and the mikania weed starts to reinvade. Consequently, it is important that the rust be established as widely as possible in the area of release, including in all habitat types, in order to give it the highest opportunity to survive the dry season.

Current Successes in the CBC of *M. micrantha* Using the Rust *P. spegazzinii*

Organizations in other countries in tropical Asia, realizing the scale and devastating potential of the invasion by *M. micrantha*, and following the lead from India and China, have developed their own release programmes for the rust. The mikania CBC programme in India has become a 'pilot project' for pathogen CBC technology in Asia and, thus, the experience gained by the various stakeholders is providing a model for other countries in the region. The successes to date, and the possible reasons for the later success in achieving establishment and spread of the rust have been summarized in Table 10.2.

Table 10.2. Releases of the rust *Puccinia spegazzinii* for the control of *Mikania micrantha*.

Location released	Organization (contact[a]) *Donor*[b]	Year of first release (origin of pathotype)	Outcome of field release	Probable reason for failure/success	Future
Jorhat, Assam, India	Assam Agricultural University (AAU) (K.C. Puzari) *DFID and Natural Resources International Ltd (NRIL), UK*	2005 (Trinidad) 2007 (Peru)	Did not establish Rust spread from initial release site, but died out after three generations	Licence restriction: permitted only to release at one site Resistance in weed populations where released Hot climatic conditions not conducive to rust spread and survival	The rust cultures have been lost at AAU No more releases are planned
Thrissur, Kerala, India	Kerala Forest Research Institute (KFRI) (K.V. Sankaran) *DFID and NRIL, UK*	2006 (Trinidad)	Did not establish Rust spread from initial release site, going through a number of generations, but did not appear to survive in the field during the dry season	Licence restriction: permitted only to release at two sites Late-season releases may have prevented adequate build-up in the rust population prior to the dry season	The rust culture has been lost at KFRI Funding being sought for resuming project
Guangdong Province, south China	Chinese Academy of Agricultural Sciences (Weidong Fu) and Guangdong Entomological Institute (Shichou Han) *Defra; Darwin Initiative, UK*	2006 (Argentina)	Probably did not establish Rust spread from initial release site, going through a number of generations	Licence restriction: permitted only to release at two sites	Rust from Ecuador under consideration for release by Wuhan Botanical Gardens (Jiangqing Ding)
Southern Taiwan, China	Taiwan National University (S.S. Tzean) *Taiwan Forestry Bureau*	2008 (Ecuador)	Rust established and spread from initial release site. Impact not yet measured	New pathotype released with better efficacy	Monitoring programme planned
Keravat and Tavilo, Papua New Guinea	National Agricultural Research Institute, Cocoa and Coconut Institute and Oil Palm Research Association (M. Day, Biosecurity Queensland, Australia) *ACIAR*	2009 (Ecuador)	Rust established and spreading. Widely released in over 550 sites in 15 provinces (over 2500 infected potted plants put out into the field). At some sites, the rust has moved 40 km in 18 months. Significant field impact recorded	New pathotype released with better efficacy Rust released at many sites with different microclimatic conditions	Monitoring and release programme ongoing Many of the remote sites are still to be monitored
Viti Levu, Fiji	Secretariat of the Pacific Community (M. Day) *ACIAR*	2009 (Ecuador)	Rust has been released at over 65 sites and has established at more than 15 sites	As above	Casual monitoring only
Vanuatu	Department of Agriculture and Fisheries, Australia (M. Day) *AusAID*	2012 (Ecuador)	Established and widespread	As above	Release programme ongoing

[a]Our thanks go to the contact people listed for their personal communications, which provided updates on the release programmes.
[a]ACIAR, the Australian Centre for International Agricultural Research; AusAID, Australian Aid; Defra: Department for Environment, Food and Rural Affairs, UK; DFID, Department for International Development, UK.

The rust was released in Taiwan in 2008, following the release in mainland China; this project was led by Taiwan National University and supported by national funding (S.S. Tzean, Taiwan National University, 2008, personal communication). In both mainland China and Taiwan, full host-range testing was repeated in-country, despite it having been undertaken twice before (in the UK and India). This demonstrates just how cautious the authorities are when it comes to a new technology – in this case, the use of pathogens for weed biocontrol. Following these releases, the Australian Centre for International Agricultural Research (ACIAR) funded a mikania-rust implementation project in Fiji and Papua New Guinea. The ministries of agriculture agreed that only a few native species required testing (Day *et al.*, 2013a) as a supplement to the extensive host-range testing data provided by CABI and published by Ellison *et al.* (2008). This should now be the way forward for other countries considering the release of the rust, and it will save time and money.

The rust has established, is spreading and is causing significant damage to mikania weed in Taiwan, Fiji, Papua New Guinea and, most recently, Vanuatu. It is also being considered for reintroduction into mainland China (J. Ding, Wuhan Botanical Gardens, 2012, personal communication) and for release in Palau (J. Miles, National Biodiversity Coordinator, 2013, personal communication), the Cook Islands and Guam (Q. Paynter, Landcare Research, New Zealand, 2014, personal communication), and Malaysia (P. Kaliannan, Department of Agriculture, 2014, personal communication).

Papua New Guinea – a success story in the making

Prior to the release of the rust in Papua New Guinea, an assessment of the distribution and socio-economic impact of mikania weed was undertaken. These baseline data were gathered to provide justification for the implementation of a CBC strategy and enable the impact of the rust to be assessed in the future (Day *et al.*, 2012). Glasshouse and field studies were undertaken as well, and these showed that the growth rate and dry weight of rust-infected mikania weed are significantly reduced compared with uninfected plants (Day *et al.*, 2013a). A simple but effective method was developed to mass produce rust-infected plants and release the rust in the field (see Fig. 10.9, formerly Plate 7). To date, the rust has been released at nearly 550 sites in all 15 provinces infested with *M. micrantha*, and has established in 12 provinces. It has established better in the wetter areas and at some of these sites has spread naturally up to 40 km in 18 months (Day *et al.*, 2013b).

A rust monitoring and evaluation programme has also been established in Papua New Guinea. Two sites in the province of East New Britain (Kerevat and Tavilo) were evaluated on a monthly basis to monitor the impact of the rust on the mikania weed populations (Day *et al.*, 2013b). Fig. 10.10 shows data from the site near Kerevat, where the growth of rusted *Mikania* plants was suppressed over the first 2 years following the release of the rust, which allowed *Glycine wightii* (a legume used as fodder or as a cover crop) to increase in abundance and further suppress mikania weed. However, in 2010, there was an extended dry season, during which the rust could not be found and the mikania population started to increase again. Following the start of the rains in early 2011, rust infection was observed again, and the mikania population began once more to decrease. At the second site under regular evaluation, at Tavilo, 18 months post-release of the rust, the mikania weed cover had been reduced from 100 to 40% (Fig. 10.11).

The host-specificity screening of the rust undertaken by Ellison *et al.* (2008) revealed that *M. cordata*, the only native species from the genus *Mikania* in Asia, is susceptible to *P. spegazzinii*. *M. cordata* occurs in Papua New Guinea and, indeed, infection of this species has been observed in the field (Ellison and Day, 2011). Moreover, the level of rust infection is lower than that on *M. micrantha*, and the damage is less significant on the native plant. In addition, the niche occupied by *M. cordata* includes

Fig. 10.9. Field release of the *Mikania micrantha* rust, *Puccinia spegazzinii*, in Vanuatu and Papua New Guinea: (a) simple but effective technology, a rust inoculation box; (b) putting out rust-infected plants into the field; (c) natural spread of the rust on to field plants, showing heavy rust infection, and an inset close-up of rust-infected leaf. Photos courtesy M.D. Day.

altitudes at which *M. micrantha* and *P. spegazzinii* do not thrive; hence, these populations will act as a reservoir of *M. cordata*. What is more, *M. micrantha* is invading and destroying the habitat where *M. cordata* occurs and is outcompeting the native species, so unless *M. micrantha* is controlled, the native plant will inevitably be negatively affected (Ellison *et al.*, 2008).

Assessment of successful releases

The successful establishment, spread and impact of *P. spegazzinii* in Papua New Guinea is likely to be attributable to the four main factors that are briefly described below: isolate selection, release strategy, and biotype and environmental compatibility (Ellison and Day, 2011).

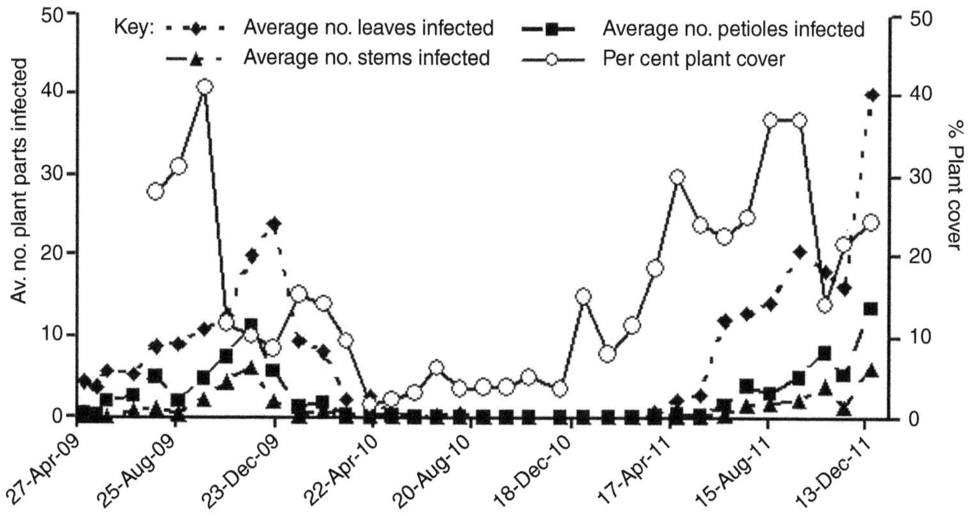

Fig. 10.10. The mean number of leaves, petioles and stems of *Mikania micrantha* infected by *Puccinia spegazzinii* and the percentage plant cover of *M. micrantha* at a release site near Kerevat in East New Britain Province, Papua New Guinea. The rust was released on 10 January 2009 and its effect monitored from April 2009 to December 2011. Reproduced from Day *et al.* (2013b).

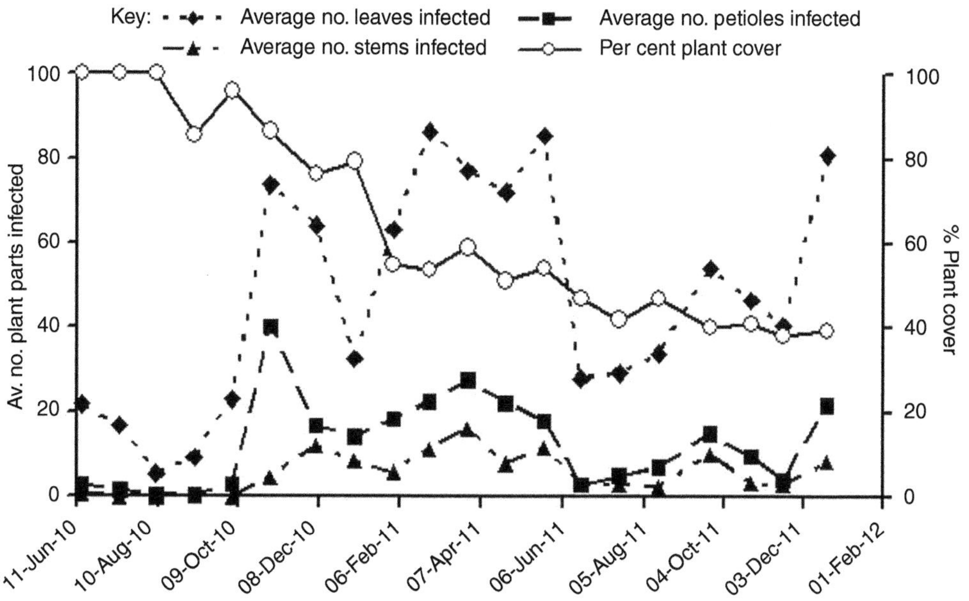

Fig. 10.11. The mean number of leaves, petioles and stems of *Mikania micrantha* infected by *Puccinia spegazzinii* and the percentage plant cover of *M. micrantha* at Tavilo, in East New Britain Province, Papua New Guinea. The rust was released on 31 March 2010 and its effect monitored from June 2010 to January 2012. Reproduced from Day *et al.* (2013b).

Rust isolate selection

A new isolate of the rust was selected for release in Taiwan, Papua New Guinea, Fiji and Vanuatu that had been collected during later surveys in South America. Field observations followed by glasshouse-based assessment indicated that this isolate might be a more effective biological control agent in the invasive range of the plant than the original isolate from Trinidad. It has the following characteristics:

- The new isolate causes canker production in the stems and petioles of *M. micrantha* (Fig. 10.12), removing resources from the plant as well as damaging the stems and petioles more severely than other isolates.
- The cankers improve the survival of the rust over extended periods by allowing it to survive in the cankers on dormant plant stems, such as during the dry season.
- This isolate is able to infect *M. micrantha* plants that have high levels of anthocyanin, which has been found to limit the infection caused by the other isolates of the rust that have been tested.

Fig. 10.12. *Mikania micrantha* infected by an isolate of *Puccinia spegazzinii* that causes hyperplasic (swollen) canker production in the stems and petioles (see arrows). Photos courtesy C.A. Ellison.

Rust release strategy

- The rust was applied at high doses; large numbers of infected plants (>2500) were placed in the field.
- Optimum microclimatic areas were selected (shaded and humid, near standing water).
- The mikania plants were young, with abundant shoot production near to the ground where the infected plants were positioned, and the meristematic tissue (in the growing points), is the most rust-susceptible part of the plant.
- The rust was released at multiple sites (550) over wide areas (15 provinces) that encompass different climatic conditions and habitats.

Plant biotype compatibility

- Fully susceptible genotypes of mikania weed were present (Day *et al.*, 2013a).

Environmental compatibility

- Much of Papua New Guinea where the rust was released has highly favourable environmental conditions for rust infection – moderate temperatures, high humidity and a short dry season.

Mikania rust is clearly adapted to the humid tropics and subtropics, and has not established in the drier areas where it has been released in either Papua New Guinea or Taiwan. The rust also favours a short to non-existent dry season. Unfortunately, mikania weed is still invasive in some of these less humid areas and the evidence suggests that the rust is unlikely to thrive in these areas. Fortunately, there are many co-evolved natural enemies of this plant in its native range still to be evaluated (Cock, 1982a; Barreto and Evans, 1995), and work now needs to be initiated to identify those that are adapted to drier environments.

Predicting the spread and impact of mikania rust on the economy

The initial monitoring and evaluation studies in Papua New Guinea are cause for optimism over the long-term efficacy of the rust as a CBC agent for mikania. By looking at similar studies on related rusts, it may be possible to gain an insight into the long-distance dispersal and likely future impact of the mikania rust. One such example was reported by Morin *et al.* (1996), who undertook a detailed assessment of the rust *Puccinia xanthii* in Australia, where it is an introduced pathogen of Noogoora burr (*Xanthium occidentale*). This rust has the same type of life cycle as *P. spegazzinii* and causes similar damage to the host plant. During the first 4 months following its 'accidental' introduction, *P. xanthii* was found to have spread 83 km from the first record of this rust in Australia, and the dispersal stage, the basidiospore, was found to be capable of surviving over long distances – 25 km on average – before landing on a host plant. This is surprising because basidiospores are considered to be a rather delicate and fragile spore type. Chippendale (1995) undertook an economic analysis of the impact of CBC on the weed. He found that only 3 years after the introduction of the rust, there were general observations from landholders that it was having a visible impact on the weed in the field. This was backed up by a significant reduction in the number of burrs (seed) in wool clippings. The economic study concluded that the total benefit from biological control was almost AU$17 million in less than 10 years. As CBC is sustainable, the savings to the farmer of this programme will continue in perpetuity.

Although a suite of natural enemies may be required to achieve significant control of some invasive alien weeds, these studies suggest that *P. spegazzinii* may well prove to be a 'silver bullet', i.e. the only biological control agent needed to achieve acceptable, sustainable control in climatic areas that have conditions conducive to rust infection. The rust spread rapidly under the environmental conditions in Papua New Guinea, and there was evidence that it was already exerting a significant impact on the abundance of *M. micrantha* in the areas of release within less than two growing seasons. In the long term, it is expected that the growth and fecundity of *M. micrantha* will be severely reduced over a significant part of its range. In some situations, such as in annual cropping systems or in newly established tree plantations, it may be necessary to combine CBC with conventional control methods; for example the sowing of cover crops, which will also help to provide the humid microclimate that will favour the establishment and spread of the rust. However, the rust is still likely to reduce the level of other control measures necessary in these situations.

Raising Awareness of the Use of CBC

For fostering public support, Warner (2012) considers the importance of improved science communication strategies that engage the public and foster public support for 'classical biological control for nature'. An important part of the mikania weed CBC programme in India involved an awareness-raising campaign on the benefits of biological control among the communities where the rust was planned to be released (Sankaran *et al.*, 2008). Staff from KFRI carried out questionnaire-based surveys on farmers, and this demonstrated that 90% of them were willing to try the biological control agent on their farms. A local-language brochure for Kerala farmers on the sustainable management of invasive alien weeds was published, and demonstrations, exhibitions and workshops were held for agricultural and forestry extension services and students from universities and schools. The media were also engaged to try to expand the audience and reach local government officials and policy makers. Local newspapers in Kerala published articles on the release of the rust, depicting it as a welcome solution to the weed problem. In collaboration with the University of Calicut, a documentary film was made: *Weeds: the Biological Invaders*, which was broadcast nationwide and won an award for the best documentary film on humanity, environment and human rights in 2004.

Discussion and Implications for Donors

Five countries (Australia, Canada, New Zealand, South Africa, the USA) account for the majority of all successful weed biological control projects (Winston *et al.*, 2014). Sometimes, these are based on the early selection of a single very effective biological control agent – the so-called 'silver bullet' mentioned above – but other projects took much longer, either because a combination of biological control agents had to be introduced before control was achieved, or because an effective agent was only discovered a number of years into the CBC programme. The great majority of programmes have focused efforts on insect natural enemies, with pathogens generally only being considered if insects have not proven to be effective. However, many new projects now consider insects and pathogens simultaneously and on their merits.

If the establishment of effective natural enemies is achieved following the first releases, it may be as little as 7 years from the start of a biological control programme until effective control of the target weed is evident, at least in the initial release areas, e.g. purple loosestrife (*Lythrum salicaria*) in North America. Other CBC programmes may last decades, yet still be ultimately successful, e.g. leafy spurge (*Euphorbia esula*) in the USA. Relatively few CBC programmes against weeds are completely unsuccessful, but when partial control is not considered adequate, potential new biological control agents may be found and evaluated over many years. For continental-scale problems, with huge, widely dispersed infestations, the process of weed suppression over all populations can take a long time (decades), even after a heavy impact is apparent locally. In these situations, control can be speeded up by implementing a redistribution programme for the biological control agents, perhaps for many years – as, for example, in New Zealand, with the suite of agents that have been released against the European gorse species, *Ulex europaeus* (Hill *et al.*, 2000).

Such a long-term time frame may present difficulties for governments and donors used to short-term research and implementation projects, especially when their own requirements are to have milestones and to show impact. It has taken many decades in the five countries listed above – which have the most extensive experience of weed CBC – to build the confidence to make such long-term investments with a reasonable expectation of return. This is not the situation for developing countries and donor agencies.

It is clear from the literature that CBC projects targeting weeds that are only a serious problem in developing countries are few and far between, apart from the extensive work that has been undertaken in South Africa, which is a country that has both developed and developing sectors (Cock *et al.*, 2000). Only four programmes have been identified that were carried out and implemented with, and on behalf of, developing countries, had reasonable resources and were reported in sufficient detail to evaluate. These are discussed below.

1. The successful biological control of Koster's curse (*Clidemia hirta*) in Fiji (Simmonds, 1933) was carried out with support from the (colonial) government of Fiji in the 1930s, along with several similarly resourced insect biological control projects (Paine, 1994). The control of Koster's curse was achieved in 3 years (Simmonds, 1933).

2. In the 1940s, the Mauritius sugar industry provided operating costs for the Commonwealth Institute of Biological Control (CIBC, now integrated within CABI) to research insect biological control agents of black sage (*Varronia curassavica*; formerly *Cordia curassavica*) in Trinidad, and this led to very successful control of the plant within 6–7 years (Greathead, 1971; Fowler *et al.*, 2000). At that time, the CIBC was core funded for scientists' salaries. Some years later, the success was repeated in Malaysia and Sri Lanka at little additional cost (Winston *et al.*, 2014).

3. In the 1960s, the CIBC started work on the insect and mite biological control agents of Siam weed (*Chromolaena odorata*) in the Neotropical region, initially with West African plantation industry support for the operating costs, but subsequently with increased support from aid agencies. This led to the introduction of several insect biological control agents, one of which, the moth *Pareuchaetes pseudoinsulata*, gave varying degrees of control in different situations, including in India and Sri Lanka (Cock and Holloway, 1982; Winston *et al.*, 2014). More recently, the ACIAR has supported renewed activity on this weed in Indonesia, and further biological control agents have been introduced with reports of significant impact (Day and McFadyen, 2004). No pathogens have yet been used, although potential agents have been identified (Barreto and Evans, 1994). This programme, which started nearly 40 years ago, is not yet over, but success has been achieved in several locations, and should be repeatable in most areas where *C. odorata* is still a problem. For instance, India has recently (2005) introduced a stem gall fly (*Cecidochares connexa*); also, interestingly, *P. pseudoinsulata*, which was released in India in 1973 and apparently did not establish at the time, has been rediscovered and is starting to increase in abundance (Varma *et al.*, 2006).

4. The fourth example is the programme against *M. micrantha*. This programme started over 30 years ago, but until now only one insect agent had been released in two countries. The release of the rust fungus in Papua New Guinea, Fiji, Taiwan and Vanuatu, as already described in this chapter, appears to be the first success against this target and it is to be hoped that it will be a critical turning point.

During the first DFID-funded mikania programme in 1978–1981 (see History of the CBC of *M. micrantha*, Insect agents, near the beginning of the chapter), other potentially effective insect biological control agents were found during the surveys (Cock 1982a), but these could not be evaluated and tested within the 3 years of the project. The original project prioritized completing the testing of one agent, so that there was potential for follow-on and uptake. A considerably longer original project, or one that proceeded in phases, would have been better, to allow testing of other (second and subsequent) agents to proceed while the first was being released. However, there was no funding for follow-up on additional biological control agents once the first was screened. At that time, the use of pathogens for weed biological control was still very much in its infancy, and so it is realistic to suggest that studies on pathogens would have been delayed for some years, as the apparently more promising insects would have been prioritized for evaluation.

Very few complete weed CBC projects (i.e. from establishing the suitability of a biological control approach, to surveys for biological control agents, to the selection and safety testing of potential biological control agents, to authorization, release, monitoring and evaluation) have been carried out entirely for the benefit of developing countries and funded by aid agencies. Mikania may be the first, though much work on *C. odorata* has proceeded in this way as well. The global mikania programme has proceeded in two main phases (insects and pathogens) supported by DFID, but other agencies, such as ACIAR, are now also supporting this work. Other weed biological control projects for developing countries either predate extensive aid-based support to independent developing countries, or have been fast-tracked utilizing earlier work by and for the five developed countries most active in CBC, which aimed to solve weed problems in their own subtropical and tropical areas. Examples include projects for the control of the salvinia water fern, water hyacinth (*Eichhornia crassipes*), water lettuce (*Pistia stratiotes*), parthenium weed (*Parthenium hysterophorus*), puncture vine (*Tribulus* spp.), and other species that have been mentioned above (Winston *et al.*, 2014).

The fast-track strategy could have (and probably has) given rise to an expectation that introducing a known, effective biological control agent is a relatively quick, cheap fix. Although this can be true, it is certainly not the case where there is no previous research on which to draw. Biological control scientists may be at least partly responsible for this misconception by promising too much, too quickly and too cheaply. Thus, it is important to clarify perspectives and plan in a realistic time frame and budget framework of at least 10 years. That way, developing countries can start to tackle their priority invasive alien weeds, and obtain the benefits of weed biological control, which are already evident in the projects of the five main countries that have undertaken them. The take-home messages for donors and governments are that:

- Classical weed biological control offers an effective means of managing many alien invasive plants, especially where the areas involved are huge, and alternative management options are inadequate, expensive and unsustainable.
- The most cost-effective way to implement biological weed control is to exploit the research and experience of developed countries and to make known biological control agents for invasive alien weeds that are available to developing countries, following appropriate testing.
- The more challenging approach is where a weed is not already a biological control target of a developed country, and no research has been done; in this case, a long-term commitment is critical to achieving success.
- By adopting a long-term strategy, not only is success more probable, but such an approach also offers considerably more opportunity for capacity building in the target (and source) countries.

This last point is not meant to suggest that sponsors should change their strategies and start committing to 10- to 20-year weed biological control programmes, but it does suggest that a realistic time frame and mechanisms for continuing funding are important, and could exist as a series of linked phases. If success is not achieved based on the first agent(s) released, the risk is that the whole programme may be delayed by funding hiccups, leading to loss of biological materials, loss of expertise and the loss of test plant collections. The development and implementation of the use of the *P. spegazzinii* rust fungus has proceeded in phases, and the DFID and NRIL are to be congratulated for their perseverance in seeing these phases of the project through to completion. This aspect is particularly relevant because there were considerable pressures on NRIL project managers to show timely and tangible results from the projects they chose to fund.

Acknowledgements

We would like to acknowledge the support given by DFID and ICAR in the undertaking of the work reported here. We would also like to acknowledge the contributions made by all members of the collaborative team, that have led to the successful conclusion of the mikania weed fungal CBC project: KFRI (particularly Dr K.V. Sankaran), AAU (particularly Dr K.C. Puzari) and PDBC (particularly Dr P. Sreerama Kumar, as well as two former directors of the PDBC, Prof. R.J. Rabindra and the late Dr S.P. Singh, who were instrumental in the development of the earlier stages of the pathogen programme), NBPGR (particularly Drs U. Dev and R.K. Khetarpal) and CABI (particularly Drs S.T. Murphy, H.C. Evans and S.E. Thomas, and Ms D.H. Djeddour).

This chapter is an output from a research project funded by DFID for the benefit of developing countries (Crop Protection Research Programme: Project No. R8502). The views expressed are not necessarily those of DFID. The rust fungi used were held and studied in the quarantine facility of the CABI UK Centre (Ascot) under licence from Defra, UK (licence no. PHL 182/4869).

References

Bale, J., van Lenteren, J. and Bigler, F. (2008). Biological control and sustainable food production. *Philosophical Transactions of the Royal Society B: Biological Sciences* 363, 761–776.

Barreto, R.W. and Evans, H.C. (1994) The mycobiota of the weed *Chromolaena odorata* in southern Brazil with particular reference to fungal pathogens for biological control. *Mycological Research* 98, 1107–1116.

Barreto, R.W. and Evans, H.C. (1995) The mycobiota of the weed *Mikania micrantha* in southern Brazil with particular reference to fungal pathogens for biological control. *Mycological Research* 99, 343–352.

Barton, J. (2012) Predictability of pathogen host range in classical biological control of weeds: an update. *BioControl* 57, 289–305.

Blossey, B. and Nötzold, R. (1995) Evolution of increased competitive ability in invasive nonindigenous plants: a hypothesis. *Journal of Ecology* 83, 887–889.

Bruzzese, E. and Hasan, S. (1983) A whole leaf clearing and staining technique for host specificity studies of rust fungi. *Plant Pathology* 32, 335–338.

CBD (2010) The Nagoya Protocol on Access to Genetic Resources and the Fair and Equitable Sharing of Benefits Arising from their Utilization to the Convention on Biological Diversity. Available at: www.cbd.int/abs/ (accessed 23 March 2016).

Charudattan, R. (2001) Biological control of weeds by means of plant pathogens: significance for integrated weed management in modern agroecology. *BioControl* 46, 229–260.

Chippendale, J.F. (1995) The biological control of Noogoora burr (*Xanthium occidentale*) in Queensland: an economic perspective. In: Delfosse, E.S. and Scott, P.R. (eds) *Proceedings of the VIII International Symposium on Biological Control of Weeds, Lincoln University, Canterbury, New Zealand, 2–7 February 1992.* Commonwealth Scientific and Industrial Research Organisation (CSIRO), Melbourne, Australia, pp. 185–192.

Clewley, G.D., Eschen, R., Shaw, R.H. and Wright, D.J. (2012) The effectiveness of classical biological control of invasive plants. *Journal of Applied Ecology* 49, 1287–1295.

Cock, M.J.W. (1982a) Potential biological control agents for *Mikania micrantha* HBK from the Neotropical region. *Tropical Pest Management* 28, 242–254.

Cock, M.J.W. (1982b) The biology and host specificity of *Liothrips mikaniae* (Priesner) (Thysanoptera: Phlaeothripidae), a potential biological control agent of *Mikania micrantha* (Compositae). *Bulletin of Entomological Research* 72, 523–533.

Cock, M.J.W. and Holloway, J.D. (1982) The history of, and prospects for, the biological control of *Chromolaena odorata* (Compositae) by *Pareuchaetes pseudoinsulata* Rego Barros and allies (Lepidoptera: Arctiidae). *Bulletin of Entomological Research* 72, 193–205.

Cock, M.J.W., Ellison, C.A., Evans, H.C. and Ooi, P.A.C. (2000) Can failure be turned into success for biological control of mile-a-minute weed (*Mikania micrantha*)? In: Spencer, N.R. (ed.) *Proceedings of the X International Symposium on Biological Control of Weeds, Bozeman, Montana, 4–14 July 1999.* USDA Forest Service, Forest Health Technology Enterprise Team, Morgantown, West Virginia, pp. 155–167.

Cock, M.J.W., van Lenteren, J.C., Brodeur, J., Barratt, B.I.P., Bigler, F., Bolckmans, K., Cônsoli, F.L., Haas, F., Mason, P.G. and Parra, J.R.P. (2010) Do new access and benefit sharing procedures under the Convention on Biological Diversity threaten the future of biological control? *BioControl* 55, 199–218.

Cullen, J.M., Kable, P.F. and Catt, M. (1973) Epidemic spread of a rust imported for biological control. *Nature* 244, 462–464.

Culliney, T.W. (2005) Benefits of classical biological control for managing invasive plants. *Critical Reviews in Plant Sciences* 24, 131–150.

Day, M.D. and McFadyen, R.E. (eds) (2004) *Chromolaena in the Asia-Pacific Region. Proceedings of the Sixth International Workshop on Biological Control and Management of Chromolaena, Cairns, Australia, 6–9 May 2003.* ACIAR Technical Reports No. 55, Australian Centre for International Agricultural Research, Canberra.

Day, M.D. and Neser, S. (2000) Factors influencing the biological control of *Lantana camara* in Australia and South Africa. In: Spencer, N.R. (ed.) *Proceedings of the X International Symposium on Biological Control of Weeds, Bozeman, Montana, 4–14 July 1999.* USDA Forest Service, Forest Health Technology Enterprise Team, Morgantown, West Virginia, pp. 897–908.

Day, M.D., Kawi, A., Kurika, K., Dewhurst, C.F., Waisale, S., Saul-Maora, J, Fidelis, J., Bokosou, J., Moxon, J., Orapa, W. and Senaratne, K.A.D. (2012) *Mikania micrantha* Kunth (Asteraceae) (mile-a-minute): its distribution and physical and socioeconomic impacts in Papua New Guinea. *Pacific Science* 66, 213–223.

Day, M.D., Kawi, A.P. and Ellison, C.A. (2013a) Assessing the potential of the rust fungus *Puccinia spegazzinii* as a classical biological control agent for the invasive weed *Mikania micrantha* in Papua New Guinea. *Biological Control* 67, 253–261.

Day, M.D., Kawi, A.P., Fidelis, J., Tunabuna, A., Orapa, W., Swamy, B., Ratutini, J., Saul-Maora, J. and Dewhurst, C.F. (2013b) Biology, field release and monitoring of the rust *Puccinia spegazzinii* de Toni (Pucciniales: Pucciniaceae), a biocontrol agent of *Mikania micrantha* Kunth (Asteraceae) in Papua New Guinea and Fiji. In: Wu, Y., Johnson, T., Sing, S., Raghu, R., Wheeler, G., Pratt, P., Warner, K., Center, T., Goolsby J. and Reardon, R. (eds) *Proceedings of the XIII International Symposium on*

Biological Control of Weeds, Waikoloa, Hawaii, 11–16 September 2011. USDA Forest Service, Forest Health Technology Enterprise Team, Morgantown, West Virginia, pp. 211–217.

Diamond, J. (2005) *Collapse. How Societies Choose to Fail or Survive*. Viking Penguin, New York and Allen Lane, London.

Ellison, C.A. (2001) Classical biological control of *Mikania micrantha*. In: Sankaran, K.V., Murphy, S.T. and Evans, H.C. (eds) *Alien Weeds in Moist Tropical Zones: Banes and Benefits. Proceedings of a Workshop, Kerala Forest Research Institute, Peechi, India, 2–4 November 1999*. Kerala Forest Research Institute, Peechi, India and CABI Bioscience, UK Centre (Ascot), Ascot, UK, pp. 131–138.

Ellison, C.A. and Day, M., (2011) Current status of releases of *Puccinia spegazzinii* for *Mikania micrantha* control. *Biocontrol News and Information* 32, 1N.

Ellison, C.A. and Murphy, S.T. (2001) *Puccinia spegazzinii* de Toni (Basidiomycetes: Urecinales) a potential biological control agent for *Mikania micrantha* Kunth. ex H.B.K. (Asteraceae) in India. Dossier for the Indian plant health authorities. CABI Bioscience, UK Centre (Ascot), Ascot, UK.

Ellison, C.A., Evans, H.C. and Ineson, J. (2004) The significance of intraspecies pathogenicity in the selection of a rust pathotype for the classical biological control of *Mikania micrantha* (mile-a-minute weed) in Southeast Asia. In: Cullen, J.M., Briese, D.T., Kriticos, D.J., Lonsdale, W.M., Morin, L. and Scott, J.K. (eds) *Proceedings of the XI International Symposium on Biological Control of Weeds, Canberra, Australia, 27 April–2 May 2003*. CSIRO (Commonwealth Scientific and Industrial Research Organisation) Entomology, Canberra, pp. 102–107.

Ellison C.A., Evans, H.C., Djeddour, D.H. and Thomas, S.E. (2008) Biology and host range of the rust fungus *Puccinia spegazzinii*: a new classical biological control agent for the invasive, alien weed *Mikania micrantha* in Asia. *Biological Control* 45, 133–145.

Evans, H.C. (1987) Fungal pathogens of some subtropical and tropical weeds and the possibilities for biological control. *Biocontrol News and Information* 8, 7–30.

Evans, H.C. (2002) Plant pathogens for biological control of weeds. In: Waller, J.M., Lenné, J.M. and Waller, S.J. (eds) *Plant Pathologist's Pocketbook*, 3rd edn. CAB International, Wallingford, UK, pp. 366–378.

Evans, H.C. (2008) The endophyte release hypothesis: implications for classical biological control and plant invasions. In: Julien, M.H., Sforza, R., Bon, M.C., Evans, H.C., Hatcher, P.E., Hinz, H.L. and Rector, B.G. (eds) *Proceedings of the XII International Symposium on Biological Control of Weeds, La Grande Motte, France, 22–27 April 2007*. CAB International, Wallingford, UK, pp. 20–25.

Evans, H.C. and Ellison, C.A. (2005) The biology and taxonomy of rust fungi associated with the Neotropical vine *Mikania micrantha*, a major invasive weed in Asia. *Mycologia* 97, 935–947.

Fowler, S.V., Ganeshan, S., Mauremootoo, J. and Mungroo, Y. (2000) Biological control of weeds in Mauritius: past successes revisited and present challenges. In: Spencer, N.R. (ed.) *Proceedings of the X International Symposium on Biological Control of Weeds, Bozeman, Montana, 4–14 July 1999*. USDA Forest Service, Forest Health Technology Enterprise Team, Morgantown, West Virginia, pp. 43–50.

Greathead, D.J. (1971) *A Review of Biological Control in the Ethiopian Region*. Technical Communication No. 5, Commonwealth Institute of Biological Control, Commonwealth Agricultural Bureau, Farnham Royal, UK [now CAB International, Wallingford, UK].

Harley, K.L.S. and Forno, I.W. (1992) *Biological Control of Weeds: A Handbook for Practitioners and Students*. Inkata Press, Melbourne, Australia.

Hasan, S. and Wapshere, A.J. (1973) The biology of *Puccinia chondrillina* a potential biological control agent of skeleton weed. *Annals of Applied Biology* 74, 325–332.

Hill, R.L., Gourlay, A.H. and Fowler, S.V. (2000) The biological control programme against gorse in New Zealand. In: Spencer, N.R. (ed.) *Proceedings of the X International Symposium on Biological Control of Weeds, Bozeman, Montana, 4–14 July 1999*. USDA Forest Service, Forest Health Technology Enterprise Team, Morgantown, West Virginia, pp. 909–917.

IPPC (2005) *ISPM 3. Guidelines for the Export, Shipment, Import and Release of Biological Control Agents and Other Beneficial Organisms*. International Standards for Phytosanitary Measures, Secretariat of the International Plant Protection Convention, Food and Agriculture Organization of the United Nations, Rome. Available at: https://www.ippc.int/en/core-activities/standards-setting/ispms/ (accessed 2 March 2017).

Kairo, M.T.K., Cock, M.J.W. and Quinlan, M.M. (2003) An assessment of the use of the code of conduct for the import and release of exotic biological control agents (ISPM No. 3) since its endorsement as an international standard. *Biocontrol News and Information* 24, 15N–27N.

Keane, R.M. and Crawley, M.J. (2002) Exotic plant invasions and the enemy release hypothesis. *Trends in Ecology and Evolution* 17, 164–170.

Labrada, R. (1996) The importance of biological control for the reduction of the incidence of major weeds in developing countries. In: Moran, V.C. and Hoffmann, J.H. (eds) *Proceedings of the IX International Symposium on Biological Control of Weeds, Stellenbosch, South Africa, 19–26 January 1996.* University of Cape Town, Rondebosch, South Africa, pp. 287–290.

Mack, R.N., Simberloff, D., Lonsdale, W.M., Evans, H., Clout, M. and Bazzaz, F.A. (2000) Biotic invasions: causes, epidemiology, global consequences and control. *Ecological Applications* 10, 689–710.

Marohasy, J. (1996) Host shifts in biological weed control: real problems, semantic difficulties or poor science? *International Journal of Pest Management* 42, 71–75.

Marsden, J.S., Martin, G.E., Parham, D.J., Ridsdill-Smith, T.J. and Johnston, B.G. (1980) *Returns on Australian Agricultural Research.* Commonwealth Scientific and Industrial Research Organisation, Canberra.

McFadyen, R.E.C. (1998) Biological control of weeds. *Annual Review of Entomology* 43, 369–393.

McFadyen, R.E.C. (2000) Successes in biological control of weeds. In: Spencer, N.R. (ed.) *Proceedings of the X International Symposium on Biological Control of Weeds, Bozeman, Montana, 4–14 July 1999.* USDA Forest Service, Forest Health Technology Enterprise Team, Morgantown, West Virginia, pp. 3–14.

McFadyen, R.E.C. (2012) Benefits from biological control of weeds in Australia. *Pakistan Journal of Weed Science Research* 18, 333–340.

Mitchell, C.E. and Power, A.G. (2003) Release of plants from fungal and viral pathogens. *Nature* 421, 625–627.

Morin, L., Auld, B.A. and Smith, H.E. (1996) Rust epidemics, climate and control of *Xanthium occidentale.* In: Moran, V.C. and Hoffmann, J.H. (eds) *Proceedings of the IX International Symposium on Biological Control of Weeds, Stellenbosch, South Africa, 19–26 January 1996.* University of Cape Town, Rondebosch, South Africa, pp. 385–391.

Morin, L., Evans, K.J. and Sheppard, A.W. (2006) Selection of pathogen agents in weed biological control: critical issues and peculiarities in relation to arthropod agents. *Australian Journal of Entomology* 45, 349–365.

Morris, M.J. (1997) Impact of the gall-forming rust fungus *Uromycladium tepperianum* on the invasive tree *Acacia saligna* in South Africa. *Biological Control* 10, 75–82.

Mortensen, K. (1986) Biological control of weeds with plant pathogens. *Canadian Journal of Plant Pathology* 8, 229–231.

Muniappan, R. and Viraktamath, C.A. (1993) Invasive alien weeds in the Western Ghats. *Current Science* 64, 555–557.

Paine, R.W. (1994) *Recollections of a Pacific Entomologist 1925–1966.* ACIAR Monograph 27, Australian Centre for International Agricultural Research, Canberra.

Parker, C. (1972) The *Mikania* problem. *Pest Articles and News Summaries (PANS)* 18, 312–315.

Pemberton, R.W. (2000) Predictable risk to native plants in weed biological control. *Oecologia* 125, 489–494.

Perrings, C., Mooney, H. and Williamson, M. (2010) *Bio-invasions and Globalization: Ecology, Economics, Management and Policy.* Oxford University Press, Oxford, UK.

Pimentel, D. (ed.) (2002) *Biological Invasions: Economic and Environmental Costs of Alien Plant, Animal, and Microbe Species.* CRC Press, Boca Raton, Florida.

Sankaran, K.V., Muraleedharan, P.K. and Anitha, V. (2001a) Integrated management of the alien invasive weed *Mikania micrantha* in the Western Ghats. KFRI Research Report No. 202, Kerala Forest Research Institute, Peechi, India.

Sankaran, K.V., Murphy, S.T. and Evans, H.C. (eds) (2001b) *Alien Weeds in Moist Tropical Zones: Banes and Benefits. Proceedings of a Workshop, Kerala Forest Research Institute, Peechi, India, 2–4 November 1999.* Kerala Forest Research Institute, Peechi, India and CABI Bioscience, UK Centre (Ascot), Ascot, UK.

Sankaran, K.V., Puzari, K.C., Ellison, C.A., Sreerama Kumar, P. and Dev, U. (2008) Field release of the rust fungus *Puccinia spegazzinii* to control *Mikania micrantha* in India: protocols and awareness raising. In: Julien, M.H., Sforza, R., Bon, M.C., Evans, H.C., Hatcher, P.E., Hinz, H.L. and Rector, B.G. (eds) *Proceedings of the XII International Symposium on Biological Control of Weeds, La Grande Motte, France, 22–27 April 2007.* CAB International, Wallingford, UK, pp. 384–389.

Sen Sarma, P.K. and Mishra, S.C. (1986) Biological control of forest weeds in India – retrospect and prospects. *Indian Forester* 112, 1088–1093.

Simmonds, H.W. (1933) The biological control of the weed *Clidemia hirta* D. Don, in Fiji. *Bulletin of Entomological Research* 24, 324–328.

Sreerama Kumar, P., Rabindra, R.J., Dev, U., Puzari, K.C., Sankaran, K.V., Khetarpal, R.K., Ellison, C.A. and Murphy, S.T. (2005) India to release the first fungal pathogen for the classical biological control of a weed. *Biocontrol News and Information* 26, 71N–72N.

Sreerama Kumar, P., Rabindra, R.J. and Ellison, C.A. (2008) Expanding classical biological control of weeds with pathogens in India: the way forward. In: Julien, M.H., Sforza, R., Bon, M.C., Evans, H.C., Hatcher, P.E., Hinz, H.L. and Rector, B.G. (eds) *Proceedings of the XII International Symposium on Biological Control of Weeds, La Grande Motte, France, 22–27 April 2007.* CAB International, Wallingford, UK, pp. 165–172.

Sreerama Kumar, P., Dev, U., Ellison, C.A., Puzari, K.C., Sankaran, K.V. and Joshi, N. (2016) Exotic rust fungus to manage the invasive mile-a-minute weed in India: pre-release evaluation and status of establishment in the field. *Indian Journal of Weed Science* 48, 206–214.

Suckling, D.M. and Sforza, F.F.H. (2014) What magnitude are observed non-target impacts from weed biocontrol? *PLoS ONE* 9(1): e84847. Available at: http://dx.doi.org/0.1371/journal.pone.0084847.

Thomas, P.A. and Room, P.M. (1986) Taxonomy and control of *Salvinia molesta. Nature* 320, 581–584.

Tomley, A.J. and Evans, H.C. (2004) Establishment and preliminary impact of the rust, *Maravalia cryptostegiae*, on the invasive alien weed, *Cryptostegia grandiflora* in Queensland, Australia. *Plant Pathology* 53, 475–484.

Varma, R.V., Shetty, A., Swaran, P.R., Paduvil, R. and Shamsudeen, R.S.M. (2006) Establishment of *Pareuchaetes pseudoinsulata* (Lepidoptera: Arctiidae), an exotic biocontrol agent of the weed, *Chromolaena odorata* (Asteraceae) in the forests of Kerala, India. *Entomon* 31, 49–51.

Wapshere, A.J. (1974) A strategy for evaluating the safety of organisms for biological weed control. *Annals of Applied Biology* 77, 201–211.

Warner, K.D. (2012) Fighting pathophobia: how to construct constructive public engagement with biocontrol for nature without augmenting public fears. *BioControl* 57, 307–317.

Waterhouse, D.F. (1994) *Biological Control of Weeds: Southeast Asian Prospects.* Australian Centre for International Agricultural Research, Canberra.

Willer, H., Lernoud, J. and Kilcher, L. (eds) (2013) *The World of Organic Agriculture. Statistics and Emerging Trends 2013.* FiBL–IFOAM Report, Research Institute of Organic Agriculture (FiBL), Frick, Switzerland and International Federation of Organic Agriculture Movements (IFOAM), Bonn, Germany.

Wilson, C.L. (1969) Use of plant pathogens in weed control. *Annual Review of Plant Pathology* 7, 411–433.

Winston, R.L., Schwarzländer, M., Hinz, H.L., Day, M.D., Cock, M.J.W. and Julien, M.H. (eds) (2014) *Biological Control of Weeds: A World Catalogue of Agents and Their Target Weeds*, 5th edn. USDA Forest Service, Forest Health Technology Enterprise Team, Morgantown, West Virginia.

Wittenberg, R. and Cock, M.J.W. (eds) (2001) *Invasive Alien Species: A Toolkit of Best Prevention and Management Practices.* CAB International, Wallingford, UK.

11 Policy Frameworks for the Implementation of a Classical Biological Control Strategy: the Chinese Experience

Jianqing Ding*

School of Life Sciences, Henan University, Kaifeng, China

Introduction

China is one of the mega-diverse countries of the world, with over 30,000 species of higher plants, 6347 species of vertebrates and 3862 species of fish (Chen, 1994; Xu *et al.*, 2000). In recent years, the conservation of China's biodiversity and protection of its environment have been critical domestic and international concerns (Liu and Diamond, 2005), and in this context invasive species have been recognized as posing an increasing threat to China's economy and ecosystems, largely due to escalating international activities and commerce (Normile, 2004).

Many approaches have been employed to control invasive species and prevent further pest introductions in the campaign against biological invasions in China and the wider world. Classical biological control through the screening, introduction and release of host-specific natural enemies of an invasive plant from its native region has been regarded as one of the more promising control approaches worldwide for more than 100 years (Winston *et al.*, 2014). Biological control may provide self-sustaining, broad-scale control of an invasive plant when one of its natural enemies, such as an insect or fungal agent, establishes a population successfully in the areas to which that plant has been introduced, owing to the ability of the natural enemy to disperse to find its host plant food resource. This is in sharp contrast to the use of manual, mechanical and chemical control measures, which typically require repeated treatment and provide control only at or near the site of application. Although an entire research project from the screening of natural enemies to successful control may be relatively expensive, a weed control programme based on biological control can be inexpensive in the long term. For example, the International Institute of Tropical Agriculture (IITA) has estimated that the biological control of water hyacinth (*Eichhornia crassipes*) following the introduction, mass rearing and release of two Neotropical *Neochetina* weevils in Benin was likely to yield a benefit:cost ratio of 124:1 over the 20 years following the releases (De Groote *et al.*, 2003). Similarly, a benefit:cost ratio of 112:1 has been demonstrated for the successful biological control of skeleton weed (*Chondrilla juncea*) in Australia (Marsden *et al.*, 1980).

The exchange of natural enemies, that is, the international import or export of potential insect/pathogen agents, is key to the implementation of a classical biological

* E-mail: dingjianqing@yahoo.com

control programme for an invasive plant. Therefore, cooperation is required between the recipient country – where the invasive plant is a problem, and the donor country – where the plant is native. An appropriate policy framework is essential to provide lawful guidance and regulation for the exchange of natural enemies between countries. Unfortunately, most developing countries lack such a policy framework, leading to often insurmountable problems in the import and export of insects/pathogens. This may have the following consequences: (i) the restriction of all international imports of natural enemies because of the unfounded fear that they may become pests themselves, or vice versa (by allowing the import of any exotic natural enemy); and (ii) forbidding all exports of natural enemies to other countries, allegedly to protect the source country's biological resources, or vice versa (by allowing any insect/pathogen to be exported). Hence, a country may either face more novel threats from invasive species if there is no policy framework to regulate and manage the import and release of potential biological control agents in place, or its biological control programmes may be jeopardized if all foreign introductions are prohibited.

In Asia, China is one of several developing countries that have implemented active classical biological control programmes and achieved successes in their campaign against invasive species. Although classical biological control is still in its early stages in China, and many aspects including the policy framework need to be greatly improved, the experience that has been gained may be of value to other Asian countries. Certainly, China shares with them many similar social, political and scientific issues, in particular in the management of invasive species. The purposes of this chapter are: (i) to review the current status and impact of, and the constraints to providing solutions for invasive plants in China; (ii) to report China's experiences in classical biological control; and (iii) to review and assess the current policy framework in terms of its role in the implementation of a classical biological control strategy.

Invasive Plants in China: Current Status, Impacts and Constraints to Providing Solutions

The first survey of invasive plants in China was reported by Ding and Wang (1999) in a Chinese government document on China's biodiversity. This survey, which was largely based on published Chinese literature, reported 58 invasive plant species in the agriculture and forestry systems in the country. In recent years, great advances in research on exotic plant species have been made as invasive species have become a growing concern in China. Qiang and Chao (2000) reported that there were 108 invasive weeds, in 76 genera and 23 families. A general survey of the exotic plant species in China conducted by the Institute of Botany and the Institute of Zoology of the Chinese Academy of Sciences showed there were about 300 exotic plants (Xie *et al.*, 2000).

The most important invasive plants in China include water hyacinth, alligator weed (*Alternanthera philoxeroides*), Crofton weed (*Ageratina adenophora*), common and giant ragweed (*Ambrosia artemisiifolia* and *A. trifida*) and mikania (*Mikania micrantha*); the latter species is also called mile-a-minute weed, but as there are two plants found in China that are commonly given the name of 'mile-a-minute' – the exotic invasive mikania and the native *Persicaria perfoliata* (which is invasive elsewhere) – the name of mile-a-minute is not used subsequently in this chapter to avoid confusion. Other important invasive plants in China are tall goldenrod (*Solidago altissima*) and the common and smooth cordgrasses (*Spartina anglica* and *S. alterniflora*) (see Box 11.1).

Invasive plants have been a critical issue in China for more than 30 years because of their negative impact on the environment and economy but, until recently, the Chinese government and public were unaware of the challenges that they present. There are no specific national acts or laws to define or regulate the introduction of invasive species, with the result that many exotic species are imported into China every year, some of which may prove to be potentially invasive. Research on developing the appropriate

Box 11.1. A roll call of invasive plants in China.

Invasive plants pose a great threat to China's ecosystems, cause economic losses in agriculture, fishery, forestry and other industries and, in some instances, have even become a problem for human health.

The water hyacinth (*Eichhornia crassipes*) has invaded 17 provinces in southern and central China, and has covered water surfaces in river courses, lakes and ponds. It alters ecosystem services and decreases soluble oxygen, leading to a decline in native aquatic biodiversity. It has aided the spread of human diseases, and has had economic impacts by impeding water flows, hindering navigation, and damaging irrigation and hydroelectricity facilities (Ding *et al.*, 1995; Lu *et al.*, 2007).

Mikania (*Mikania micrantha*), a more recent invasive climbing plant from South America, competes aggressively with native plants in the Neilingding National Reserve in Guangdong Province, southern China, and this has resulted in large areas of forests being killed, land degradation and habitat loss for protected animals such as rhesus macaques (*Macaca mulatta*) (Zhang *et al.*, 2004). At its peak, the plant covered some 40–60% of shrub and woodland in Neilingding (Feng *et al.*, 2002).

- In south-western China, Crofton weed (*Ageratina adenophora*) has driven many native plant species, such as *Persicaria perfoliata*, to local extinction (Ding Jianqing, Chinese Academy of Sciences, unpublished data).
- Pollen from ragweed (*Ambrosia artemisiifolia*) causes hay fever in susceptible people in many invaded areas in China, as they are allergic to its pollen (Li *et al.*, 2009).
- The biological and ecological characteristics of alligator weed (*Alternanthera philoxeroides*) have been well studied owing to its threat to biodiversity and ecosystem functions (Pan *et al.*, 2007). Although the use of a biological control agent has successfully suppressed its growth and reproduction in aquatic habitats in many provinces in southern China, it is still problematic in terrestrial habitats. The invasion of fields by alligator weed can cause 19–63% yield losses in five crops: rice, wheat, maize, sweet potato and lettuce (Tan, 1994).
- In the provinces of Zhejiang and Fujian in south-eastern China, the common and smooth cordgrasses (*Spartina anglica* and *S. alterniflora*) have invaded and occupied large coastal areas, leading to a dramatic decrease in aquaculture yields, as many of the crabs, mussels and fish have died (Huang, 1990). Alteration of the habitat structure and food resources has also led to a dramatic decline in avian species richness (Gan *et al.*, 2010). However, a 16 year chronosequence study has indicated that smooth cordgrass is being inhibited by an accumulation of habitat changes created by the two *Spartina* species themselves (Tang *et al.*, 2012).

technology for early warning and prediction systems to prevent future invasions is at an early stage, particularly at the national level.

Control approaches for most of the major invasive plants, e.g. Crofton weed and water hyacinth, involve manual removal and the application of chemical herbicides, but these methods are often implemented with little skill or knowledge because of poor levels of training in the technologies and lack of funding support. In addition, conflicts of interest may restrict the management of invasive plants in many areas, as individuals and different sectors of society often have differing opinions about whether an exotic plant is harmful or beneficial (Ding and Wan, 1993). For example, water hyacinth has been regarded as an ideal resource to make fertilizer, paper, pig feed and even food for human consumption, by many researchers in China (Ding *et al.*, 1995), even though its negative impacts on the environment are well known.

Classical Biological Control in China

Classical biological control was initiated in the mid-1980s as a novel strategy and technology against invasive plants in China, when the chrysomelid flea beetle *Agasicles hygrophila* was introduced from the USA to control alligator weed in southern China. This project, like almost all of China's classical biological control programmes, was led by the Institute of Biological Control (now the Institute of (Agro-)Environment and Sustainable Development, IEDA) of the

Chinese Academy of Agricultural Sciences (CAAS), in collaboration with foreign partners and domestic collaborators. By 2011, a total of 14 species of insects and pathogens had been imported from foreign countries or spread from a neighbouring country (in the case of Crofton weed) to control six invasive plant species in China (see Table 11.1).

Biological control of the alligator weed was implemented in the USA in the late 1950s by the US Army Corps of Engineers and the USDA-ARS (US Department of Agriculture – Agricultural Research Service). The US scientists successfully screened the South American beetle *A. hygrophila* and released it in the western and southern USA between 1964 and 1979 (Buckingham, 2002). Following successful control of alligator weed in the USA, the same beetle was introduced into China in 1986, where it underwent host-range tests on some 50 native plant species before being released in southern China in 1988. By 2001, the beetle was providing significant suppression of the weed in 14 provinces, particularly in aquatic habitats, and alligator weed has ceased to be dominant in many water bodies in southern China in recent years (Ding *et al.*, 2006).

Following this success, to control the common and giant ragweeds, another chrysomelid leaf beetle, *Zygogramma suturalis*, was introduced from North America via what was then the Soviet Union in the late 1980s, and a tortricid moth, *Epiblema strenuana*, was introduced from Mexico via Australia in the 1990s (Wan *et al.*, 1995). Immediately after their introduction into China and before they were released in the field, host-range tests were conducted to ensure their host specificity. Both were subsequently released. The leaf beetle failed to establish a population because of the presence of native predators, but the moth has successfully established and its impact on the ragweeds is under evaluation.

Biological control of the water hyacinth in China began in 1995 with the introduction of two *Neochetina* weevils sourced from cultures in Florida, USA, that were originally from Argentina. After host-range tests on 46 native plant species had confirmed their specificity, the weevils were released in Zhejiang and Fujian Provinces in south-eastern China in 1996–2000 (Fig 11.2). They successfully established and overwintered at Wenzhou in Zhejiang Province, and significantly suppressed water hyacinth at some release sites (Ding *et al.*, 2006).

Biological control of mikania became the hot project in classical weed biological control against an invasive plant in China in 2003, when CABI's centre in the UK developed a biological control project for China in collaboration with the Chinese Institute of Biological Control and the Guangdong Entomology Research Institute. In 2004, a highly host-specific Neotropical rust fungus, *Puccinia spegazzinii*, was introduced from CABI, and established in culture in quarantine in Beijing. Host-range tests were successfully completed prior to field release in 2006 (Fu *et al.*, 2006; Li *et al.*, 2007). Although the rust established in the field, there is no evidence that it has spread and still persists (Ellison and Day, 2011). More recently, a different isolate of the rust, from Ecuador, was introduced into quarantine in China at the Invasion Ecology and Biocontrol Laboratory, Wuhan Botanical Garden (CAAS), via Papua New Guinea. This isolate is proving to be an effective agent in Taiwan and Fiji, as well as in Papua New Guinea. Preliminary results from Wuhan look promising, though further evaluation is still needed (See Chapter 10, this volume).

Providing natural enemies to other countries

While China introduces biological control agents from other countries, it also helps other countries by exporting natural enemies to them to control invasive plants that are native to China. By 2014, a total of 34 species of insects and pathogens had been exported from China to foreign countries to control 15 invasive plant species (Table 11.2).

Hydrilla verticillata is a submersed aquatic plant with a broad native region that includes China that has invaded the USA, where it first arrived in Florida. To control it,

Table 11.1. Agents introduced for biological control of weeds in China (see also Fig. 11.1).

Target weed (Family)	Biological control agent	Source country/ies	Status	Reference
Ageratina adenophora (syn. *Eupatorium adenophorum*) (Asteraceae) (Fig. 11.1a)	*Passalora ageratinae* (syn. *Phaeoramularia eupatorii-odorati*, *Cercospora eupatorii*) (Fig. 11.1b)	Probably Mexico (natural spread of pathogen from Nepal probably carried by adult flies of *Procecidochares utilis*)	Unknown	Ma *et al.*, 2003a
	Procecidochares utilis (Fig. 11.1c)	Probably Mexico (spread from Nepal to China, then intentionally redistributed)	Unknown	Ma et al., 2003a
Alternanthera philoxeroides (Amaranthaceae)	*Agasicles hygrophila*	USA	Successful control in aquatic habitats in southern China; established populations found in terrestrial habitats but not suppressing the weed	Ma *et al.*, 2003b
Ambrosia artemisiifolia and *A. trifida* (Asteraceae) (Fig. 11.1d)	*Epiblema strenuana* (Fig. 11.1e)	Australia	Released, established populations in the provinces of Fujian, Hunan, Hubei and Jiangxi; impact of control under evaluation	Wan *et al.*, 1995
	Euaresta bella	Canada	Not released, introduced into quarantine but no established population	Wan *et al.*, 1995
	Euaresta festiva	Canada	Not released, introduced into quarantine but no established population	Wan *et al.*, 1995
	Liothrips sp.	Canada	Not released, introduced into quarantine but no established population	Wan *et al.*, 1995
	Tarachidia candefacta	Canada	Not released, introduced into quarantine but no established population	Wan *et al.*, 1995
	Zygogramma suturalis	Canada and the USA via (the then) Soviet Union	Released, did not establish populations in nature	Wan *et al.*, 1995

continued

Table 11.1. *continued*

Target weed (Family)	Biological control agent	Source country/ies	Status	Reference
Eichhornia crassipes (Pontederiaceae)	*Eccritotarsus catarinensis*	Brazil via South Africa	Released in Guangxi and Zhejiang provinces; did not establish	Ding *et al*., 2001b
	Neochetina bruchi	Argentina via the USA	Released, established populations in Fujian and Zhejiang provinces; established at Wenzhou, Zhejiang, contributing to significant control at some sites	Ding *et al*., 2001a,b 2006
	Neochetina eichhorniae	Argentina via the USA	Released, established populations in Fujian and Zhejiang provinces; established at Wenzhou, Zhejiang, contributing to significant control at some sites	Ding *et al*., 2001a,b, 2006
Mikania micrantha (Asteraceae)	*Actinote* spp.	Costa Rica/Brazil via Indonesia	Released in Guangdong Province, augmentative release needed for establishment	Li *et al*., 2004, Liu *et al*., 2007
	Puccinia spegazzinii	Argentina via the UK	Released in Guangdong Province in 2006, but failed to establish	Ellison and Day, 2011
		Ecuador via Papua New Guinea/ UK	Waiting for approval for release in nature	J. Ding, unpublished data

Fig. 11.1. Biological control of invasive weeds in China. (a) an infestation by young plants of *Ageratina adenophora*, with a negative impact on biodiversity; (b) a leaf of *A. adenophora* infected with the leaf spot pathogen *Passalora ageratinae*, also showing a young gall of *Procecidochares utilis* on the stem (arrows); (c) a mature gall of *P. utilis* gall on a stem of *A. adenophora*; (d) an infestation by *Ambrosia artemisiifolia* of a recently ploughed agricultural field; (e) damage by the ragweed borer *Epiblema strenuana* on leaves of *A. artemisiifolia* (arrows show beetle pupae). Photos courtesy C.A. Ellison.

Fig. 11.2. Dr Jianqing Ding releasing the water hyacinth weevil *Neochetina eichhorniae* for the biological control of water hyacinth (*Eichhornia crassipes*) in Wenzhou, Zhejiang Province, southern China. Photo courtesy Zhongnan Fan.

the first releases of the leaf-mining fly *Hydrellia pakistanae* were made in Florida in 1987 from material collected in India and Pakistan. Later, a strain of the fly from China was also released in the USA by the USDA, in cooperation with Chinese partners. In 1989, the congener *Hydrellia sarahae sarahae* was sent to Florida from China, but no releases were made because host-specificity testing indicated that it would have had a potentially broad host range in the USA (Balciunas *et al.*, 2002; Ding *et al.*, 2006).

The salt cedars *Tamarix ramosissima* and *T. chinensis*, are deciduous shrubs or small trees that are native in the Old World (Gaskin and Schaal, 2002) but invasive in the western USA, where they cause damage to riparian areas by displacing the native communities, degrading wildlife habitats and reducing water flow and groundwater levels (DeLoach *et al.*, 2000). A chrysomelid leaf beetle, *Diorhabda carinulata* (previously identified as *D. elongata deserticola*) was found to inflict substantial damage to salt cedar in its native range. A strain of the beetle imported from Xinjiang in north-western China, together with a second strain from Chilik in Kazakhstan, were approved for introduction and release following successful host-range testing and risk assessments (DeLoach *et al.*, 2003). The weevils were released in 2001. Spread in areas north of 38° latitude was rapid and the impact often dramatic. The beetles are now widespread in Nevada, Utah, Colorado and Wyoming, where they cause extensive defoliation of *T. ramosissima*, leading to dieback and even death after some years (DeLoach *et al.*, 2004; Ding *et al.*, 2006; Dudley and Bean, 2012).

Persicaria perfoliata is an annual or perennial herb native to Asia that has invaded the north-eastern USA. Already present in an area stretching from North Carolina to Massachusetts and westwards to Ohio, it is continuing to spread. More than 100 arthropod species were recorded from the plant during field surveys in China that were initiated in 1996, but only one, the

Table 11.2. Natural enemies exported from China as potential biological control agents in other parts of the world.

Target weed	Potential biological control agents	Recipient country	Status	References
Ailanthus altissima	*Eucryptorrhynchus brandti*	USA	Imported into quarantine in the USA, under evaluation	Ding *et al.*, 2006
	Eucryptorrhynchus chinensis	USA	Imported into quarantine in the USA, under evaluation	Ding *et al.*, 2006
Cirsium arvense	*Altica carduorum*	Canada	Unknown	J. Ding, unpublished data
	Cleonus sp.	Canada	Unknown	J. Ding, unpublished data
	Lixus sp.	Canada	Unknown	J. Ding, unpublished data
	Pustula spinulosa	USA via UK	Under evaluation in China and UK (imported into quarantine)	Wan *et al.*, 2014
	Thamnargus spp.	Canada	Unknown	J. Ding, unpublished data
Dioscorea bulbifera	*Lilioceris cheni*	USA	Released and established	Center *et al.*, 2013
	Lilioceris egena	USA	Imported into quarantine in the USA, under testing	Center *et al.*, 2013
Euphorbia esula	*Aphthona chinchihi*	USA, possibly via France	Unknown	Pemberton and Wang, 1989
	Chamaesphecia sp.	Canada	Rejected due to larval poor development	A. Gassmann, CABI, personal communication
	Hyles euphorbiae	Canada, USA	Unknown	J. Ding, unpublished data
	Oberea doncelii	Canada	Rejected due to broad host range	A. Gassmann, CABI, personal communication
	Puccinia spp.	USA	Unknown	Ma *et al.*, 2003a
Fallopia japonica	*Gallerucida bifasciata*	USA	Needs further evaluation	Wang *et al.*, 2008
Hydrilla verticillata	*Hydrellia pakistanae*	USA	Released and established in Florida and other parts of south-eastern USA	Center *et al.*, 1997; Balciunas *et al.*, 2002
	Hydrellia sarahae sarahae	USA	Host-specificity testing indicated potentially broad host range; not released	Balciunas *et al.*, 2002; J. J.K. Balciunas, personal communication
	Macroplea sp.	USA	Unable to rear adults from quarantine; additional field information needed	Balciunas *et al.*, 2002
	Mycoleptodiscus terrestris	USA	Under evaluation	Balciunas *et al.*, 2002; Shearer and Jackson, 2006

continued

Table 11.2. *continued*

Target weed	Potential biological control agents	Recipient country	Status	References
Ligustrum sinense	*Argopistes tsekooni*	USA	Needs further evaluation.	Zhang *et al.*, 2008
Myriophyllum spicatum	*Eubrychius* sp.	USA	Unknown	Ma *et al.*, 2003a
	Phytobius sp.	USA	Unknown	Ma *et al.*, 2003a
Persicaria perfoliata (syn. *Polygonum perfoliatum*)	*Rhinoncomimus latipes*	USA	Released in 10 states from 2004, promising impact/control	Ding *et al.*, 2004; Lake *et al.*, 2011; Hough-Goldstein *et al.*, 2012; Smith and Goldstein, 2014
	Timandra griseata	USA	Rejected due to broad host range	Price *et al.*, 2003
Pueraria montana var. *lobata*	*Arges* sp.	USA	Under evaluation	Ding *et al.*, 2006
	Gonioctena tredecimmaculata	USA	Rejected as not safe for release	Frye *et al.*, 2007
	Ornatalcides trifidus (syn. *Mesalcidodes trifidus*)	USA	Rejected as not safe for release	Frye *et al.*, 2007
Rubus ellipticus	*Epiblema tetragonana*	USA, Hawaii	Under evaluation	Wu *et al.*, 2013, 2014
	Epinotia ustulana	USA, Hawaii	Under evaluation	Wu *et al.*, 2013, 2014
Tamarix chinensis and *T. ramosissima*	*Diorhabda carinulata* (syn. *D. elongata deserticola*)	USA	Released in western USA in 2001, significant impact/control north of latitude 38°N	Deloach *et al.*, 2000; Ding *et al.*, 2006; Dudley and Bean, 2012
Trapa natans	*Galerucella birmanica*	USA	Imported into quarantine in the USA, under evaluation	Ding *et al.*, 2006
Triadica sebifera (syn. *Sapium sebiferum*)	*Bikasha collaris*	USA	Imported into quarantine in the USA, under evaluation	Huang *et al.*, 2011
	Gadirtha inexacta	USA	Imported into USA quarantine, under evaluation	Wang *et al.*, 2012
	Heterapoderopsis bicallosicollis	USA	Rejected as a threat to other species	Steininger *et al.*, 2013

curculionid weevil *Rhinoncomimus latipes*, proved sufficiently host specific to meet US regulatory requirements. A population in Hunan Province proved most damaging of the three populations tested, and weevils of this strain were mass reared and released in Delaware and New Jersey in 2004; Fig. 11.3 shows the larval stage. They have since been released in ten states, and impact assessments indicate that the beetles are having a measurable and at times rapid impact on *P. perfoliata* (Ding *et al.*, 2004, 2006; Hough-Goldstein *et al.*, 2012).

In addition to the control of the above species, China has helped Canada and USA with the biological control of leafy spurge (*Euphorbia esula*) and Canada thistle (*Cirsium arvense*) through insect surveys conducted in north-western China. Most of the natural enemies found for these two plants were shipped to the CABI centre at Delémont in Switzerland during 1991–1995.

Policy Frameworks: Challenges and Perspectives

Appropriate policy frameworks are the law-based cornerstones for conducting and implementing classical biological control. In China, the import and export of natural enemies have been processed according to the Law of the People's Republic of China on the Entry and Exit Animal and Plant Quarantine, 1991, although currently there is no specific national law for classical biological control. The purpose of the legislation and implementation of this quarantine law is to protect domestic animals and plants in China from direct introduction to diseases, pest insects and weeds, and also harmful organisms carried by introduced animals and plants. The quarantine law also prevents the export of diseased animals or plants and disease- or pest-carrying animals and plants from China.

The official enforcing agency charged with implementing the quarantine law is the National Bureau of Inspection and Quarantine of Imported and Exported Animals and Plants. This Bureau operates from its headquarters in Beijing and has branches in all provinces and at some key airport and seaport cities. All applications for import (or export) of living insects and pathogens from (or to) foreign countries must be submitted to either the headquarters or its affiliated branches. A permit for the import of natural enemies may be issued once the insect or pathogen is evaluated by the Bureau's officials and judged to be a no-risk or low-risk organism.

In the case of introductions for biological control, once the application for import is approved, the applicants are first required to maintain and rear the natural enemies in an isolated quarantine facility for a specific time (usually 60–90 days) to ensure that no unapproved (and potential or actual pest) organism has been inadvertently introduced. Even if the host range of the natural enemy has been tested, and has been reported as host specific to the target weed in previous experiments in other countries, complementary host-range testing with some of China's native plants is also required before field release is permitted. All the seven previously released insect species and one intentionally released pathogen (see Table 11.1) imported from foreign countries for weed control in China were retested in this way before field release in China (Wang and Wang, 1988; Wan *et al.*, 1995; Ding *et al.*, 2002; Fu *et al.*, 2006).

Fig. 11.3. Larva of the weevil *Rhinoncomimus latipes*, which has been introduced from China to the USA for the biological control of the invasive plant *Persicaria perfoliata*. Photo courtesy the author.

Classical biological control has been conducted and implemented in China since the mid-1980s (Wang, 1989). Many Chinese national and provincial governmental agencies, such as the Ministry of Science and Technology, the Ministry of Agriculture and the Ministry of Forestry, as well as China's Natural Scientific Foundation, have supported classical biological control financially since the 1980s. The achievements of several biological control projects, such as the successful control of alligator weed, were well recognized by the government by the granting of awards to the personnel who implemented the research and extension of the project.

However, the policy frameworks to facilitate the implementation of a comprehensive biological control programme remain ambiguous and even deficient. Gaps in national, regional and various industry systems exist and more may arise in the near future, given the recent increasing effort against invasive species. There is often a lack of communication between biological control scientists and quarantine officials on the import and export of insects or pathogens, as there are no specific laws and regulations to follow. The biological control workers may believe that the insects or pathogens they want to introduce will contribute to the suppression of an invasive plant, but the officials from the Inspection and Quarantine Bureau may be more concerned with the potential risk that these agents pose to China's ecosystems than their impact on the invasive plant, and this could ultimately hamper the introduction. Equally, a local branch office may issue an import permit for an insect or pathogen to control a plant that is a weed in the region, although the putative agent could spread to neighbouring provinces where it may attack other plants. There is currently no inter-province or national-scale communications network to deal with the import and export of biological control agents.

China is facing a great challenge and has a need to pass a specific law to facilitate successful biological control programmes. This need is becoming more pressing as the impacts of invasive plants become of increasing concern and many other control approaches fail to provide viable solutions. A national biological control Act or Regulation is necessary to provide a legal framework within which biological control workers can perform research and conduct the necessary international exchange of natural enemies. The functioning of this law would not only benefit the biological control of invasive plants, but should also help to prevent the introduction of potentially invasive species in the future through the strict regulation of imports. The law should also include an appropriate framework for assessing whether permission should be given for a putative beneficial insect to be released in nature after host-range tests; at present, there is no official documentation to guide this process.

Acknowledgements

I thank Carol Ellison and Sean Murphy for their kind invitation to write this chapter. I am also grateful for Carol Ellison and Rebecca Murphy for their comments that improved earlier versions of this chapter.

References

Balciunas, J.K., Grodowitz, M.J., Cofrancesco, A.F. and Shearer, J.F. (2002) *Hydrilla*. In: van Driesche, R., Blossey, B., Hoddle, M., Lyon, S. and Reardon, R. (eds) *Biological Control of Invasive Plants in the Eastern United States.* USDA Forest Service, Forest Health Technology Enterprise Team, Morgantown, West Virginia, pp. 91–114.

Buckingham, G.R. (2002) Alligatorweed. In: van Driesche, R., Blossey, B., Hoddle, M., Lyon, S. and Reardon, R. (eds) *Biological Control of Invasive Plants in the Eastern United States.* USDA Forest Service, Forest Health Technology Enterprise Team, Morgantown, West Virginia, pp. 5–15.

Center, T.D., Grodowitz, M.J., Cofrancesco, A.F., Jubinsky, G., Snoddy, E. and Freedman, J.E. (1997) Establishment of *Hydrellia pakistanae* (Diptera: Ephydridae) for the biological control of the submersed aquatic plant *Hydrilla verticillata* (Hydrocharitaceae) in the southeastern United States. *Biological Control* 8, 65–73.

Center, T.D., Rayamajhi, M., Dray, F.A., Madeira, P.M., Witkus, G., Rohrig, E., Mattison, E., Lake, E., Smith, M., Zhang J.L. *et al.* (2013) Host range validation, molecular identification and release and establishment of a Chinese biotype of the Asian leaf beetle *Lilioceris cheni* (Coleoptera: Chrysomelidae: Criocerinae) for control of *Dioscorea bulbifera* L. in the southern United States. *Biocontrol Science and Technology* 23, 735–755.

Chen, L. (1994) Conservation in biodiversity and countermeasures. In: Qian, Y. and Ma, K. (eds) *Principles and Methods of the Researches on Biodiversity.* China Science and Technology Press, Beijing, pp. 13–35.

De Groote, H., Ajuonu, O., Attignon, S., Djessou, R. and Neuenschwander, P. (2003) Economic impact of biological control of water hyacinth in southern Benin. *Ecological Economics* 45, 105–117.

DeLoach, C.J., Carruthers, R.I., Lovich, J.E., Dudley, T.L. and Smith, S.D. (2000) Ecological interactions in the biological control of saltcedar (*Tamarix* spp.) in the United States: toward a new understanding. In: Spencer, N.R. (ed.) *Proceedings of the X International Symposium on Biological Control Weeds, Bozeman, Montana, 4–14 July 1999.* USDA Forest Service, Forest Health Technology Enterprise Team, Morgantown, West Virginia, pp. 819–873.

DeLoach, C.J., Lewis, P.A., Carruthers, R.I., Herr, J.C., Tracy, J.L. and Johnson, J. (2003) Host specificity of a leafbeetle, *Diorhabda elongate deserticola* (Coleoptera: Chrysomelidae) from Asia, for biological control of saltcedars (*Tamarix*: Tamaricaceae) in the western United States. *Biological Control* 27, 117–147.

DeLoach, C.J., Carruthers, R., Dudley, T., Eberts, D., Kazmer, D., Knutson, A., Bean, D., Knight, J., Lewis, P., Tracy, J. *et al.* (2004) First results for control of saltcedar (*Tamarix* spp.) in the open field in the western United States. In: Cullen, J.M., Briese, D.T., Kriticos, D.J., Lonsdale, W.M., Morin, L. and Scott, J.K. (eds) *Proceedings of the XI International Symposium on Biological Control of Weeds, Canberra, Australia, 27 April–2 May 2003.* Commonwealth Scientific and Industrial Research Organisation (CSIRO), Canberra, pp. 506–513.

Ding, J. and Wan, F. (1993) Conflict of interests in biological control in weeds. In: Hu, T. (ed.) *Insect Ecology.* China Science and Technological Press, Beijing, China, pp. 102–106. [In Chinese.]

Ding, J. and Wang, R. (1999) Invasive alien species and their impact on biodiversity in China. In: Chen, C.D. (ed.) *China's Biodiversity: A Country Study.* Chinese Environmental Press, Beijing, pp. 72–75.

Ding, J., Wang, R., Fan, Z., Chen, Z. and Fu, W. (1995) The distribution, damage and control strategy of [the] aquatic weed, water hyacinth (*Eichhornia crassipes*) in China. *Chinese Journal of Weed Science* 9, 49–52. [In Chinese, English abstract.]

Ding, J., Chen, Z.-Q., Fu, W.-D., Wang, R., Zhang, G.L., Fan, Z., Fang, Y. and Xu, L. (2001a) Control [of] *Eichhornia crassipes*, an invasive aquatic weed in south China with *Neochetina eichhorniae. Chinese Journal of Biological Control* 17, 97–100. [In Chinese, English abstract.]

Ding, J., Wang, R., Fu, W.-D. and Zhang, G.-L. (2001b) Water hyacinth in China: its distribution, problems and control status. In: Julien, M.H., Hill, M.P., Center, T.D. and Ding, J. (eds) *Biological and Integrated Control of Water Hyacinth. Proceedings of the Second Meeting of the Global Working Group for the Biological and Integrated Control of Water Hyacinth, Beijing, China, 9–12 October 2000.* ACIAR Proceedings No. 102, Australian Centre for International Agricultural Research, Canberra, pp. 29–32.

Ding, J., Chen, Z.-Q., Fu, W.-D. and Wang, R. (2002) Biology and host range of [the] water hyacinth weevil, *Neochetina eichhorniae. Chinese Journal of Biological Control* 18, 153–157. [In Chinese, English abstract.]

Ding, J., Fu, W.[-D.], Wu, Y., Reardon, R. and Zhang, G.[-L.] (2004) Exploratory survey in China for potential insect biocontrol agents of mile-a-minute weed, *Polygonum perfoliatum* L., in eastern USA. *Biological Control* 30, 487–495.

Ding J., Reardon, R., Wu, Y., Zheng, H., Fu, W.-D. (2006) Biological control of invasive plants through collaboration between China and the United States of America: a perspective. *Biological Invasions* 8, 1439–1450.

Dudley, T. and Bean, D. (2012) *Tamarix* biocontrol programme enters a new and uncertain phase. *Biocontrol News and Information* 33, 10N–13N.

Ellison, C. and Day, M. (2011) Current status of releases of *Puccinia spegazzinii* for *Mikania micrantha* control. *Biocontrol News and Information* 32, 1N.

Feng, H.-L., Cao, H.-L., Liang, X.-D., Zhou, X. and Ye, W.-H. (2002) The distribution and harmful effect of *Mikania micrantha* in Guangdong. *Journal of Tropical and Subtropical Botany* 10, 263–270. [In Chinese, English abstract.]

Frye, M.J., Hough-Goldstein, J. and Sun, J.H. (2007) Biology and preliminary host range assessment of two potential kudzu biological control agents. *Environmental Entomology* 36, 1430–1440.

Fu, W.-D, Yang, M.-L. and Ding, J. (2006) Biology and host specificity of *Puccinia spegazzinii*, a potential biocontrol agent for *Mikania micrantha*. *Chinese Journal of Biological Control* 22, 67–72. [In Chinese, English abstract.]

Gan, X.-J., Choi, C.-Y., Wang, Y., Ma, Z.-J., Chen, J.-K. and Li, B. (2010) Alteration of habitat structure and food resources by invasive smooth cordgrass affects habitat use by wintering saltmarsh birds at Chongming Dongtan, east China. *The Auk* 127, 317–327.

Gaskin, J.F. and Schaal, B.A. (2002) Hybrid *Tamarix* widespread in US invasion and undetected in native Asian range. *Proceedings of the National Academy of Sciences of the United States of America* 99, 11256–11259.

Hough-Goldstein, J., Lake, E. and Reardon, R. (2012) Status of an ongoing biological control program for the invasive vine *Persicaria perfoliata* in eastern North America. *BioControl* 57, 181–189.

Huang, W., Wheeler, G.S., Purcell, M.F and Ding, J. (2011) The host range and impact of *Bikasha collaris* (Coleoptera: Chrysomelidae), a promising candidate agent for biological control of Chinese tallow, *Triadica sebifera* (Euphorbiaceae) in the United States. *Biological Control* 56, 230–238.

Huang, Z. (1990) Common cordgrass turns into big disaster – appeal from Dong Wu Yang Sea. *Newspaper of Agriculture and Fishery of China*, Beijing. [No further publication details available.]

Lake, E.C., Hough-Goldstein, J., Shropshire, K.J. and D'Amico, V. (2011) Establishment and dispersal of the biological control weevil *Rhinoncomimus latipes* on mile-a-minute weed, *Persicaria perfoliata*. *Biological Control* 58, 294–301.

Li, J., Sun, B., Huang, Y., Lin, X., Zhao, D., Tan, G., Wu, J., Zhao, H., Cao, L. and Zhong, N. (2009) A multicentre study assessing the prevalence of sensitizations in patients with asthma and/or rhinitis in China. *Allergy* 64, 1083–1092.

Li, Z.-G., Han, S.-C., Guo, M.-F., Luo, L.-F., Li, L.-Y., de Chenon, R.D., Day, M.D. and McFadyen, R.E. (2004) Rearing *Actinote thalia pyrrha* (Fabricius) and *Actinote anteas* (Doubleday and Hewitson) with cutting[s of] and potted *Mikania micrantha* Kunth. In: Day, M.D. and McFadyen, R.E. (eds) *Chromolaena in the Asia-Pacific Region. Proceedings of the Sixth International Workshop on Biological Control and Management of Chromolaena, Cairns, Australia, 6–9 May 2003.* Australian Centre for International Agricultural Research, Canberra, pp. 36–38.

Li, Z.-G., Han, S.-C., Li, L.-Y., Li, J. and Lu, J.-W. (2007) Pathogenicity of *Puccinia spegazzinii* in South China. *China Journal of Biological Control* 23(suppl.), 57–59.

Liu, J. and Diamond, J. (2005) China's environment in a globalizing world. *Nature* 435, 1179–1186.

Liu, X.-L., Han, S.-C. and Zeng, L. (2007) Life table analysis of natural population of *Actinote anteas* (Nymphalidae), a biocontrol agent of *Mikania micrantha* (Compositae). *Chinese Journal of Biological Control* 23, 127–132. [In Chinese, English abstract.]

Lu, J.-B., Wu, J.-G., Fu, Z.-H. and Zhu, L. (2007) Water hyacinth in China: a sustainability science-based management framework. *Environmental Management* 40, 823–830.

Ma, R., Wang, R. and Ding, J. (2003a) Classical biological control of exotic weeds. *Acta Ecologica Sinica* 23, 2677–2688. [In Chinese.]

Ma, R., Ding, J. and Wang, R. (2003b) Population adaptability of *Agasicles hygrophila* on different ecotypes of alligatorweed. *Chinese Journal of Biological Control* 19, 54–58. [In Chinese, English abstract.]

Marsden, J.S., Martin, G.E., Parham, D.J., Ridsdill-Smith, T.J. and Johnston, B.G. (1980) Skeleton weed control. In: *Returns on Australian Agricultural Research.* Commonwealth Scientific and Industrial Research Organisation (CSIRO) Division of Entomology, Canberra, pp. 84–93.

Normile, D. (2004) Expanding trade with China creates ecological backlash. *Science* 306, 968–969.

Pan, X.-Y., Geng, Y.-P., Sosa, A., Zhang, W.-J., Li, B. and Chen, J.-K. (2007) Invasive *Alternanthera philoxeroides*: biology, ecology and management. *Acta Phytotaxonomica Sinica* 45, 884–900.

Pemberton, R.W. and Wang, R. (1989) Survey for natural enemies of *Euphorbia esula* L. in northern China and Inner Mongolia. *Chinese Journal of Biological Control* 5, 64–67.

Price, D.L., Hough-Goldstein, J. and Smith, M.T. (2003) Biology, rearing, and preliminary evaluation of host range of two potential biological control agents for mile-a-minute weed, *Polygonum perfoliatum* L. *Environmental Entomology* 32, 229–236.

Qiang, S. and Chao, X.-Z. (2000) Survey and analysis of exotic weeds in China. *Journal of Plant Resources and Environment* 9, 34–38. [In Chinese, English abstract.]

Shearer, J.F. and Jackson, M.A. (2006) Liquid culturing of microsclerotia of *Mycoleptodiscus terrestris*, a potential biological control agent for the management of hydrilla. *Biological Control* 38, 298–306.

Smith, J.R. and Hough-Goldstein, J. (2014) Impact of herbivory on mile-a-minute weed (*Persicaria perfoliata*) seed production and viability. *Biological Control* 76, 60–64.

Steininger, M.S., Wright, S.A., Ding, J. and Wheeler, G.S. (2013) Biology and host range of *Heterapoderopsis bicallosicollis*: a potential biological control agent for Chinese tallow *Triadica sebifera*. *Biocontrol Science and Technology* 23, 816–828.

Tan, W. (1994) Evaluation of loss in several crops caused by alligatorweed. *Chinese Journal of Weed Science* 8, 28–31.

Tang, L., Gao, Y., Wang, C.-H., Zhao, B. and Li, B. (2012) A plant invader declines through its modification to habitats: a case study of a 16-year chronosequence of *Spartina alterniflora* invasion in a salt marsh. *Ecological Engineering* 49, 181–185.

Wan, F., Wang, R. and Ding, J. (1995) Biological control of *Ambrosia artemisiifolia* with potential insect agents, *Zygogramma suturalis* and *Epiblema strenuana*. In: Delfosse, E. and Scott, R.R. (eds) *Proceedings of the VIII International Symposium on Biological Control of Weeds, Canterbury, New Zealand, 2–7 February 1992*. Commonwealth Scientific and Industrial Research Organisation (CSIRO), Melbourne, Australia, pp. 193–200.

Wan, H.H., Ellison, C., Li, H.M., Evans, H., Liu, T.G., Zhang, Y.J., Hinz, H. and Zhang, F. (2014) Is there now hope for the biological control of *Cirsium arvense*? In: Impson, F.A.C., Kleinjan, C.A. and Hoffmann, J.H. (eds) *Proceedings of the XIV International Symposium on Biological Control of Weeds, Kruger National Park, South Africa, 2–7 March 2014*. University of Cape Town, Rondebosch, South Africa, p. 17. [Abstract.]

Wang, R. (1989) Biological control of weeds in China: a status report. In: Delfosse, E. (ed.) *Proceedings of the VII International Symposium on Biological Control of Weeds, Rome, Italy, 6–11 March 1988*. Istituto Sperimentale per la Patologia Vegetale, Ministero dell'Agricoltura e delle Foreste, Rome, pp. 689–693.

Wang, R. and Wang, Y. (1988) Host specificity tests for *Agasicles hygrophila*, a biological control agent of alligatorweed. *Chinese Journal of Biological Control* 4, 14–17. [In Chinese, English abstract.]

Wang, Y., Ding, J. and Zhang, G.-P. (2008) *Gallerucida bifasciata* (Coleoptera: Chrysomelidae), a potential biological control agent for Japanese knotweed (*Fallopia japonica*). *Biocontrol Science and Technology* 18, 59–74.

Wang, Y., Zhu, L., Gu, X., Wheeler, G.S., Purcell, M. and Ding, J. (2012) Pre-release assessment of *Gadirtha inexacta*, a proposed biological control agent of Chinese tallow (*Triadica sebifera*) in the United States. *Biological Control* 63, 304–309.

Winston, R.L., Schwarzländer, M., Hinz, H.L., Day, M.D., Cock, M.J.W. and Julien, M.H. (eds) (2014) *Biological Control of Weeds: A World Catalogue of Agents and Their Target Weeds*, 5th edn. USDA Forest Service, Forest Health Technology Enterprise Team, Morgantown, West Virginia.

Wu, K., Center, T.D., Yang, C.-H., Zhang, J., Zhang, J.-L. and Ding, J. (2013) Potential classical biological control of invasive Himalayan yellow raspberry, *Rubus ellipticus* (Rosaceae). *Pacific Science* 67, 59–80.

Wu, K., Zhang J., Zhang G.-A. and Ding J. (2014) *Epiblema tetragonana* and *Epinotia ustulana* (Lepidoptera: Tortricidae), two potential biological control agents for the invasive plant, *Rubus ellipticus*. *Biological Control* 77, 51–58.

Xie, Y., Li, Z., Gregg, W.P. and Li, D. (2000) Invasive species in China – an overview. *Biodiversity and Conservation* 10, 1317–1341.

Xu, H.-G., Wang, D.-H. and Sun, X.-F. (2000) Biodiversity clearing-house mechanism in China: present status and future needs. *Biodiversity and Conservation* 9, 361–378.

Zhang, L.Y., Ye, W.H., Cao, H.L. and Feng, H.L. (2004) *Mikania micrantha* H.B.K. in China – an overview. *Weed Research* 44, 42–49.

Zhang, Y.-Z., Hanula, J.L. and Sun, J.-H. (2008) Host specificity *of Argopistes tsekooni* (Coleoptera: Chrysomelidae), a potential biological control agent of Chinese privet. *Journal of Economic Entomology* 101, 1146–1151.

12 Policy Frameworks for the Implementation of a Classical Biological Control Strategy: the Indian Experience

R.J. Rabindra, P. Sreerama Kumar* and Abraham Verghese

ICAR (Indian Council of Agricultural Research)-National Bureau of Agricultural Insect Resources, Bengaluru, India

Introduction

The onset of the Green Revolution in India ushered in an era of food security. Modernization of agriculture through the use of improved seeds and chemical fertilizers, coupled with good agricultural practices, resulted in quantum increases, particularly in the yields of wheat and rice. However, the country continues to face the pressure of increasing food production to meet the needs of its growing human population. With no further opportunities for bringing additional areas into cultivation on the one hand, and the shrinking amount of available cultivable land due to invasion by housing and industry in peri-urban areas on the other, there is enormous pressure on the existing land for food production. Intensive agriculture involving chemical inputs, particularly pesticides for the management of pests, diseases and weeds, has been found to be unsustainable and to have serious undesirable ecological consequences.

Therefore, the Government of India has adopted integrated pest management (IPM) as a national policy. Consequently, biological control receives significant attention from the Indian Council of Agricultural Research (ICAR). The National Bureau of Agricultural Insect Resources (NBAIR), one of ICAR's constituent units, focuses its attention on harnessing natural biological resources for the eco-friendly management of pests, diseases and weeds for enhancing the productivity and welfare of farmers, through an All-India Coordinated Research Project (AICRP).

This chapter is intended to share the Indian experience on importing and utilizing a rust pathogen for the classical biological control of a weed – the first ever attempt to introduce a plant pathogen for classical biological control of a weed by any country in the region. As a prelude to this, the chapter first highlights the overall problem of alien weeds in India, the use of classical biological control for these weeds and the existing frameworks and guidelines that enable the introduction of host-specific natural enemies to manage invasive pests in India. Lastly, the chapter discusses India as a provider of weed biological control agents.

The Problem of Alien Weeds in India

Highly intensified land-use for crop production, as seen in many parts of the country, has resulted in the depletion of underground

* Corresponding author. E-mail: psreeramakumar@yahoo.co.in

water sources, land degradation, salinization, nutrient depletion and desertification, with serious consequences for the livelihood of the communities that are dependent on the different agroecosystems involved. Unsustainable cultivation practices and land use, deforestation, excessive grazing and monocropping have upset the ecological balance.

Over the last century, vegetational modification has occurred owing to biological invasion by alien plants following their either deliberate or unintentional introduction into India. Their inherent properties of efficient nutrient uptake and use (as exemplified by *Mikania micrantha* in Chapter 2, this volume) facilitate the invasion of disturbed land by many exotic weeds, which, in turn, leads to loss of biodiversity. A few aquatic weeds, including water hyacinth (*Eichhornia crassipes*) and water fern (*Salvinia molesta*), have also invaded India. As well as to choking navigation and irrigation canals, these weeds have increased evaporation losses from precious water resources through plant transpiration (Van der Weert and Kamerling, 1974). The pernicious Siam weed (*Chromolaena odorata*) is a serious threat to the regeneration of natural forests because of its inhibitory (allelopathic) effects on the germination and survival of native plants. In addition, it depletes the soil nutrients and competes with native plants for moisture and nutrients. Invasion by *C. odorata* has created a shortage of fodder and fuel (Doddamani *et al.*, 1998) and affected the growth of teak, bamboo, eucalyptus, coconut and areca nut (Swaminath and Shivanna, 2001). Similarly, mikania (*M. micrantha*) is affecting plantations of banana, coffee, tea, teak, rubber and oil palm, and crops of pineapple, ginger and tapioca in Kerala (Sankaran and Sreenivasan, 2001) and tea and coffee in the northeastern states (Gogoi, 2001). Parthenium weed (*Parthenium hysterophorus*), also called Congress grass in India, has invaded several agroecosystems, resulting in changes in plant diversity, in addition to causing a number of health problems in humans and animals. *Lantana camara*, *Ageratum conyzoides*, *Ageratina adenophora* and *Mimosa diplotricha* var. *diplotricha* are among the major alien weeds of the tropical moist forest zones.

The huge losses of revenue, as well as the biodiversity losses, caused by invasive alien weed species prompted the Government of India to notify specified quarantine weed species under Schedule VIII of the Plant Quarantine (Regulation of Import into India) Order, 2003 (see Appendix 12.1) to encourage all stakeholders to watch out for such plants, with the aim of preventing their entry and spread in India.

Classical Biological Control in India

The classical biological control of alien weeds is not new to India, although the first case was accidental: in the late 18th century, successful control of the prickly pear (*Opuntia vulgaris*) was achieved in 5–6 years in northern India by the scale insect *Dactylopius ceylonicus*, following its inadvertent introduction from Brazil; the intended introduction was *D. coccus*, for use in commercial cochineal dye production. Attempts to redeploy *D. ceylonicus* against *O. stricta* var. *dillenii* in southern India were unsuccessful, however. The first successful intentional introduction of a weed biocontrol agent to India took place in 1926, when another *Dactylopius* species, *D. opuntiae*, of North American origin was introduced from Sri Lanka and controlled *O. stricta* and the closely related *O. elatior* on more than 40,000 ha (Sreerama Kumar *et al.*, 2008). Since then, a number of phytophagous arthropods and other agents have been introduced against invasive terrestrial and aquatic weeds in India with varying degrees of success (see Table 12.1). The spectacular success of the control of *Salvinia molesta* with the insect *Cyrtobagous salviniae* in Kerala gave a boost to the biocontrol approach in India (Joy *et al.*, 1987), but there have been few introductions in recent years.

In the post-independence era, classical biological control of weeds and insect pests became more systematic and scientific when specific programmes began at the then

Table 12.1. Organisms introduced into India for the biological control of weeds.[a]

Target weed	Introduced organism[b]	Year	Origin of biocontrol agent	Result
Terrestrial weeds				
Ageratina adenophora	*Procecidochares utilis*	1963	Mexico via Hawaii/Australia/New Zealand	Established, widespread but little impact
Chromolaena odorata	*Apion brunneonigrum*[c,d]	1972, 1976, 1982/83	Trinidad	Not established
	Pareuchaetes pseudoinsulata	1973	Trinidad	Established, resurgence recorded in Kerala
		1984	Trinidad via India/Sri Lanka	Established, localized initial impact
	Phestinia costella (as *Mescinia parvula*)[d]	1986	Trinidad	Quarantine culture failed, not released
	Cecidochares connexa	2005	Colombia via Indonesia	Established, significant impact in wet, low-elevation areas
Lantana camara	*Ophiomyia lantanae*	1921	Mexico via Hawaii	Established, widespread but no impact
	Orthezia insignis	1921	Mexico via Hawaii	Established, widespread but no target impact (some non-target attack)
	Teleonemia scrupulosa	1941	Mexico via Hawaii/Fiji/Australia	Quarantine culture destroyed but agent escaped. Established, widespread but little impact
	Diastema tigris	1971	Trinidad	Not established
	Salbia haemorrhoidalis	1971	Thailand	Not established
	Uroplata girardi	1972	Brazil via Hawaii/Australia	Established, no impact
	Octotoma scabripennis	1972	Mexico via Hawaii/Australia	Established, no impact
Mikania micrantha	*Puccinia spegazzinii* (rust pathogen)	2005	Trinidad	Not established
		2007	Peru	Not established
Opuntia elatior	*Dactylopius ceylonicus*	post-1863	Brazil, Mexico	Not established
	Dactylopius opuntiae	1926/27	USA via Australia/Sri Lanka	Established, widespread, heavy impact, complete control
Opuntia monacantha (as *O. vulgaris*)	*Dactylopius ceylonicus* (as *D. coccus*)	1795, 1821	Brazil	Established, widespread, heavy impact, complete control
	Dactylopius confusus	1836, 1838	South America via Germany/South Africa	Not established

Opuntia stricta (as var. *dillenii*)	*Dactylopius ceylonicus*	post-1863	Brazil, Mexico	Not established
	Dactylopius opuntiae	1926/27	USA via Australia/Sri Lanka	Established, widespread, heavy impact, complete control
Orobanche sp.	*Phytomyza orobanchia*[e]	1982	Yugoslavia	Rearing failed, not released
Parthenium hysterophorus	*Smicronyx lutulentus*[d]	1983	Mexico	Quarantine culture failed, not released
	Zygogramma bicolorata	1984	Mexico	Established, variable impact, population reduction in some regions
	Epiblema strenuana[f]	1983	Mexico	Not host specific, destroyed in quarantine
Aquatic weeds				
Eichhornia crassipes	*Neochetina eichhorniae*	1983	Argentina via USA/Australia	Established, initial impact and good control in combination with *N. bruchi*; resurgence since, variable impact
	Neochetina bruchi	1984	Argentina via USA	Established, initial impact and good control in combination with *N. eichhorniae*; resurgence since, variable impact
	Orthogalumna terebrantis	1986	Argentina via USA	Established, no independent impact
Salvinia molesta	*Paulinia acuminata*	1974	Trinidad	Not established
	Cyrtobagous salviniae	1983	Brazil via Australia	Established, widespread, heavy impact, good control
Various species[g]	*Osphronemus goramy* (fish)	Early 1800s	Java (Indonesia)	Stocked at Calcutta Botanical Gardens, but perished by 1941
		1865	Mauritius	Breeding occurred, but condition of stock not satisfactory
		1916	Java (Indonesia) and Mauritius	Well established in Tamil Nadu and distributed to other areas
	Tilapia mossambica (fish)	1953	Africa	Limited control of soft-leaved weeds in fish ponds
	Hypophthalmichthys molitrix (fish)	1959	?	Established, not very effective
	Ctenopharyngodon idella (fish)	1959, 1962	China–Hong Kong	Widely established and successful in fish ponds, tanks, etc.

[a]Arthropod/pathogen introduction/biocontrol information from Winston *et al.* (2014) with additional references where indicated; [b]Arthropod unless otherwise stated; [c]Chacko and Narasimhan (1988); [d]Singh (1998); [e]Sreerama Kumar *et al.* (2008); [f]Singh (1997); [g]Singh (1989).

Indian Station of the Commonwealth Institute of Biological Control based in Bangalore (now Bengaluru). From 1977, country-specific programmes were implemented under the auspices of the AICRP on the 'Biological control of crop pests and weeds'. This subsequently became part of the remit of the Project Directorate of Biological Control (PDBC, now NBAIR), which was created in 1993 under ICAR, together with its research programmes initiated at the Indian Institute of Horticultural Research (IIHR), Bangaluru. In addition, the World Bank funded ICAR National Agricultural Technology Project (NATP) supported an ICAR–CABI work plan, under which a workshop on quarantine procedures and facilities for biological control agents was organized at PDBC during May 2002 (Ramani *et al.*, 2004). This, together with the ICAR–CABI collaborative project (funded by the UK Department for International Development, DFID) on 'Classical biological control of *Mikania micrantha*', are both a clear reflection ICAR's policy to support classical biological control in India. In the case of mikania, the rust fungus *Puccinia spegazzinii* was tested for host specificity in the quarantine facility of the National Bureau of Plant Genetic Resources (NBPGR) and released in Assam and Kerala (see Chapter 10, this volume).

Other recent Indian programmes on the classical biological control of weeds are (see also Table 12.1):

- *Chromolaena odorata:* the tephritid gall fly *Cecidochares connexa* was imported from Indonesia and quarantined in the laboratory to eliminate any parasitoids or pathogens. Extensive host-specificity tests revealed the specificity of the insect to *C. odorata* and the flies were released in Bengaluru. Several hundred galls were produced on *C. odorata* plants there and plant growth was suppressed by up to 30%. The gall fly is now multiplying quickly on the west coast of India and more suppression of the weed can be expected in the years to come.
- *Parthenium hysterophorus*: the Department of Biotechnology of the Ministry of Science and Technology, Government of India had approved a project for the PDBC to introduce and evaluate the seed weevil *Smicronyx lutulentus* for the biological control of parthenium weed. Although this work could not be progressed at that time, expansion of the classical biological control of parthenium weed is possible in the future.
- *Mimosa diplotricha* var. *diplotricha*: the psyllid *Heteropsylla spinulosa* is being considered for the control of *M. diplotricha* var. *diplotricha*.

Policy Framework Development

Classical biological control is one of the activities of NBAIR and, under the current framework of regulation, the importation of natural enemies from their native range for the biological control of pests and invasive alien plants will continue. ICAR, with IPM as one of its cardinal national policies, will continue to encourage and support research and development activities for the biological control of pests and weeds in agricultural, horticultural and agroforestry ecosystems. Biological control is a national priority, and so a whole range of other institutes in India have a clear policy to support research, development and extension activities to enhance the uptake of biocontrol technologies for sustainable crop production, including:

- the Department of Agriculture Cooperation and Farmers Welfare of the Ministry of Agriculture and Farmers Welfare;
- the Department of Biotechnology of the Ministry of Science and Technology;
- the Department of Environment, Forest and Climate Change of the Government of India; and
- various state agricultural universities, departments of agriculture and horticulture and institutes, such as the Kerala Forest Research Institute.

Because biological control is a priority area, the planning commission in India has allocated an adequate budget for research projects in biological control during the current planning period, as in previous years. There has never been a situation of lack of funds for biocontrol research and development in India. The Department of Agriculture Cooperation and Farmers Welfare encourages the production and use of biocontrol agents through centrally sponsored schemes. Subsidies are extended to farmers for procuring biocontrol agents. Through the AICRP, biocontrol agents are multiplied and distributed to farmers. Attempts are, therefore, in full gear to make bioagent production an agricultural enterprise.

Importation of natural enemies of invasive alien weeds

As per the policies of the Government of India, ICAR and the Department of Agriculture Cooperation and Farmers Welfare will continue to permit the import of beneficial arthropods and pathogens for the classical biological control of pests and weeds, provided that all of the procedures for the import, post-entry quarantine (to ensure that all biocontrol agent cultures are pure and no contaminants have been accidentally introduced), safety testing and evaluation (in both the laboratory and the field) are followed. A workshop and consultancy organized by the then PDBC under the ICAR–CABI work plan funded under the ICAR–NATP project (see previous section) was part of strengthening this important activity. The general procedures for the introduction of biocontrol agents for weed control using arthropods and fungi, based on both international standards (Kairo *et al.*, 2003; IPPC, 2005) and unpublished technical support documents provided by CABI, were laid down during the workshop (Bhumannavar, 2004; Ellison, 2004).

A consultation, provided by experts from CABI in November/December 2002, addressed the design of a state-of-the-art, international-standard quarantine facility at PDBC, including the structure of the building, its facilities and specifications, and protocols for the handling of introduced biocontrol agents. The proceedings of these two activities are covered by Ramani *et al.* (2004).

As a result of these activities and in accordance with the policy of enhancing biological control in India, ICAR has established a quarantine facility of international standard at NBAIR that can handle both arthropod and microbial biocontrol agents. It includes a quarantine area with an air-handling system with HEPA (high-efficiency particulate air) filters and airtight doors to prevent the accidental escape of organisms into the environment. It has a glasshouse with a shatterproof polycarbonate roof for the containment of weed-feeding insects and fungal pathogens. This facility will cater for the needs of importation of biocontrol agents into India. It may also be used for 'third-country quarantine' of biocontrol agents on behalf of other countries that do not have adequate quarantine facilities.

Guidelines for the importation of natural enemies of invasive alien weeds

NBAIR now is identified as the nodal agency for the import, quarantine and field release of all biocontrol agents for pest, disease and weed management. With effect from 1 April 2004, a new order, the Plant Quarantine (Regulation of Import into India) Order, 2003, came into effect by a notification in the Gazette of India (S.O.1322 CE) dated 18 November 2003 by the then Ministry of Agriculture. This exercises the powers conferred by subsection (I) of Section 3 of the Destructive Insects and Pests Act, 1914 (2 of 1914). The revised guidelines pertaining to the import of biocontrol agents are as follows:

1. No consignment of live insects, microbial cultures or biocontrol agents shall be permitted into India without a valid import permit issued by the Plant Protection Adviser.

2. Every application for a permit to import insects or microbial cultures, including algae or biocontrol agents, shall be made using PQ Form 12 (Appendix 12.2), which shall be sent at least 2 months in advance to the Plant Protection Adviser along with a fee of Indian Rs. 200/- towards registration.

3. The Plant Protection Adviser shall issue the permit in the form of PQ Form 13 (Appendix 12.3) in triplicate, if satisfied of the purpose for which import is being made and subject to such conditions imposed thereon. A blue/violet colour tag or label in the form of PQ Form 14 (Appendix 12.4) shall be issued which shall be affixed on the parcel at the time of export.

4. All the consignments of insects, microbial cultures and biocontrol agents shall be permitted only through specified ports of entry. The consignment of beneficial insects shall be accompanied by a certificate issued by the national plant protection organization in the country of origin with additional declarations of its freedom from specified parasites, parasitoids and hyperparasites. The consignment of beneficial insects/biocontrol agents shall be subjected to post-entry quarantine as may be prescribed by the Plant Protection Adviser.

As a consequence of the Plant Quarantine Order, 2003, the following orders have been repealed:

- Rules for regulating the import of insects into India notified under F-193/40 dated 3 February 1941; and
- Rules regulating the import of fungi into India notified under F.16-S(I)/43A dated 10 May 1943.

The National Biodiversity Authority (NBA) was established in 2003. It is a statutory, autonomous body and it performs facilitative, regulatory and advisory functions on issues of conservation, sustainable use of biological resources and the fair and equitable sharing of benefits arising from the use of biological resources. With the enactment of the Biological Diversity Act (BDA), 2002 and the framing of the Biological Diversity Rules in 2004 (Government of India, 2004), further revision of the guidelines for the importation of biocontrol agents is being contemplated.

In September 2012, NBAIR (then the National Bureau of Agriculturally Important Insects, NBAII) was designated as a repository for agriculturally important insects, mites and spiders under the BDA. In accordance with subsection (2) of Section 39 of the Act, NBAIR shall also keep in safe custody the representative samples as voucher specimens of the biological materials accessed in accordance with the provisions of Section 19 of the Act. Other relevant information related to the material, such as DNA fingerprints, shall be available if required by NBA. In addition, NBAIR shall keep in safe custody the type specimen deposited by any person who discovers a new taxon, in accordance with sub-section (3) of Section 39 of the Act.

The Mikania Experience

The ICAR–CABI project on 'Classical biological control of *Mikania micrantha*' enabled a collaboration of stakeholders as varied as research scientists, growers, tea associations and forest administrators, and brought them into the ambit of biological control for the first time in India. In Kerala, the Forest Department and farmers gave support; in Assam, the tea industry extended full cooperation to the biological control effort.

The initial investigations of the rust fungus *P. spegazzinii* for the classical biological control of mikania were carried out in the quarantine facility at CABI's centre in the UK. The screening of several isolates of the rust against the weed resulted in the identification of a strain of *P. spegazzinii* from Trinidad in the West Indies, strain IMI 393067, which was infective to several populations of *M. micrantha* from India (Ellison *et al.*, 2004). Initial host-specificity testing at CABI had shown this strain to be highly host specific (Chapter 10, this volume).

Based on the import permit issued by the Plant Protection Advisor to the Government of India, the PDBC imported the rust into India in bare-rooted plants of mikania and established the rust on the Kerala population of mikania in the quarantine facility of NBPGR, New Delhi.

Policy on host-specificity tests

As per the policy decision of ICAR, PDBC and NBPGR, in association with Kerala Forest Research Institute (KFRI) and Assam Agricultural University (AAU) conducted host-specificity tests on 74 species of host plants, including some of those tested at CABI, to confirm the specificity and safety of the rust. Test plant selection was based on the centrifugal phylogenetic test protocol (Wapshere, 1974). The list of host plants to be tested was finalized in consultation with several rust specialists as well as systematic botanists in India; economically important plants from Kerala and Assam were included because the rust was to be released in these two states. Because the rust produced chlorotic spots on sunflower, several cultivated sunflower varieties and hybrids were screened. Careful observations revealed that the rust did not invade the plant tissues and hence could not complete its life cycle. Plant growth, flowering and seed setting were normal on all of the test plants. Thus, these results showed that the rust did not infect any of the test plants, including sunflower (which it affected, but upon which it could not complete its life cycle).

Policy on field release

While issuing the permit for release of the rust in the field, as per the policy of the Government of India, the Plant Protection Advisor approved the release of the rust at two specific locations in each of the states of Kerala and Assam. Rust propagation units were set up in KFRI and AAU under the ICAR–CABI project for propagating the large quantities of the rust needed for the release strategy. These units could be used if a planned follow-on project is successfully funded in the near future (see Chapter 10, this volume).

India as a Provider of Weed Biological Control Agents

Like the exotic weeds that have invaded India from elsewhere, Indian native plants have also been moved around the world as potential floricultural or agricultural crops, and there are a number of examples where these species have become invasive aliens in their introduced range. India is now providing support internationally to develop a classical biological control strategy to manage these weeds. In this section, some examples are given of this international collaboration between CABI and India, showing how government policy is supporting this research.

Prior to 2012, the responsibility for the export of plant, microbe and animal material lay with the head of plant protection at NBPGR. However, in 2013 a Government Office Order (F No 8(2)/2011-Coord (Tech) of 19/10/12) came into effect that divided responsibility for the management of genetic resources among three Indian institutes – NBPGR, NBAIR and NBAIM (the National Bureau of Agriculturally Important Microorganisms). Table 12.2 lists these both these institutes and the other Indian institutions that are involved in the export of biological resources for furthering biological control of weeds originating in India, with a brief outline of their roles and the export procedures that they use.

Impatiens glandulifera: biological control in the UK

Impatiens glandulifera (Himalayan balsam) is a riverine weed in many parts of Europe. Under a consortium of UK donors, a project was initiated in 2006 to find a suitable biocontrol agent to release in the UK. In collaboration with NBPGR and KFRI, and following NBA notification, surveys were conducted in the Indian Himalaya, and a rust fungus (*Puccinia komarovii* var. *glanduliferae*) identified as a suitable biocontrol agent (Tanner *et al.*, 2015). Export of biological control agents was at the time dealt with by the NBPGR following approval by the Department of Agricultural Research and Education (DARE) of the then Ministry of Agriculture. Trial releases of the rust were undertaken in the UK in 2014.

Table 12.2. Role of Indian institutes in export of biological resources.

Institute	Role	Export procedure
ICAR-National Bureau of Plant Genetic Resources (NBPGR), New Delhi (www.nbpgr.ernet.in/)	Nodal institute for acquisition and management of indigenous and exotic plant genetic resources for food and agriculture	Authorizes export of plant genetic material following approval from the National Biodiversity Authority (DARE)
ICAR-National Bureau of Agricultural Insect Resources (NBAIR), (formerly National Bureau of Agriculturally Important Insects, NBAII), Bengaluru, Karnataka (www.nbair.res.in/)	Nodal agency for collection, characterization, documentation, conservation, exchange and utilization of agriculturally important insect resources (including mites, spiders and related arthropods) for sustainable agriculture	Authorizes export of insects for research purposes based on a Memorandum of Understanding (MoU) following approval from DARE, with intimation to the National Biodiversity Authority (NBA)
ICAR-National Bureau of Agriculturally Important Microorganisms (NBAIM), Mau Nath Bhanjan, Uttar Pradesh (www.nbaim.org.in/)	Nodal institute for acquisition and management of indigenous and exotic microbial genetic resources for food and agriculture	Authorizes export of fungi (and other microorganisms) following approval from DARE
Department of Agricultural Research and Education (DARE), Ministry of Agriculture and Farmers Welfare, New Delhi (www.dare.nic.in/)	Responsible for international cooperation in agricultural research. Authorizes Memoranda of Understanding (MoU) and Material Transfer Agreements (MTA) under which biological material can be exported following review by an export facilitation committee	Processes proposals for exchange of germplasm under MoU/MTA signed between foreign governments/ organizations and finally conveying the Department's approval for export to relevant institutes for executing the transaction
National Biodiversity Authority (NBA), Chennai, Tamil Nadu (www.nbaindia.org/)	Autonomous body that performs facilitative, regulatory and advisory functions for the Government of India, including the issue of sustainable use of biological resources and fair equitable sharing of benefits of use. Responsible for implementation of the Convention on Biological Diversity (CBD) and its Nagoya Protocol	NBA requires to be notified of all surveys undertaken to collect biological control agents. Permission to be sought for export of biological resources for commercial purposes. If under research collaboration, permission from the designated ministry to be sought and NBA to be notified

Hedychium spp.: biological control in Hawaii and New Zealand

The wild ginger species complex of *Hedychium gardnerianum*, *H. flavescens* and *H. coronarium* have escaped ornamental cultivation and are now invasive in many countries around the world. In New Zealand and Hawaii, these aggressive colonizers invade native forests, displace native species and pose a continuing and growing threat to native biodiversity. A review of the scientific and botanical literature identified the Eastern Himalaya as the centre of origin of these *Hedychium* spp. Surveys conducted in the Indian part of this region (initially assisted by KFRI) from 2008 to the present have prioritized a chloropid fly and a *Tetratopus* weevil for study. Following the deposition of voucher specimens at NBPGR and the Indian Agricultural Research Institute (IARI), review by an export facilitation committee and subsequent approval by DARE, these two species have been exported to the UK each year since 2011. Following the Government Office Order in 2013, all voucher specimens associated with the project were transferred from NBPGR to NBAIR and export facilitation of the insect species was, accordingly, coordinated through the latter institute. The work is being complemented through collaboration with Sikkim University, with research undertaken by an Indian student in Sikkim and surveys facilitated by the Department of Forests, Environment and Wildlife Management (Sikkim Government). The research is ongoing.

Rubus spp.: biological control in Hawaii and the Galapagos Islands

Yellow Himalayan raspberry (*Rubus ellipticus* var. *obcordatus*) is regarded as one of the world's 100 worst invasive species by the International Union for the Conservation of Nature (IUCN). Native to tropical and subtropical India, it is now highly invasive in natural areas of Hawaii. A related species, *R. niveus* (mora) was introduced into research organizations around the world from India to assess its value as an agricultural plant, due to the delicious berries it produces. Unfortunately, the plant escaped cultivation in a number of regions, including East Africa, Ecuador (including the Galapagos Islands) and Hawaii. It has become a serious weed in both agricultural and natural systems as it forms an impenetrable spiny thicket that shades out other plant species and alters the habitat of native animals. A project was initiated in 2012 with surveys to investigate the natural enemy diversity of both of these *Rubus* spp. in the Indian Himalaya. To date, herbarium specimens of the plant have been deposited at NBPGR, pinned specimens of insect natural enemies at the NBAIR and dried leaves bearing fungi at NBAIM for identification. The next stage will be to apply for export of living natural enemies for assessment in the UK in the CABI quarantine facility.

Conclusions

The biological control of pests, diseases and weeds as a component of IPM is accepted as a national policy by ICAR, and classical biological control of alien weeds is one of the priorities. In-country host-specificity tests for weed biocontrol agents in the Indian quarantine facilities are obligatory before the agent can be approved for release. The Indian quarantine facility for biocontrol agents at NBAIR has the capability to include pathogens. The facility may also be used for third-country quarantine.

Several weed-feeding insects have already been introduced into India and a few have established successfully, the latest being the tephritid fly *Cecidochares connexa* for the management of *Chromolaena odorata*. The import of the rust fungus *P. spegazzinii* into India is a landmark in the development of classical biological control of weeds using pathogens. It has paved the way for the use of fungal pathogens as biocontrol agents for invasive alien weeds in India.

Acknowledgements

The authors are grateful to Djami Djeddour and Carol Ellison for valuable inputs on CABI's recent international projects in India, and to Chandish R. Ballal for supportive information on export procedures.

References

Bhumannavar, B.S. (2004) General procedures for introduction of biocontrol agents: current and proposed guidelines for weed control arthropods. In: Ramani, S., Bhumannavar, B.S. and Rabindra, R.J. (eds) *Quarantine Procedure and Facilities for Biological Control Agents.* Technical Document No. 54, Project Directorate of Biological Control, Bangaluru, India, pp. 47–51.

Chacko, M.J. and Narasimham, A.U. (1988) Biocontrol attempts against *Chromolaena odorata* in India – a review. In: Muniappan, R. (ed.) *Proceedings of the First International Workshop on Biological Control of* Chromolaena odorata, *Bangkok, Thailand, 29 February–4 March 1988.* Agricultural Experiment Station, University of Guam, Mangilao, Guam, pp. 65–79.

Doddamani, M.B., Chetti, M.B., Koti, R.V. and Patil, S.A. (1998) Distribution of chromolaena in different parts of Karnataka. In: Ferrar, P., Muniappan, R. and Jayanth, K.P. (eds) *Proceedings of the Fourth International Workshop on Biological Control and Management of* Chromolaena odorata, *Bangalore, India, 14–18 October 1996.* Agricultural Experiment Station, University of Guam, Mangilao, Guam.

Ellison, C.A. (2004). A review of the guidelines, protocols and containment facilities used for the introduction of fungal biocontrol agents. In: Ramani, S., Bhumannavar, B.S. and Rabindra, R.J. (eds) *Quarantine Procedures and Facilities for Biological Control Agents.* Technical Document No. 54, Project Directorate of Biological Control, Bangaluru, India, pp. 17–21.

Ellison, C.A., Evans, H.C. and Ineson, J. (2004) The significance of intraspecies pathogenicity in the selection of a rust pathotype for the classical biological control of *Mikania micrantha* (mile-a-minute weed) in Southeast Asia. In: Cullen, J.M., Briese, D.T., Kriticos, D.J., Lonsdale, W.M., Morin, L. and Scott, J.K. (eds) *Proceedings of the XI International Symposium on Biological Control of Weeds, Canberra, Australia, 27 April – 2 May 2003.* CSIRO (Commonwealth Scientific and Industrial Research Organisation) Entomology, Canberra, pp. 102–107.

Gogoi, A.K. (2001) Status of mikania infestation in northeastern India: management options and future research thrust. In: Sankaran, K.V., Murphy, S.T. and Evans, H.C. (eds) *Alien Weeds in Moist Tropical Zones: Banes and Benefits. Proceedings of a Workshop, Kerala Forest Research Institute, Peechi, India, 2–4 November 1999.* Kerala Forest Research Institute, Peechi, India and CABI Bioscience, UK Centre (Ascot), Ascot, UK, pp. 77–79.

Government of India (2004) *Biological Diversity Rules, 2004.* National Biodiversity Authority, Ministry of Environment and Forests, Chennai, India.

IPPC (2005) *ISPM 3. Guidelines for the Export, Shipment, Import and Release of Biological Control Agents and Other Beneficial Organisms.* International Standards for Phytosanitary Measures, Secretariat of the International Plant Protection Convention, Food and Agriculture Organization of the United Nations, Rome. Available at: https://www.ippc.int/en/core-activities/standards-setting/ispms/ (accessed 2 March 2017).

Joy, P.J., Satheesan, N.V. and Lyla, K.R. (1987) Biological control of weeds in Kerala. In: Joseph, K.J. and Abdurahiman, U.C. (eds) *Advances in Biological Control Research in India. Proceedings of the First National Seminar on Entomophagous Insects and other Arthropods and their Potential in Biological Control, Calicut University, Calicut, India, 9–11 October 1985.* University of Calicut, Calicut, India, pp. 247–251.

Kairo, M.T.K., Cock, M.J.W. and Quinlan, M.M. (2003) An assessment of the use of the code of conduct for the import and release of exotic biological control agents (ISPM No. 3) since its endorsement as an international standard. *Biocontrol News and Information* 24, 15N–27N.

Ramani, S., Bhumannavar, B.S. and Rabindra, R.J. (eds) (2004) *Quarantine Procedures and Facilities for Biological Control Agents. Proceedings of the ICAR-CABI Workshop and Consultancy on Development of Quarantine Techniques for Biological Control Agents.* Technical Document No. 54, Project Directorate of Biological Control, Bangaluru, India.

Sankaran, K.V. and Sreenivasan, M.A. (2001) Status of mikania infestation in the Western Ghats. In: Sankaran, K.V., Murphy, S.T. and Evans, H.C. (eds) *Alien Weeds in Moist Tropical Zones: Banes and Benefits. Proceedings of a Workshop, Kerala Forest Research Institute, Peechi, India, 2–4 November 1999.* Kerala Forest Research Institute, Peechi, India and CABI Bioscience, UK Centre (Ascot), Ascot, UK, pp. 67–76.

Singh, S.P. (1989) Biological Suppression of Weeds. Technical Bulletin No. 1, Biological Control Centre, National Centre for Integrated Pest Management, Bangaluru, India.

Singh, S.P. (1997) Perspectives in biological control of parthenium in India. In: Mahadevappa, M. and Patil, V.C. (eds) *Proceedings of the First International Conference on Parthenium Management, Dharwad, India, 6–8 October 1997.* University of Agricultural Sciences, Dharwad, India, pp. 22–32.

Singh, S.P. (1998) A review of biological suppression of *Chromolaena odorata* (Linnaeus) King and Robinson in India. In: Ferrar, P., Muniappan, R. and Jayanth, K.P. (eds) *Proceedings of the Fourth International Workshop on Biological Control and Management of* Chromolaena odorata, *Bangalore, India, 14–18 October 1996.* Agricultural Experiment Station, University of Guam, Mangilao, Guam, pp. 86–92.

Sreerama Kumar, P., Rabindra, R.J. and Ellison, C.A. (2008) Expanding classical biological control of weeds with pathogens in India: the way forward. In: Julien, M.H., Sforza, R., Bon, M.C., Evans, H.C., Hatcher, P.E., Hinz, H.L. and Rector, B.G. (eds) *Proceedings of the XII International Symposium on Biological Control of Weeds, La Grande Motte, France, 22–27 April 2007.* CAB International, Wallingford, UK, pp. 165–172.

Swaminath, M.H. and Shivanna, M. (2001) The ecological impact of *Chromolaena odorata* in the Western Ghats forests of Karnataka and the management strategies to minimize the impact. In: Sankaran, K.V., Murphy, S.T. and Evans, H.C. (eds) *Alien Weeds in Moist Tropical Zones: Banes and Benefits. Proceedings of a Workshop, Kerala Forest Research Institute, Peechi, India, 2–4 November 1999.* Kerala Forest Research Institute, Peechi, India and CABI Bioscience, UK Centre (Ascot), Ascot, UK, pp. 112–114.

Tanner, R.A., Pollard, K.M., Varia, S., Evans, H.C. and Ellison C.A. (2015) First release of a fungal classical biocontrol agent against an invasive alien weed in Europe: biology of the rust, *Puccinia komarovii* var. *glanduliferae*. *Plant Pathology* 64, 1130–1139.

Van der Weert, R. and Kamerling, G.E. (1974) Evapotranspiration of water hyacinth (*Eichhornia crassipes*). *Journal of Hydrology* 22, 201–212.

Wapshere, A.J. (1974) A strategy for evaluating the safety of organisms for biological weed control. *Annals of Applied Biology* 77, 201–211.

Winston, R.L., Schwarzländer, M., Hinz, H.L., Day, M.D., Cock, M.J.W. and Julien, M.H. (eds) (2014) *Biological Control of Weeds: A World Catalogue of Agents and Their Target Weeds*, 5th edn. USDA Forest Service, Forest Health Technology Enterprise Team, Morgantown, West Virginia.

Appendices

The documents in Appendices 12.1–12.4 that follow are available at: http://plantquarantineindia.nic.in

Appendix 12.1: Plant Quarantine Import Regulations (India) – List of Quarantine Weed Species

Plant Quarantine (Regulation of Import into India) Order, 2003
SCHEDULE VIII
[See Clauses 3 (12) and 8 (1)]
List of Quarantine Weed Species

1.	*Abutilon theophrasti*	31.	*Echinochloa crus-pavonis*
2.	*Agrostemma githago*	32.	*Echium plantagineum*
3.	*Alectra sp.*	33.	*Emex australis*
4.	*Allium vineale*	34.	*Emex spinosa*
5.	*Ambrosia artemisiifolia*	35.	*Froelichia floridana*
6.	*Ambrosia maritima*	36.	*Helianthus californicus*
7.	*Ambrosia psilostachya*	37.	*H. ciliaris*
8.	*Ambrosia trifida*	38.	*H. petiolaris*
9.	*Ammi visnaga*	39.	*H. scaberrimus*
10.	*Apera spica-venti*	40.	*Heliotropium amplexicaule*
11.	*Arceuthobium oxycedri*	41.	*Ipomoea coccinea*
12.	*Avena sterilis*	42.	*Leersia japonica*
13.	*Baccharis halimifolia*	43.	*Lolium rigidum*
14.	*Bromus diandrus*	44.	*Matricaria perforatum*
15.	*Bromus rigidus*	45.	*Mimosa pigra*
16.	*Bromus secalinus*	46.	*Orobanche cumana*
17.	*Cardus pycnocephalus*	47.	*Phalaris paradoxa*
18.	*Cenchrus tribuloides*	48.	*Fallopia japonica (Polygonum cuspidatum)*
19.	*Centaurea diffusa*	49.	*Persicaria perfoliata (P. perfoliatum)*
20.	*C. maculosa*	50.	*Proboscidea lovisianica*
21.	*C. melitensis*	51.	*Raphanus raphanistrum*
22.	*C. solstitialis*	52.	*Rumex crispus*
23.	*Chondrilla juncea*	53.	*Salsola vermiculata*
24.	*Cichorium endivia*	54.	*Senecio jacobaea*
25.	*C. pumilum*	55.	*Solanum carolinense*
26.	*C. spinosum*	56.	*Striga hermonthica*
27.	*Cordia curassavica*	57.	*Thesium australe*
28.	*Cuscuta australis*	58.	*T. humiale*
29.	*Cynoglossum officinale*	59.	*Vicia villosa*
30.	*Desmodium tortuosum*	60.	*Viola arvensis*
		61.	*Xanthium spinosum*

Appendix 12.2: Plant Quarantine Form 12

PQ Form 12

Application for permit to import live insects/mites/nematodes/microbial cultures including algae/bio-control agents

To The Plant Protection Adviser to the Government of India, Directorate of Plant Protection, Quarantine & Storage, NMV-IV, Faridabad (Haryana)-121001	

I/ We hereby make an application, in accordance with provision of Clause 7 of Plant Quarantine Regulation of Import Order, 2003, made under Sub-section (1) of the Section 3 of the Destructive Insects & Pests Act, (2 of 1914) for a permission to import of following insects/mites/nematodes/ microbial cultures. biocontrol agents for research/experimental purpose as detailed below:

1.	Description of insects/mites/nematodes/ microbial cultures/biocontrol agents intended to import (common/scientific names)	
2.	Taxon (class/order/family/sub-family tribe/races or strains)	
3.	Stages of the organism	
4.	Number of specimens or units	
5.	Host species, if any (common/Scientific Name)	
6.	Mode of packing & no. of packages and distinguishing marks, if any	
7.	Country of origin & foreign port of shipment	
8.	Mode of shipment & point of entry	
9.	Name & address of importer	
10.	Name & address of exporter	
11.	Approximate date of import	
12.	Purpose of import	

Declaration

I/ We hereby undertake to abide by the instructions/guidelines issued by the Plant Protection Adviser to the Govt. of India from time to time in this regard.

Date :
Place :

(Seal)

(Signature of Applicant)

Appendix 12.3: Plant Quarantine Form 13

PQ Form 13

(Emblem)
Government of India
Ministry of Agriculture
Department of Agriculture & Cooperation
Directorate of Plant Protection, Quarantine & Storage
NH-IV, Faridabad (Haryana-121001)
Permit For Import of live insects/mites/nematodes/microbial cultures

Including algae/bio-control agents

Permit No.____________________ Date of issue ____________

Valid up to _____________

In accordance with the provision of clause 7 (3) of the Plant Quarantine (Regulation of Import into India) Order 2003 issued under Sub-section (1) of Section 3 of the Destructive Insects & Pests At, 1914,I hereby grant Permission for import of following insects/mites/nematodes/microbial cultures/ Biocontrol agents as detailed below:

1.	Name and address of importer			2.	Name and address of exporter				
3.	Country of origin			4.	Point of Entry				
5.	Description of organism (Common/ Scientific Name)	6.	Taxon (Class/ family/ order etc.)	7.	Stage of organism, host species, if any	8.	No of specimens/ units	9.	Mode of packing and distinguishing

10.	The above permission is granted subject to following conditions :-
(1)	No substitute is permitted for the kind or organism permitted for import under this permit.
(2)	The consignment shall be accompanied by an official certificate issued by an appropriate authority in the country of origin for freedom from: (a)________________________________ (b)________________________________
(3)	The consignment of bio-control agents shall be held under post-entry quarantine at ______________________________ (Name of Institute/Organization) for a period of _______________ before release for fields trials.
(4)	The permit shall intimate the Plant Protection Adviser of any change of address and comply with his instructions.

	Place : Date: Place:	Seal	Name & (Signature of issuing authority) Stamp of Organization

Appendix 12.4: Plant Quarantine Form 14 – Label

PQ Form 14

Face of label

BLUE/VIOLET LABEL

Permit No.________________ Valid up to ___________________

This package contains: Live insects/mites/nematodes/microbial cultures/bio-control agents.
Do not open except in the presence of plant quarantine authority

RUSH AND DELIVER TO
Officer-in-charge
Plant Quarantine Station
at________________

Reverse of label

Directions for mailing live insects/mites/nematodes/microbial cultures including algae/biocontrol agents

Under this label only material covered under this permit should be shipped and any other material be denied entry.

Place within the package the Consignee's name and address and Invoice.

Paste securely the Blue/violet label on the face of each package.

Do not write anything on this label.

Do not place any delivery address outside package.
Place on outside of package name and address of foreign shipper.

Index

Page numbers in **bold** type refer to figures, tables and boxed text. 'IAS' throughout the index stands for 'Invasive Alien Species'.